Time and the Philosophy of Action

W0018552

Although scholarship in philosophy of action has grown in recent years, there has been little work explicitly dealing with the role of time in agency—a role with great significance for the study of action theory. As the articles in this collection demonstrate, virtually every fundamental issue in the philosophy of action involves considerations of time. The four sections of this volume address the metaphysics of action, diachronic practical rationality, the relation between deliberation and action, and the phenomenology of agency, providing an overview of the central developments in each area with an emphasis on the role of temporality. Including contributions by established, rising, and new voices in the field, *Time and the Philosophy of Action* brings analytic work in philosophy of action together with contributions from continental philosophy and cognitive science to elaborate the central thesis that agency not only develops in time but is shaped by it at every level.

Roman Altshuler is Assistant Professor at Kutztown University, USA

Michael J. Sigrist is Professorial Lecturer at George Washington University, USA

Routledge Studies in Contemporary Philosophy

For a full list of titles in this series, please visit www.routledge.com

Time and the Philosophy of Action

Edited by
Roman Altshuler and
Michael J. Sigrist

Routledge
Taylor & Francis Group

LONDON AND NEW YORK

First published 2016 by Routledge

2 Park Square, Milton Park, Abingdon, Oxfordshire OX14 4RN
52 Vanderbilt Avenue, New York, NY 10017

Routledge is an imprint of the Taylor & Francis Group, an informa business

First issued in paperback 2018

Library of Congress Cataloging-in-Publication Data
Names: Altshuler, Roman, editor.
Title: Time and the philosophy of action / edited by Roman Altshuler and Michael J. Sigrist.
Description: 1 [edition]. | New York : Routledge, 2016. | Series: Routledge studies in contemporary philosophy ; 83 | Includes bibliographical references and index.
Identifiers: LCCN 2016005447 | ISBN 9780415735247 (hardback : alk. paper)
Subjects: LCSH: Act (Philosophy) | Time.
Classification: LCC B105.A35 T56 2016 | DDC 128/.4—dc23
LC record available at http://lccn.loc.gov/2016005447

ISBN: 978-0-415-73524-7 (hbk)
ISBN: 978-0-367-14414-2 (pbk)

Typeset in Sabon
by Apex CoVantage, LLC

Contents

Acknowledgments

Many of the papers collected in this volume were first presented at the Time and Agency conference at the George Washington University on November 18–19, 2011, with the support of the George Washington University Department of Philosophy, the Columbian College of Arts and Sciences, and the George Washington University Philosophy Club. The editors would like to thank those who helped make the conference a success, including several presenters whose work did not make it to this volume—Jennifer Morton, Chauncey Maher, Lynne Rudder Baker, and Patrick Fleming—as well as the philosophers whose comments at the conference helped initiate discussion and revision of the papers—Kyle Fruh, Michael Brownstein, Irene Bucelli, Lior Levy, Alex Madva, Steven Gross, C. Hans Pedersen, Carl Hammer, Michael Brent, and John Schwenkler. Gail Weiss, former chair of the Department of Philosophy, also deserves special mention for her efforts and assistance.

The editors would also like to thank those who provided advice and support at various stages of this project, especially the editorial team at Routledge, as well as Andrei Buckareff, Patrick Stokes, Jonathan Webber, David Velleman, and three referees for Routledge. Roman would like especially to thank Lauren Trainor for her support, advice, and putting up with his perpetual late nights. And Michael would like to thank Clare Berke for her care and patience during this long process.

1   Introduction

Roman Altshuler and Michael J. Sigrist

We do things in time. Philosophy of action can capture this phenomenon in at least two ways. On one hand, it might focus on the way that temporal preferences and long-term temporal horizons affect the rationality of decisions in the present (see, e.g., Parfit 1984; Rawls 1971). Such work may focus on the way we discount the distant future, for example, or prioritize the future over the past. Approaches of this kind treat time as, in a sense, something external to agency; they set various constraints on what we may rationally intend, for example. But if temporal considerations can constrain agency, it follows that they can also structure it internally, and if they can do so, it seems likely that this is possible because agency is already temporally structured. This volume focuses on the way that agency is temporally structured from within, such that time is an ineliminable constituent of agency.

This approach is supported by reflection on some ordinary facts about action and agency. Since actions take time, we might wonder whether they have temporal parts and whether those parts must have the same temporal structure as the broader actions they constitute. Since actions undergo change as they unfold, we might wonder whether they are events or processes. Since actions often involve prior intentions, we might wonder how our acting selves are coordinated with our earlier intending selves, what content those intentions must have in order to avoid committing our later selves to courses of action that may become inadvisable, what psychological mechanisms we rely on in forming intentions for future actions, and how those intentions are shaped by long-term commitments. Since rational actions seem to require deliberation, we might wonder whether the deliberation must occur just prior to intention, or whether it may come long before, or even follow upon the action. Since we find ourselves acting out of a certain background of habits and prior commitments, we might wonder how those temporal phenomena interact with the temporality of actions.

Common to these questions is the idea that time structures agency from within. But the diversity of these questions gives rise to a further suggestion: that the various structures involved in agency—action, intention, deliberation, commitment—each operate along different timescales, which require both independent and joint elaboration. The chapters in this volume examine both the temporalities operative at these different levels of agency individually and the interactions involved in their reciprocal manifestation.

1. The Metaphysics of Action

On one common line of thought, we do some things by means of doing other things. I type "things," for example, by typing *t*, and I type *t* by moving my finger. But what would happen if we did *all* things by doing other things? In that case, I could only perform any action A by means of B, which in turn I could only perform by means of C, and a regress would result. The performance of any action would require me to first perform an infinite number of actions as means to it. But since, as defenders of basic action reason following Arthur Danto's initial line of thought (Danto 1963, 1965), it is impossible for finite agents to perform an infinite number of actions, if I am to ever get anything done, it must be possible for me to perform some basic actions: actions that I perform directly, without doing anything else as a means to performing them. Such actions are typically called instrumentally or teleologically basic.

There is another line of thought seemingly leading to the same conclusion. If I put down my whiskey glass absentmindedly, having forgotten to take a sip, I have still done something: I brought the glass up to my mouth. And even if I didn't get that far, I at least raised it off the table. But what if, as I begin to raise the glass off the table, I am struck by an oncoming train? I have not in fact raised the glass off the table, and yet it seems I have done *something*. Whatever it is I have done, the thinking goes, must be more basic than what I had set out to do; there must, then, be basic actions: actions that cannot fail if interrupted.

It is tempting to think that the two lines of thought must lead to the same conclusion. If I am interrupted in my performance of A having completed only a part of it—B, say—then it seems natural to assume that I must have been aiming to do A by means of B; why else would I have done B in the course of attempting A? This consideration leads us to question what might otherwise be a reasonable thought, namely that two distinct concepts of basicness are at work. However, neither line of thought is entirely convincing in its own right, and thus each has its critics. What's worse, the two lines of thought do not seem to mesh neatly together. The first implies that basic actions are simple bodily movements. The second, on the other hand, leads to the thought that, as all bodily action is subject to failure, basic actions cannot be bodily actions at all but must instead be tryings or volitions.

We are thus faced with a question of how to accommodate both lines of thought. One could, of course, reject one or both. Critics of basic action, as we will see later, may play on the uneasy fit between them. But other alternatives are available: to reject the letter of one line of thought while preserving its spirit, for example, or to respond to the concerns raised by both—as some of the authors in this volume suggest—by distinguishing between different senses in which agents *do* things. A reconciliation of the two, however, could be aided by seeing them as operating along different dimensions of thinking about the temporality of agency. The first line of thought stems from the fact that actions, typically, reflect a process, explicit or implicit, of diachronic reasoning. The second stems from the fact that actions unfold over time. Thus, while the second leads to the conclusion that basic actions must not take time at all, the first instead seeks to construct our doings out of simple, but temporally extended, parts. This leaves it open to us to deny, for example, that what we *do* must have the sort of temporality that admits of interruption and division, or to deny that an intended action must have intentional units at its core.

Attempts to define basic action largely died down in the 1990s (for an exception, see Ruben 2003). But the idea of basic action did not go away; it has continued to be presupposed in much of the recent theorizing about action. In any case, basic action remains an important problem. If basic actions are, indeed, the essential building blocks of all other actions, then arguably we cannot know what actions are without knowing what basic actions are. The question should, therefore, be at the forefront of thinking about action. One could, of course, attempt to escape the logic of this position by denying that there is anything special about basic action. Perhaps basic actions are just like other actions, only simpler (or, perhaps, shorter). If that's the case, we need not understand the parts to grasp the whole; we can simply go about studying action, assuming that whatever it turns out to be, basic action is a miniature version of that. If this assumption is behind action theory's relative neglect of the question, however, the assumption has first to be proven. More importantly, it seems unlikely to be true. Basic actions, on traditional views, have a curious property other actions don't share: they are immune to failure. If that's the case, it seems hard to believe that questions about the nature of basic action can be quarantined from other questions about agency.

First, the possibility of failure seems to be a persistent feature of our agency in the world. The possibility of action with natural immunity may have relevance to far grander issues. We find a concern with failure-proof action going back as far as the Stoics, who counseled that we worry about controlling our cognitions rather than the failure-prone bodily movements that stem from them (a consideration clearly behind the second line of thought, discussed earlier). In more recent times, H. A. Prichard (1949) advanced the Kantian-tinged suggestion that our duty is confined to intention, since we cannot have duties to do something that, like all bodily movement, is not within our control. Second, if basic actions have unique properties, and if all other actions decompose into them, it is unlikely that questions about the nature of basicness can avoid spilling over into questions about action. In short, the thought that the question of basic action can be set aside in action theory rests on a truncated conception of basic action and its relation to non-basic actions, a conception that demands further examination.

The thought that our conception of basic action must inform our conception of action as such has fueled two recent skeptical attacks on the concept. Douglas Lavin and Michael Thompson see basic action at the foundation of the currently dominant view: the causal theory of action. On this view, actions are essentially events involving bodily happenings, caused by appropriate mental states of the agent such as beliefs, desires, or intentions. The idea of basic action seemingly allows this picture to work: if complex actions are, at bottom, simple events strung together by mental events internal to the agent, they can fit easily into the framework of the causal theory. Lavin thus envisions basic action as presenting us with a picture of arational bodily movement, unfolding in isolation from the agent's causal activity of planning such unfoldings in advance via means-end rationality. Lavin (2013) rightly rejects this picture as alienating agents from movements that can hardly be described as actions at all, and for good measure attacks the coherence of basic action as such, employing an argument similar to Thompson's. The latter (Thompson 2008, pt. 2), playing the two lines of thought against each other, argues that since every action is temporally extended, it can be broken down into an infinite number of shorter segments. Any arbitrary segment of a larger action is ostensibly an intentional action in its own right, and yet each such action can itself be further decomposed. Thus, no actions are basic; at best, some are more basic than others, and yet less basic than their own arbitrarily chosen constituent segments. The first three chapters in Part I represent critical responses to Lavin and Thompson's objections.

Central to this skeptical strategy is the thought that the boundaries of a basic action are arbitrary, such that any part of an action, however basic that action may appear, can itself be taken as a more basic action. In Chapter 2, Santiago Amaya proposes a novel criterion for making those boundaries non-arbitrary: instead of thinking of basic actions as movements that cannot be interrupted, he argues that we should take them to be ones that agents cannot slip in performing. Taking his cue from linguistics, Amaya notes that while verbal slips are universal to language users, which slips are possible for a particular speaker will depend on the speaker's competences; while it seems speakers can utter certain sounds directly, and thus cannot slip in these utterances, other utterances are vulnerable to such slips. Amaya proposes extending the analysis to actions in general: when an agent can perform an action without the possibility of slipping, she can perform it directly, and the action is thus basic for her. What sets the boundaries of basic action on this account is not the possibility of interruption by external circumstances, but rather the agent's competence in performing it without the possibility of internal failure.

In Chapter 4, Ben Wolfson further develops this latter possibility in crafting his response to Lavin. It is a mistake, he argues, to class all actions into only two categories: complex actions structured by means-end rationality on one hand, and arational bodily movements that hardly qualify as actions on the other. There are, he points out, things we do that fall into an intermediate category in which we clearly exercise rationality and yet do so directly without recourse to means-end reasoning: understanding a sentence in a language we know, for example. Wolfson thus defends basic actions as involving second-natural rational capacities, such that we know how to do them without doing anything else; that our exercise of such capacities may be interrupted by external factors is, then, irrelevant to the question of whether the exercise of the capacities is basic as such.

In Chapter 3, Kim Frost explores a seeming antinomy in discussions of basic action. On one hand, we have defenders of basic action, who argue that basic actions are necessary if agents are to act at all. On the other hand, we have Thompson's critique, on which every action is divisible into an infinite number of actions. In his rejection of Thompson's argument, Frost shows that the increasingly microscopic movements Thompson takes to be constitutive of any action may themselves be *aspects* of actions rather than intentional actions in their own right. But if the argument thus fails, neither does this show that the other side of the antinomy is right. In some cases, actions do not exhaustively decompose into a finite set of basic actions. What's more, if agents can perform larger actions without intentionally performing the individual aspects of those actions—if, that is, they can perform actions by engaging in non-intentional *activities*—then the pressure to posit basic actions in every case dissolves. Agents can engage in the world directly, via engagement in activities, without having to first do many more basic things.

Here we have a new possibility: that we do not have to choose between a picture of complex actions as events consisting of smaller events, on the one hand, and on the other the counterfactual possibility that agents cannot engage with the world directly at all. In Chapter 5, Helen Steward argues that the mistake lies in thinking of actions as events in the first place; if we understand them as processes, the agent's direct involvement with the world becomes clearer. Events do not persist through time and they do not undergo changes; rather, they are constituted by changes. Processes, on the other hand, can change: my typing becomes more furious when I think I am on to a thought, and it slows down when I am fishing for my next one. Since events do not change, in asking what causes them we are trying to find only what sets them off. Processes, on the other hand, have causes that also sustain them and bring about the changes in their unfolding; unlike events, which have a single cause at their beginning, processes require ongoing causal input over time. In thinking of actions as events, then, we leave out the agent's contribution, taking the initial cause to be merely an event or chain of events in the agent's mind. If actions are processes, however, agents are needed to explain why they continue and undergo change. I typed this, perhaps, because I wanted to write an introduction, but the process of its typing required far more than this initial desire from me.

2. Diachronic Practical Rationality

We can imagine agents whose only actions are basic actions. These would still be agents, in the strict sense, since, when they moved their bodies, the question "Why?" would apply to them, thus satisfying Anscombe's widely accepted minimal criterion for agency (Anscombe 2000). However, such agents would not perform actions for the sake of some other action. They would be minimal or atomic agents (see Setiya 2014 for this suggestion). We may assume, however, that humans are typically not agents of *this* sort. We string our basic actions together into complex ones, sometimes ranging over long periods of time, and with an eye to accomplishing tasks we could not via basic actions alone. Stringing actions together toward a further end, however, requires planning: we need our intentions to be consistent with each other, so that we do not constantly undertake self-defeating actions. The current understanding of intentions as components of plans, and thus best understood via the functional role they play in such plans, stems from seminal work by Michael Bratman (1987). On Bratman's account, since intentions play a role in intrapersonal coordination over time (as well as coordination with others), they must exhibit stability. If intentions were not stable, if they didn't bind us to courses of action, they would be useless for planning purposes.

Forming an intention puts an end to deliberation. This means that intentions commit us to action in the absence of further deliberation. If the action follows immediately, this is unproblematic, but it raises an important difficulty when the action is meant to occur at some future point and thus future-directed intentions—particularly ones aimed at a remote future—come into play. Timmy may, for example, deliberate about whether he should ask Molly or Milly to the dance when school resumes in two months. Timmy decides that Milly's lightheartedness and blonde curls are reason enough to ask her. When the time to ask comes around, however, Timmy discovers that Milly has already settled on a date for the dance and, furthermore, that Molly's sense of humor, which he had previously overlooked, is subtle but wicked. Surely at this point it would be irrational for Timmy to stick to his original intention: Molly, not Milly, is clearly the person to ask in these new circumstances. Intentions must be both conditional and partial if they are to do their work. They must be conditional, since it must be possible for the agent to revise them in light of unfavorable circumstances, such as Milly's already having a date. And they must also be partial: if Timmy originally intended to ask by phone, but overhears Molly complaining that her parents monitor her phone calls, it would be silly for him to abandon the intention to ask altogether; far better to have initially left the method of asking vague, to be filled in later as circumstances allow.

Despite their partiality and conditionality, however, intentions need to remain stable. Balancing these features may be a tricky task, as shown by Gregory Kavka's well-known Toxin Puzzle (Kavka 1983). In the Toxin

Puzzle, you are offered a million dollars to intend to drink a toxin the next day, but with a caveat: if you successfully form the intention, the money will be wired to your account *before* the time to drink comes. Since you know that drinking the toxin will make you ill, and that by that point there will be no reason to drink it, it seems you cannot even form the intention to drink. But why? It seems as if stability is constitutive of intention: while intentions must be conditional, so that under certain circumstances we must be ready to abandon them, it seems that we cannot even form them given the knowledge that they *will* have to be abandoned. This brings out an important feature of intending: it brings three temporal selves into a relation with each other. First, there is my self at the moment I form the intention. Second, there is my future self, which must carry out that intention, trusting its previous self to have intended well. And finally, there is an even further future self, which, having completed the action, now reflects on whether its previous incarnation was right to trust the earlier one. Intentions must somehow unify these three selves across time, else they cannot be formed at all.

Bratman (1999) proposes that regret avoidance is constitutive of intention: to intend something, we must expect that we will not regret so intending upon carrying out the plan. In Chapter 6, Edward Hinchman argues that this is a mistake. We can form intentions even if we expect that our actual distant future selves will regret them; what matters to intending is, rather, that the distant future self one projects at the time of intending not regret having trusted the intending self in carrying out the intention. Hinchman's account has two important upshots. First, he analyzes the relation of trust-regret in the intrapersonal case as partially analogous to its interpersonal variants: in intending, my present self must project two future selves, the acting self that will follow through on the intention and the still later self that will assess the follow-through, and aim at an intention that will be satisfactory to the latter, if not necessarily the former. Second, the focus on avoidance of projected trust-regret allows for a more plausible account of agential authority. It is not clear why the reactive attitudes of my expected future self should play any role in giving authority to my currently intending self: I may expect to suffer a catastrophic reduction in my ability to appreciate modern art in the near future, but this hardly robs my intention to visit the Museum of Modern Art of its agential authority. On the other hand, such authority is undermined if the self I currently project would regret following through on the intention.

Returning to the curious case of Timmy, we may observe that he has put himself in a bind. Having intended to ask Milly to the dance, he spent the summer thinking about her: how best to approach her, how best to assure an affirmative answer, and how lovely it will be if she accepts. Now that the plan has fallen through, he is in trouble: aside from the intense mental anguish of having his dreams crushed, Timmy finds himself without a prepared strategy for asking Molly. But the clock is ticking (Molly has been seen talking to Jeremy from second period), and Timmy must act now or lose his chance. It's too late for Timmy now, but he could have made a better plan months ago: he could have intended to ask Molly or Milly and left it at that. That is, he could have formed what Luca Ferrero (Chapter 7) calls a *pro-tempore* disjunctive intention. Importantly, such a disjunctive intention would have given him greater stability, because the chances that he would have to abandon the intention would be lowered: should it turn out that Milly already has a date, Timmy would not need to abandon his intention entirely and switch to a new one, but would simply pursue the other disjunct of the intention. Aside from providing an extra strategy for ensuring the stability of his intention, Ferrero argues, Timmy's *pro-tempore* disjunctive intention would have important practical benefits: not only would he be less heartbroken when one disjunct of his intention collapsed, but he would also have spent the summer productively preparing to ask both Milly and Molly, and so would not be unprepared for the last-minute change in circumstances.

Despite the strength of the planning model, human agents are more complex than this model alone would suggest. We do not simply coordinate our actions via plans that link our intentions together. Those plans themselves must be coordinated with each other so that they match the underlying practical identity to be expressed in our actions. The issue here isn't simply that plans must be coordinated so that they do not undermine each other (in the way plans coordinate intentions), but that our plans should not clash with the commitments by means of which we give structure and meaning to our lives. These commitments are what Bernard Williams (1973a, 1981) called projects. Projects, for Williams, include a wide range of commitments, from career success to friendship. In Williams's writing, such projects give us reasons to live and provide us with a path to happiness. And since sticking by our projects—having integrity—seems to have some kind of normative force, Williams deploys the importance of projects against both utilitarian and deontological moral theories, which he argues have the potential to clash with our projects and thus cannot be overriding (see Altshuler 2013 for discussion of projects in Williams).

Williams, like Harry Frankfurt who has made similar arguments in his analyses of love and caring (Frankfurt 1998, 2006b), is certainly right that we structure our lives, at least to some extent, by means of commitments, and that these commitments give range to a wide array of reasons that lend unity to our identities over time. But as we have just seen, these projects also have normative force: they give us reasons for action, we are open to criticism for violating those reasons, and those reasons are at least potentially strong enough to override moral demands. If my friend, hiding in my basement, is being sought by the police for public indecency, there is at least a question about whether I should turn him in; because he is my friend, I am open to criticism for turning him over too readily. My commitment to celebrating "Movember" is itself a reason to continue cultivating my moustache over my wife's objections. But it is unclear how our evaluative commitments, as Monika Betzler (Chapter 8) labels the category including projects and personal relationships, can themselves exert normative pressure, seemingly independently of the value of the objects of those commitments. Betzler defends a self-referential account of evaluative commitments, on which the reasons such commitments give us gain their normative force from the fact that valuing something itself has value. The normativity does not stem from the value of the object, but from the value of our commitment to it.

But how are our plans and evaluative commitments brought together? We make plans partially on the basis of our evaluative commitments. But we do so in light of our past experiences and with a sense of the shape our lives may take, our aspirations, our limitations (both physical and temporal), and our values. Following work by Alisdair MacIntyre (2007), Marya Schechtman (1996), and Paul Ricoeur (1990), there has been a growing interest in the idea that our practical identities (on some views, our personal identities as well) are constituted by narratives, and a great deal of discussion has focused on the ways narrative seems particularly well suited to bringing together so many diverse elements, on its ability to provide temporal unity, and on its central role in our mental lives. Peter Goldie (2012), for example, has argued that narratives have a significant role to play in Bratman's planning agency. By their nature, narratives allow for dramatic irony, by allowing for divergence between the emotional valences of their subject, the narrator, and the audience—they can present a sad scene in a funny way, for example. Thus, we can reflect on past actions, even ones we at the time viewed with glee, with regret, and we can construct narratives in which such actions are avoided in the future, cementing such narratives into action-guiding plans or, in extreme cases, into evaluative commitments.

In Chapter 9, Daniel Hutto and Patrick McGivern argue that our narrative capacities are central to our capacity for mental time travel. Mental time travel, as it has come to be called, is the ability to be aware of oneself in past and future times. As such, it is central both to first-personal episodic memory and to the ability to imagine ourselves in the future, capacities that underlie our planning agency. A growing range of work on mental time travel strongly suggests that episodic memory and first-personal imagination have a common underlying mechanism, which makes it reasonable to study mental time travel as a unified capacity. There is debate, however, over how such a capacity is to be understood. It seems to be closely connected to theory of mind—our ability to understand the mental states of others— which should not be surprising, given that mental time travel requires us to understand the mental states of our past and future selves. According to classic theory theory, mental time travel is to be understood as a capacity of applying psychological laws to particular cases; simulation theory, on the other hand, posits that it involves projecting ourselves into the mind of another person. Hutto and McGivern argue instead that our ability to understand and use narratives is necessary to both the formation and continued exercise of our capacity for mental time travel.

3. Deliberation, Motivation, and Agency

Several scholars have pointed to an equivocation that stands out in Aristotle's account of deliberation in the third and sixth books of the *Nicomachean Ethics* (see Allan 1955; Ross 1960). Is deliberation a process by which one *arrives* at practical conclusions, or one by which one could justify them, perhaps later, should the need arise? A surface reading suggests the former. For example, Aristotle compares deliberation to geometrical analysis, suggesting that deliberation is a process by which one resolves on a course of action in the same way that one resolves on an answer to a geometrical problem. Discussing this equivocation, John Cooper points out that Aristotle's discussion about the importance of speed and timing in deliberation makes little sense if it is mainly about justifying or explaining an action after the fact (Aristotle 2002, 1142b; Cooper 1975, 6). But there are good reasons to question this reading of Aristotle's account if it is intended as a general account of deliberative agency (rather than a restricted subset thereof), since most actions are not preceded by a period of concentrated deliberation that could be modeled on geometrical problem solving. Aristotle is particularly interested in ethical actions, and many of these (such as when I refrain from participating in gossip, or spontaneously return a wallet that the man in front of me has dropped) are done automatically, as it were, and yet Aristotle insists that every ethical action, *qua* ethical, proceeds from choice (Aristotle 2002, 1105a30; Cooper 1975, 7). One way of attempting to resolve the equivocation is to grant both readings: that while deliberation is the process by means of which one arrives at choice, deliberation may occur long before, or even after, action.

This equivocation trades on an important fact about agency and practical reflection, namely, that the thinking that goes into action—the planning, assessing, weighing, deciding, evaluating, gauging—often happens at some time other than the time of action itself. For instance, I might plan an itinerary for a trip long in advance of the trip itself, so that while I am actually traveling I hardly deliberate at all about where to go next. That is an example of planning for a particular occasion, but there are also, as it were, pre-formed deliberations that we put into action on general occasions. I might be so used to making risotto that I measure out the right amount of rice, broth, butter, and wine without an explicit thought to any of it, and yet if asked later, "Did I intend to put in just those amounts?", I confidently answer yes. In both cases, the deliberations that best explain my actions happen at times other than the time of action itself. This might seem to conflict with assumptions about how deliberations *cause* actions. We typically assume that causes must satisfy two conditions: they must be *prior* to and *continuous* with the event that they cause. Faulty brakes can only cause a car crash if they are faulty *prior to* the crash itself. Similarly, a prior event—my mechanic skipping breakfast, say—can only cause the crash if it is continuous in some way with the crash itself. Perhaps my mechanic skipped breakfast that morning, and so became hungry in the afternoon, and so affixed the brakes incorrectly, leading me to crash when I attempt to use the faulty brakes. Here one might reasonably say that skipping breakfast caused the crash, but only because it is part of a continuous series of events leading to the crash, each of which makes a difference to the event that follows it. There is no such continuity, however, between planning my itinerary and acting on it. There is of course a series of events that takes place between the planning and the traveling, but none of these makes a difference to the one or the other. The explanation leading from my planning deliberations to acting on them is—despite the temporal gap—immediate. In such cases, my intention serves the role of that temporal bridge; while the previous section examined the function of intentions in coordinating our agency, the question of *how* they can do so remains open. Even trickier, however, is the case of ingredient measurement, where we cannot even appeal to intentions, and where deliberate action kicks in not at a pre-specified time, but simply whenever appropriate—whenever, that is, I am called on to make a risotto.

The first two chapters in Part III—by David Velleman and John Drummond—argue, in effect, that the *priority* and *continuity* conditions need not apply in those cases where deliberation is responsible for action. That is to say, the deliberations that explain, and in some sense cause, my actions do not have to be prior to or continuous with that action itself. Velleman (Chapter 10) reviews a long line of work in philosophy of action that assumes—mostly implicitly—that practical reasoning always precedes action and takes time in the sense that it consists of occurrent thoughts that the agent consciously entertains. He argues that both temporal features of this model are wrong and misconstrue what nearly all of our thoughtful actions are like. When we attend to the experience of thoughtful action, what we find, Velleman argues, is that rather than driving our actions, reason exercises a supervisory role. For the most part, our behavior is a function of standing desires and solicitations from the environment. For instance, I might be sitting down on a hot day and decide that I want a cool drink. In that case what happens is that my attention, due to my thirst, is drawn to the refrigerator where I know the ice water is kept, and I get out of my seat to walk and fetch a glass. During all of this I'm hardly explicitly planning, thinking through, or reasoning about what to do. Instead, the rising thirst, the perception of the fridge, the disposition to quench that thirst all together simply draw me out of my seat, as it were, and lead me to the refrigerator. It would be wrong to characterize any of this as happening *to* me—I know exactly what I'm doing and in that sense I do intentionally rise and walk to the refrigerator. But I'm not deliberating in any explicit sense. Whatever reasoning is involved here—and there is reasoning, Velleman insists, which is why human agency is different from automatic behavior—is concurrent with, often even following after, rather than prior to, the behavior it rationalizes.

In Chapter 11, Drummond organizes his contribution around an antinomy of sorts that concerns the phenomenon of blame. On the one hand, we are likely to assign more blame or praise to actions that proceed from deliberation (a premeditated murder, for example, deserves harsher punishment than a crime of passion). On the other hand, an agent's need to deliberate before taking action is often treated as betraying a lack of settled character and lessens the merit (or baseness) of an action. So it seems that sometimes deliberation augments the merit or baseness of an action, while at other times it reduces them. Drummond attempts to unravel this antinomy by uncoupling the time involved in deliberation and the time of action. Deliberation can inform action even if it is not concurrent with it. It does so through the formation of habits and, ultimately, character.

Drummond agrees with Velleman that explicit cogitation—determining facts about oneself and the environment, making explicit inferences on those bases, and following through with a plan of action—rarely precedes most of our actions. Acting is mostly a perceptual, emotional, and evaluative affair, involving immediate responses directly solicited by the environment, rather than issuing from a constant series of mental conclusions. And yet, it is clearly the case that many situations evoke responses and actions that cannot but have been the result of reflection and deliberation. Imagine a baseball player reaching out to catch a line drive. In one sense it's perfectly true that he is *thinking* of catching the ball, but in another, there is no suggestion that some actual thought, "I need to catch the ball" or "to catch the ball I'll need to jump" crosses his mind. At the same time, his seeing how to catch the ball is not *simply* elicited by the situation. Many years of careful practice—deliberately learning where to hold the glove, plant his feet, gauge the path of the ball—precede that moment. For moral situations, it is often the case that this earlier preparation involves weighing complex values. In a crisis situation trained medical experts often need to ignore certain patients who are clearly suffering and in need. This is because they have given special thought beforehand to how to prioritize time and resources. That prior preparation and reflection eventually forms habits. Such habits, according to Drummond, are not automatic or merely reactive responses; they are thoughtful responses to a situation even if the thinking in play occurred much earlier than the situation itself.

Habituation is a central topic in Chapter 12 by Jonathan Webber as well. Like Drummond, Webber observes that we often act on the basis of habits that are in effect standing desires, and that these desires are themselves often the product of prior deliberations. Webber uses the idea of a career to illustrate his point: we may have decided long ago to pursue a particular career, and now, although we rarely reflect upon the reasons for that decision, it continues to inform our daily decisions and behavior at work. Unlike Drummond, Webber grounds most of his conclusions not in direct phenomenological experience but in recent work in empirical psychology. Drawing on this research, Webber examines how beliefs, not just desires or behaviors, exhibit habitual tendencies. Philosophers have not traditionally associated beliefs with any type of habit. This is because habits can persist in the face of countervailing reason and evidence. I may conclude, on the basis of evidence, that my habit of drinking a stiff martini every night is bad for my health, but nonetheless continue to desire one. Beliefs do not seem to be like that. If I question my belief about, say, a tax proposal, and after reasoning conclude that it is not fair, it seems to go against the nature of a belief to wonder whether, all the same, I continue to believe in its fairness anyway. Beliefs that are *not* so susceptible to evidence are sometimes not considered to be beliefs at all (see Gendler 2008). Webber argues that this conclusion is too quick, and that a phenomenon like the persistent effect of implicit bias in certain actions shows that beliefs, too, can have a habitual nature.

The first three chapters in Part III show that some standard assumptions about the timing of deliberation and action—that deliberation must happen prior to and as leading into an action—are limited in scope and do not hold for a general model. Velleman and Drummond show that those assumptions are not verified by our own experiential awareness, while Webber shows that they are not verified by some important results in empirical psychology. However, there is a further question of what it is like for an agent not only to deliberate *in* time, but to be aware of its actions as inherently temporal. These three chapters address that issue peripherally, but it is the central question in Shaun Gallagher's contribution.

In Chapter 13, Shaun Gallagher draws equally from classic research in the phenomenology of temporal perception and contemporary empirical psychology. In particular, he works to establish an explanatory thesis about the temporality of motor control on the basis of a Husserlian model of temporal experience (Dainton 2000; de Warren 2012; Gallagher and Zahavi 2012; Husserl 1991). According to an early model of temporal experience developed by Husserl, temporal experience is organized around a direct perceptual component wed to a short-term retention of an immediately preceding perception and the anticipation of a similar, future one (a "protention"). For reasons Gallagher explains, that model was later replaced by one in which the perceptual moment is absorbed, as it were, into the retentional and protentional moments themselves. This model stresses a dynamic interplay between primary memory and anticipation in the temporal profile of any experience, such that the content of the experience undergoes constant updating through that interplay. Gallagher argues that with these amendations to the original theory, the model of temporal experience for perception applies equally well to actions. He then shows how this analysis scales: the same temporal dynamic that is at play in the constitution of basic motor intentional control can be found higher up in conscious, deliberate actions, and even in the integration of whole periods of life at a still higher scale. Gallagher's analysis, in other words, shows how a similar temporal form is not only common among but functionally integrates behavior at the neurological and motor intentional scales with those broadest actions—like making a career, starting a family, preparing for retirement—that take place over long periods of time.

4. Phenomenology and the Temporality of Agency

Part IV in this volume takes up the theme of agency from the phenomenological tradition, looking especially at the way that time structures agency in ways other than purposive, deliberate action. B. Scot Rousse and David Ciavatta examine how time shapes agency at scales both larger and smaller than at the level of individual actions. Micah Tillman argues that Aristotle's metaphysics of change is a better framework for understanding action than causality. Henry Somers-Hall argues that historical change in the understanding of action, illustrated by Shakespeare's *Hamlet*, follows from a change in the way that time is thought to inform action.

In Chapter 14, Rousse focuses on the concept of care. Care is at the center of both Martin Heidegger's and Harry Frankfurt's accounts of agency, and by elaborating upon this connection, Rousse, in line with some other recent studies (see Altshuler 2010; Crowell 2007; Okrent 1988; Wrathall 2015), seeks to integrate Heidegger's work more closely with the mainstream literature on action and agency. Care (*Sorge*) names the way that agents orient themselves in their world through implicit, practical assessments of priority and importance. We select from among countless possibilities open to us those particular tasks and projects that we choose to perform on any given day, yet we rarely find ourselves explicitly weighing these possibilities as such. A father might, on any given morning, decide to ignore his crying child, skip work, and go to a baseball game, but when he does not do these things it is almost never because he considered and rejected those options, but because those options do not occur to him in the first place. To get around in the world at a practical level one must have priorities that structure one's daily actions and routines. We do not at any point in our lives confront a world of radically open and indeterminate possibility to find ourselves faced with the task of having to make up what to do from whole cloth. A central claim in Harry Frankfurt's work—and also, Rousse argues, Martin Heidegger's—is that these priorities or care-structures are grounded in one's practical identity. Because being a father is typically core to one's identity, the possibility of abandoning one's child to go to a baseball game is not one that most fathers explicitly reject because it is not one that most fathers would consider at all.

Practical identity grounds agency insofar as it determines the sorts of possibilities presented by the world as candidates for action. Rousse focuses on the fact that Frankfurt and Heidegger uniquely emphasize the role played by mortality in this connection and, further, why that connection implicates time. Frankfurt argues that agency involves a standing commitment to projects and values that persist into an ongoing future, and at the basis of that commitment is some care that one's life goes on at all. On Frankfurt's theory, then, a desire to continue to work on the projects that constitute one's identity is grounded in a desire to continue living, a desire that Frankfurt claims is primitive and biologically embedded (Frankfurt

2006b). Rousse, citing Bernard Williams, argues that this has it backwards. We do not first want to go on living and then find projects that will realize that desire; rather, we have projects—or, using William's term, "categorical desires" (Williams 1973b)—that themselves give us reasons for wanting to continue to live in the face of our mortality. But if it is these projects that ground the desire to live, then that desire cannot be strictly biological or natural. This is where Heidegger is useful: rather than basing the temporally articulated care structure in a biological desire to continue living, he grounds it in an "original temporality" constituted in part through an awareness of one's mortality.

While Rousse's article examines care as a sort of superstructure conditioning agency at the broadest level, David Ciavatta (Chapter 15), drawing on the work of Maurice Merleau-Ponty, looks to how intentional action is conditioned by processes that occur at lower, pre-reflective levels of agency. Some of our behaviors—breathing and blinking, for example—take place along the repetitive, cyclical time horizon that characterizes many natural events (think of the phases of the moon, the rotation of days, and the tides). This natural, cyclical time is characterized by a sort of indifference to individuation, where each recurrence is in effect indistinguishable from its predecessors. These cycles may go on forever while no particular iteration makes any difference. Historical time, by contrast—the time involved in intentional agency—occurs within a dimension in which the content and meaning of any particular moment is shaped by what has come before and comes after, so that each moment makes some sort of difference to the whole. Something as simple as stepping off an airplane can be a radically different action depending upon its temporal context—has she finally arrived on vacation, successfully fled extradition, made the biggest mistake of her life, and so forth? Philosophers of action, especially in the phenomenological tradition, have tended to focus on the historical rather than natural level of agency, but it is undeniable that we are just as much natural, bodily beings as we are cultural, self-conscious beings, and that many of our actions can be understood simultaneously along both dimensions. We need to understand how the two are related. Ciavatta argues that a "generalized agency" mediates between those two levels. At the level of historical agency, actions are discrete, but there is also a level of generalized agency that can be characterized as the way that one does things. Ciavatta argues that this generalized agency is rooted in the ongoing, cyclical processes of the body, and shows how that rootedness plays out at the historical, intentional level.

In Chapter 16, Micah Tillman also looks to the phenomenological tradition in order to draw out a theory of agency. According to that tradition, the best account of agency should descriptively match what it is like to be an agent, and for this, Tillman argues, the Aristotelian analysis of change fits better than the notion of causation that has predominated since the seventeenth century. Most importantly, the concept of change, as opposed to cause, can capture in a more comprehensive and less reductive manner the several levels of intentional and pre-intentional directedness that are at work simultaneously whenever one acts. Tillman's chapter works out a theory of agency as self-directed change based upon that analysis. He addresses themes also taken up by Ciavatta, arguing that agency is operative along several different dimensions, from the natural, to the habitual, to the rational, each with their own temporal profile. A further benefit of adopting an Aristotelian theory of change is that it makes clear how human agency is well integrated with change throughout nature. Tillman argues that this Aristotelian approach does better than the alternative causal accounts, since Aristotle's theory of change goes beyond efficient causality and locomotion, and is better suited to understand the nested and reflexive nature of structures like meaning and repetition in agency. There is an interesting disagreement between Ciavatta and Tillman on the relation between habitual and rational actions, since Tillman, unlike Ciavatta, argues that rational action must constantly react against habitual responses if it is to count as fully autonomous.

In Chapter 17, Henry Somers-Hall examines the temporality of action by looking at the representation of action in drama. Somers-Hall finds radically different senses of agency depending upon how one conceives of the place of action in the temporal order of the world. He illustrates this through a reading of Shakespeare's *Hamlet* while drawing on the work of Gilles Deleuze. *Hamlet* represents a fundamental break with Aristotle's model of the way that action is represented in tragedy. Aristotle argues in the *Poetics* that happiness or failure is the first and final object of tragedy, and character is included only in order to make sense of the action. In this depiction of agency, time as something that passes, that has an existence of its own that might interfere with the unfolding of the plot, has no place. To the extent that time figures at all in dramatic depictions of action as Aristotle understands them, it is only in the establishment of ordering (before and after). As Martha Nussbaum (2001) has observed, the actions of Greek heroes emanate directly from their character in service of the plot, and this means that, while we find extensive declarations expressing the intentions behind one action or another, deliberation has hardly any place in the surviving Greek tragedies of the fifth century. Something entirely different is found by 1603. Hamlet's actions are not pushed along by the inexorable logic of the plot. Instead, we are privy to Hamlet's inner deliberations, and sense, along with him, that it is not at all certain what he will do next. Moreover, Hamlet must reckon with time itself as a factor that plays a role in the events of the play beyond mere sequencing. The relentless succession of time forces a need to act on Hamlet, and conditions the anxieties and hesitations he experiences in the face of that pressure. Somers-Hall argues that these changes in the dramatic representation of time and action are echoed in the philosophical differences between Plato and Kant, and the embedded assumptions about time and agency that characterize their respective worlds.

5. Conclusion

While the chapters collected here address a wide range of topics and certainly do not converge on a single position, jointly they demonstrate how consideration of the temporality of agency complicates our standard picture. It may seem as if action is unidimensional and linear: we rationally deliberate about how best to satisfy our desires on the basis of our beliefs, form an intention by reaching a decision, and carry out that intention by stringing together smaller actions into a complex structure. Much of this picture, however, may need to be significantly revised on the basis of temporal considerations.

References

Allan, D.J. (1955). The Practical Syllogism. In *Autour d'Aristote* (pp. 325–340). Louvain: Publications Universitaires de Louvain.

Altshuler, R. (2010). An Unconditioned Will: The Role of Temporality in Freedom and Agency. Doctoral dissertation, SUNY Stony Brook.

—— (2013). Practical Necessity and the Constitution of Character. In C.D. Herrera & A. Perry (Eds.), *The Moral Philosophy of Bernard Williams* (pp. 40–53). Newcastle: Cambridge Scholars.

Anscombe, E. (2000). *Intention*. Cambridge, MA: Harvard University Press.

Aristotle. (2002). *Nicomachean Ethics* (S. Broadie & C. Rowe, Eds.). Oxford: Oxford University Press.

Bratman, M. (1987). *Intention, Plans, and Practical Reason*. Cambridge, MA: Harvard University Press.

—— (1999). Toxin, Temptation, and the Stability of Intention. In *Faces of Intention* (pp. 58–90). Cambridge: Cambridge University Press.

Cooper, J.M. (1975). *Reason and Human Good in Aristotle*. Cambridge, MA: Harvard University Press.

Crowell, S. (2007). Sorge or Selbstbewußtsein? Heidegger and Korsgaard on the Sources of Normativity. *European Journal of Philosophy, 15*, 315–333.

Dainton, B.F. (2000). *Stream of Consciousness: Unity and Continuity in Conscious Experience*. New York: Routledge.

Danto, A.C. (1963). What We Can Do. *Journal of Philosophy, 60*(15), 435–445.

—— (1965). Basic Actions. *American Philosophical Quarterly, 2*(2), 141–148.

De Warren, N. (2012). *Husserl and the Promise of Time*. Cambridge: Cambridge University Press.

Frankfurt, H. (1998). *Necessity, Volition, and Love*. Cambridge: Cambridge University Press.

—— (2006a). *The Reasons of Love*. Princeton, NJ: Princeton University Press.

—— (2006b). *Taking Ourselves Seriously & Getting It Right*. Stanford, CA: Stanford University Press.

Gallagher, S., & Zahavi, D. (2012). *The Phenomenological Mind* (2nd ed.). New York: Routledge.

Gendler, T.S. (2008). Alief and Belief. *Journal of Philosophy, 105*(10), 634–663.

Goldie, P. (2012). *The Mess Inside: Narrative, Emotion, and the Mind*. Oxford: Oxford University Press.

Husserl, E. G. (1991). *On the Phenomenology of the Consciousness of Internal Time* (J. B. Brough, Trans.). Dordrecht: Kluwer.

Kavka, G. S. (1983). The Toxin Puzzle. *Analysis, 43*(1), 33–36.

Lavin, D. (2013). Must There Be Basic Action? *Noûs, 47*(2), 273–301.

MacIntyre, A. C. (2007). *After Virtue: A Study in Moral Theory.* Notre Dame, IN: University of Notre Dame Press.

Nussbaum, M. C. (2001). *The Fragility of Goodness: Luck and Ethics in Greek Tragedy and Philosophy.* Cambridge: Cambridge University Press.

Okrent, M. (1988). *Heidegger's Pragmatism: Understanding, Being, and the Critique of Metaphysics.* Ithaca, NY: Cornell University Press.

Parfit, D. (1984). *Reasons and Persons.* Oxford: Oxford University Press.

Prichard, H. A. (1949). Acting, Willing, Desiring. In *Moral Obligation* (pp. 187–198). Oxford: Clarendon Press.

Rawls, J. (1971). *A Theory of Justice.* Cambridge, MA: Harvard University Press.

Ricoeur, P. (1990). *Time and Narrative, Volume 3* (K. Blamey & D. Pellauer, Trans.). Chicago: University of Chicago Press.

Ross, W. D. (1960). *Aristotle: A Complete Exposition of His Works & Thought.* New York: Meridian Books.

Ruben, D.-H. (2003). *Action and Its Explanation.* Oxford: Clarendon Press.

Schechtman, M. (1996). *The Constitution of Selves.* Ithaca, NY: Cornell University Press.

Setiya, K. (2014). Intention, Plans, and Ethical Rationalism. In M. Vargas & G. Yaffe (Eds.), *Rational and Social Agency: The Philosophy of Michael Bratman* (pp. 56–82). Oxford: Oxford University Press.

Thompson, M. (2008). *Life and Action: Elementary Structures of Practice and Practical Thought.* Cambridge, MA: Harvard University Press.

Williams, B. (1973a). A Critique of Utilitarianism. In B. Williams and J. J. C. Smart, *Utilitarianism: For and Against* (pp. 77–150). New York: Cambridge University Press.

——— (1973b). The Makropulos Case: Reflections on the Tedium of Immortality. In *Problems of the Self* (pp. 82–100). New York: Cambridge University Press.

——— (1981). Internal and External Reasons. In *Moral Luck: Philosophical Papers, 1973–1980* (pp. 101–113). New York: Cambridge University Press.

Wrathall, M. A. (2015). Autonomy, Authenticity, and the Self. In D. McManus (Ed.), *Heidegger, Authenticity and the Self* (pp. 193–214). New York: Routledge.

Part I
The Metaphysics of Action

2 Slip-Proof Actions

Santiago Amaya

Most human actions are complex, but some of them are basic. Which are these? Here, I address the question by invoking the slip, a common mistake. The proposal is this: *an action is basic for an agent if and only if the agent cannot slip in performing it.*

The argument presents some well-established results from psycholinguistics. These are generalized in the context of a philosophical theory of action. The guiding hypothesis is that speaking is a paradigmatic form of acting and, hence, thinking about speaking affords insights about human agency. The resulting criterion has the advantage of being operationalized: there are observable markers that guide its application. Thus, it gives an empirical standpoint for evaluating claims regarding basic actions.

I begin by introducing basic actions and by sketching the proposal. Then, I present the argument for it. I start with some observations regarding verbal slips, show how these observations extend to the nonverbal domain, and explain why lessons about the structure of human actions can be inferred from them. With this in hand, I discuss some widespread views on basic actions. First, I address some reasons for skepticism about their existence. Next, I examine the view that basic actions are those of which agents have basic knowledge. Finally, based on the present proposal, I criticize the thesis that all basic actions occur within the limits of the agent's body.

1. Basic Actions

Type a word in your computer. You are not God, who supposedly can do everything in one sweep. At the very least, you need to type each of the letters in the word. Now compare typing a word with clenching a fist, or raising your arm (Danto 1965; Brand 1968). According to many theorists, the latter are some of your basic actions. You don't need to do other actions *as means* to perform them. But by virtue of cobbling them up you manage to do other things.

In thinking about actions, different senses of "basic" might be at issue. The focus here will be on *teleological basicness* (Hornsby 1980, 84; Ruben 2003, 66). The idea is that because agents have a limited set of abilities, they normally face the problem of how to pool them together in order to achieve their desired results. Teleologically basic actions are those for which this kind of problem doesn't arise. As an agent, one doesn't have to figure out how, or by which means, to perform them.

Many philosophers have argued that the structure of language parallels the structure of thought.[1] Likewise, the means-end structure of action is a function of the structure of the agent's mind. This is why identifying basic actions matters. To the extent that they are the building blocks of agency, their identification provides a window into the constituents of the psychology that makes human performances count as actions. Basic actions are the behavioral manifestations of the most primitive ways in which agents think about what they do.

To illustrate, many action theorists believe there is a hierarchy of cognitive states that guides action. The hierarchy starts with high-level representations such as general policies and large-scale intentions, and bottoms out with representations in charge of the fine details of motor control (Mele 1992, 220–223; Pacherie 2006; Butterfill and Sinigaglia 2014). Within this scheme, basic actions play an important role. Identifying them helps distinguish the representations by which agents exert guidance from those by which sub-personal mechanisms exercise it for them. Being the simplest kinds of behaviors that the agent *directly* guides, basic actions mark the lower bounds of agency.

Aristotle famously held that deliberation had to stop somewhere (*EN* 1113a2). For him, the stopping point was the minor premise of the practical syllogism. The *dictum* is useful because it narrows down where to look for basic actions. But it doesn't settle the matter. You might need to do something as a means to achieve something you intend and yet not deliberate about how to execute your intention. Skilled typists, say, don't deliberate about how to type words or sentences. But typing a word or a sentence seems too "large" to count as basic for them.[2]

Arthur Danto, who introduced the topic to contemporary theorists, claimed that every agent intuitively knows which are her basic actions (1965, 145). This was a mistake. Even if agents have privileged knowledge of their actions, this privilege need not extend to the *structure* of their actions. An agent, therefore, might claim that she has the ability to do something basically and be wrong about it. Consider a parallel case. Certain epistemologies might rightly assert that some beliefs are basic in the sense of not originating from other beliefs. Yet, given that the origins of one's beliefs are not always transparent, it wouldn't follow that every thinker knows which are her basic beliefs.

Since Danto, various criteria for basic actions have been proposed. But the proposals have proven too controversial. Attempts to arrive at a criterion by induction from "intuitive" cases have been problematic because these intuitions hardly form a unified basis. Attempts to derive a solution from "first principles" quickly revert to long-standing debates about the nature of events, causation, and so forth before any significant progress is made (Baier 1971; Sandis 2010).

The criterion advanced here is meant to sidestep these problems. Being grounded on work in psycholinguistics, it offers an empirical standpoint to evaluate intuitions regarding basic actions. In addition, the criterion is neutral regarding the metaphysical positions that dominate the discussion.[3] Metaphysicians, therefore, can adopt it as a naturalistic hypothesis while they make progress on the more abstract aspects of their accounts.

2. The Proposal

The *slip* is a familiar kind of mistake. You set out to do something, but what you do by way of doing it is inappropriate, given what you intended and believed then. Think about spoonerisms. When Reverend Spooner spoke, language (not garble) came out of his mouth. The problem was that the things he said were usually in the wrong order.

Spoonerisms are just one example. In fact, there is a whole variety of slips. Some of them are lexical, such as calling your partner by the wrong name. Others, as we shall see later, are nonverbal, and even non-linguistic. All of them, however, exhibit the same general pattern. You mess up by doing some subordinate action in ways that are inappropriate in the light of your beliefs and of the overarching intention with which you act.

Upon casual inspection, it would seem that slips are random malfunctions. But the truth is that they happen in patterns shaped by the abilities of each individual agent. One important consequence, as we shall see, is that not all actions allow for slips. For each agent there are some slip-proof actions: actions such that she cannot slip in doing them. The actions are slip-proof because the agent doesn't need to do another *action* as means to perform them. Hence, there is no chance to mess up in the way characteristic of slips.

Slip-proof actions are, in this regard, basic. Or so I propose. In the slip, the intended action is broken down into its constituents. So, if you cannot slip in doing some action, a reasonable hypothesis is that it has no constituents. It is, therefore, basic for you.

3. Verbal Slips

Everyone makes slips. As a competent speaker, you intend to say one thing but wind up saying something else. It happens to Westerners and non-Westerners, and in spoken and signed languages. It has happened in ancient and modern times. Being a speaker, in short, is being a slipster.[4]

Slips come in different guises. Some involve substitutions, such as calling someone by somebody else's name. Others involve transpositions, for example, Spooner's memorable "Queer old Dean" (instead of "Dear old Queen"). And yet other slips involve omissions, as the Christian who asks God to "lead us into temptation." The possibilities would seem endless.

It would be a mistake, however, to think that verbal slips occur in random patterns. In fact, they are highly regular mistakes. Substitution slips, for instance, have lexical biases: you typically substitute intended words with real words. Further, the substituting words tend to be semantically related and of the same syntactic category. Consequently, most slips result in grammatically correct and meaningful utterances.

Importantly, some slips *never* happen. As an English speaker, you might slip and talk about a "Freudian flip," perseverating on the initial *f*. Hesitating whether to use "past" or "by," you might announce that you will "drop py" a friend's house. However, you will never be caught talking about a "Sreudian slip" (anticipating the *s*), or announcing that you will "drop bst" your friend's house (blending the two words). There is an intuitive explanation for this. Although anticipations and blends are common, initial /sr/ and /bs/ are not phonetically permissible in English.[5]

Psycholinguist Victoria Fromkin (1971, 1973) famously capitalized on these observations. Despite its fluent nature, she argued, speech is constituted by discrete linguistic units, which are sometimes substituted, transposed, or omitted. Some of them, like syntactic phrases or words, can be broken down, and some of their constituents rearranged or altogether omitted. Others, in particular phonemic segments, are basic.

At first glance, Fromkin's argument would seem simply to establish the well-known conclusion that language can be *analyzed* as having a recursive structure. But it goes deeper. It shows on the basis of behavioral data that speech acts are *produced* recursively from a set of atomic constituents. In speaking, as in language comprehension, there are psychologically real primitives.

4. Speaking Is Acting

Speaking is acting. Yet, judging by how seldom action theorists talk about speech, this would seem a forgotten truth. Arm-raising might be pervasive. But it does not beat speaking to being the paradigmatic case of agency.[6]

Of course, there are differences between verbal and nonverbal action. Speaking, for instance, is regimented by constraints absent in other forms of acting, say, the syntax of the language. Also, although agents differ vastly in the skills they have, there is nothing in the non-linguistic domain quite like the division of speakers of different languages. Likely, there is no "action organ."

Nevertheless, there are *deep* similarities. In the nonverbal domain there are slips too. You intend to do something and act on that intention. Yet, without changing your mind, you wind up doing something else. It happens to everyone: the young and the old. It happens in all sorts of places. Sometimes, slips occur in unusual circumstances. But, for the most part, they happen in circumstances where agents otherwise behave impeccably. Nonverbal slips are normal mistakes among competent agents.

Actually, these slips reveal the same kind of structure observed in their verbal counterparts. Some of them involve substitutions. You mess up a check in early January 2016 by dating it with 2015. Others involve transpositions. Rushing to clear the breakfast table, you store the milk in the cupboard and the cereal in the fridge. Yet other slips involve omissions, such as jumping into the shower with the glasses still on.[7]

Like their verbal counterparts, nonverbal slips also show a significant degree of regularity. Obviously, regularity is not measured here by rule compliance, but by compliance with some of the agent's *know-how* or *procedural knowledge*. Slips are, in this regard, displays of misdirected competence, not signs of incompetence. They rarely result in clumsy movements; typically they result in actions within one's behavioral repertoire.

Finally, rather than being brute accidents, slips are somewhat reasonable in the light of their overarching intention. This is evidently true in cases of omissions: you do everything you intended except for one thing. But it also holds when the mistake involves a substitution or a transposition. To wit, writing January 2015 on the check was not completely off the mark. Storing the milk in the cupboard and the cereal in the fridge was one way to clear the breakfast table.

5. Planning Mistakes

Consider a version of Wittgenstein's question: what is left over if I subtract the fact that my words are spoken from the fact that I spoke my words? A plausible answer is that after the subtraction at least two things are left: an intention to say something and a plan of how (or by which means) to say it.

The answer is part of a widely accepted view of the nature of action. On it, an action is the execution of an intention guided by a plan (Brand 1984; Mele and Moser 1994).[8] The execution need not be successful. A failed murder attempt might still count as an action. Nor does the plan need to be thought out in advance. Even with intentions formed ahead of time, the plans to execute them often involve a significant amount of improvisation. Think about having a conversation. You might know ahead of time the gist of what you intend to say. But in order to engage in real dialogue, you need to improvise along the way.

Plans guide the execution of intentions by assembling subordinate actions into larger wholes. Thus, they give actions the *means-ends structure* without which they would be mere sequences (Ferrero 2009; Bratman 2010). In doing so, plans serve as solutions to various practical problems that agents normally face having adopted an intention. One of them is the problem of *indeterminacy*: there is often more than one open course of action to achieve a given goal. Another is the problem of *serial ordering*: achieving a goal often involves sequencing and coordinating several actions.

Sometimes agents make mistakes solving these problems. Not all plans end up being properly aligned with the intentions they subserve. This is how slips come into existence. The plan that guides one's execution winds up not being suitable to achieve what one intends given one's beliefs. It has an incorrect step, the right steps are laid out in an incorrect order, or some of them are missing. Therefore, by acting on it, you wind up substituting, transposing or omitting one of the things you intended to do.

The mistakes happen because execution plans are often *cognitively underspecified*. Only a portion of the relevant information possessed by the agent at the time is actively brought to bear (Stemberger 1991b; Reason 1992). It is easy to see why this can happen. If you have too many things in mind or are acting under heavy time constraints, you might not be able to deliberate about all the details of your performance. Instead, you might have to follow certain automatic routines to fill in the details.

In the linguistic case, these routines are shaped by semantic and phonetic associations. This explains why slips involve words in the same semantic neighborhood and utterances that sound very much alike. In the non-linguistic domain, there are habits and well-rehearsed behavioral patterns, which is why slips look like competent performances. In general, both domains instantiate the same architectural feature of the human mind. In intention execution, exhaustive deliberation tends to be the exception. Automatic routines are the default rule.[9]

The downside is that such routines are geared towards efficiency. Semantic associations are a rough guide for lexical decision; habits are a decent stand-in for deliberation. Occasionally, however, the circumstances of action are such that the result of relying on these rough guides is behavior that fails to be a good match for one's intentions. Not being able to attend to all the aspects of one's performance, the mistake goes unedited.

6. Slip-Proof

Most everyday actions are complex: they are constituted by subordinate actions, orchestrated and assembled by plans. Therefore, their agents are susceptible to slip while performing them. Under the right circumstances, they can substitute, transpose, or omit one of the subordinate actions. This means that if an action is planned, the agent can slip in performing it. Put in the contrapositive, if an agent is not susceptible to slip, the action is not planned.

On the other hand, if an action is not planned, it is basic. Basic actions are those that an agent does directly, not by virtue of doing anything else. So, when it comes to them, there is no difference between intending the action and being settled on how to do it. There is no need to select from various open courses of action, or to sequence subordinate actions. The action, *qua* action, does not have an internal structure.

Thus, the criterion for basic actions follows. If the agent cannot slip, the action isn't planned and, therefore, is *basic for her*. Obviously, the action can figure in an agent's larger plan; basic actions are rarely done in isolation.[10] And the agent can slip performing some of the overarching actions of that plan. Yet, in performing something that is basic, there won't be a slip. Basic actions are slip-proof.

Clearly, the parallel argument does not hold for other kinds of planning mistakes, say, those due to ignorance or irrationality. Whereas there are no restrictions on how wrong beliefs can be, or principled limits on how much irrationality a mind tolerates, there are observed limitations to the ways humans slip.[11] This, at least, is what the evidence from psycholinguistics indicates.

Importantly, there is a rationale behind the evidence. One well-known characteristic of cognitively under-specified processes is their tendency to default to *contextually appropriate and well-rehearsed behaviors* (Reason 1990, 1992).[12] It is not surprising, then, that slips do not result in phonemically impermissible sounds. Articulating the phonemic segments of one's language is among the most rehearsed actions there is. The rehearsal starts even before one becomes a speaker: around their tenth month, infants start babbling in the language of their community (De Bysson-Bardies et al. 1984; Stoel-Gammon 1985).

Likewise, in the case of non-linguistic action, some action patterns are so well rehearsed that they are candidates for being slip-proof. The adult stride is likely one of them. After years of practice, raising one foot as soon the other hits the ground becomes an action unit. You don't do it by virtue of doing any other action. Nor do you need to solve the problem of how to do it. Thus, you might slip and walk to the fridge when you intended to walk to the cupboard to fetch a snack. Or you try to walk having "forgotten" that your foot is immobilized. But you won't slip by not alternating your stride. Walking, in this sense, is likely slip-proof.[13]

There is another (non-technical) sense in which walking is *not* slip-proof. If the floor is wet, if the last cocktail hit you hard, you might literally slip and fall. Internal or external conditions might genuinely prevent an agent from exercising her basic competences and skills. Similarly, speakers of a language might violate the phonemic constraints of their language, say, if they are extremely tired, or if they are asked to repeat very quickly a string of nonsense syllables (Mowrey and MacKay 1990; Pouplier 2003). But this is as it should be. Under normal conditions, basic actions are protected from certain kinds of mistakes but not from *all* sorts of failures.

What is true of uttering the phonemes of one's language might also be true about fist-clenching and other common examples. Or maybe not. One of the advantages of the criterion is that it can be *operationalized*: behaviorally speaking, slips are actions that their agents quickly recognize as falling short of what they intended. The recognition is typically accompanied by surprise. And the mistake can easily be corrected upon being noticed (Baars 1992; Poulisse 1999, ch. 2).

7. Skepticism

Given that actions theorists tend to countenance basic actions, an existence proof would seem unnecessary. Yet, Michael Thompson (2008, 107–112) denies their existence on the grounds that every action can be indefinitely divided and that each of its parts qualifies as an action in and of itself. Imagine an agent moving from point A to point C. Because the action takes time and occurs in space, Thompson argues, it could be divided into the movement from A to B and from B to C, where each of these movings is a means for the larger action and, hence, itself an action. Insofar as this kind of process of division can be iterated indefinitely, he concludes, there are no basic actions.

The argument, however, does not go through. It is, perhaps, true that every action can be indefinitely divided as Thompson envisions. Yet, from the fact that an action can be thus divided, it does *not* follow that it is so divided. Moving from A to B and from B to C *could* each be means for moving from A to C. But they *need not* be. There could be an agent for whom, because of her skills or training, the latter has become an *action unit*. In moving from A to C, she would obviously move through B. But *for her* the latter would not be a means for moving from A to C.

Consider again the case of speech production. Uttering a phoneme is evidently an action with parts: moving one's tongue, letting air out in coordination with the vibration of the chords, and so forth. In turn, these parts can be subdivided: the tongue is moved, say, beginning from the roof of the mouth to the soft palate to pronounce the German name *Bach*. Now, as anyone who has learned a new language late in life knows, one sometimes needs to do some of these things as means to pronounce new phonemes.

The situation of competent speakers, however, is not like this.[14] Although one could take these steps, one does not need to do it. At least, this is what the evidence from psycholinguistics reviewed here would seem to indicate. Speakers do not slip in uttering phonemes of their language, because for them uttering each of them is an action unit. From their perspective as fluent speakers, there are no means by which they utter them.[15]

The point can be put in a slightly different way. If an action is basic then, as far as the agent is concerned, there is no further action she needs to perform as a means. Thompson, however, objects that in the process of dividing the action of moving from A to C no *minimum sensibile* (his terminology) can be found: no point at which the division ceases to reflect how the agent structures her action. There is no such point or, if there is one, its location can only be arbitrarily settled. Either way, skepticism about basic actions ensues.

However, in the process of dividing speech acts into their constituent parts, there seems to be a natural stopping point. Under normal conditions, competent speakers do *not* slip in assembling individual phonemes out of the possible sounds they can otherwise voice. Due to years of intense rehearsal, they do not face the problem of figuring out how to cobble articulatory movements into those phonemes and, hence, they do not make the corresponding planning mistakes. In contrast, when they face the problem of assembling phonemes into larger wholes, there is the possibility of slipping. At supra-phonemic levels, speakers regularly substitute, transpose, or omit parts of the intended utterance.

8. Practical Knowledge

Many theorists appeal to the agent's practical knowledge to determine which parts of her actions can be regarded as means to perform them. Thus, they count as basic those actions such that their agents' knowledge of how to do them is basic. According to this view, knowing how to do an action is basic if it does not depend on the agent's knowing how to do some other action (Goldman 1970; Enc 2003; Smith 2010).

In a way, this is compatible with what I have been arguing so far. Slip-proof actions are normally represented by the agent in a plan for an overarching action but they are *never* the object of *further* planning. So, if knowing in a basic way how to do an action is knowledge without a further plan, the proposals would seem to agree.

The problem with the knowledge criterion is that it is unclear how to identify when a piece of knowledge is in fact basic. To illustrate, consider an argument by Jennifer Hornsby (2005). According to her, contrary to what I've urged here, voicing your thoughts in a sentence is normally basic for speakers. To do it, you do not need to exercise knowledge regarding how the vehicles that express those thoughts go together. Instead, once you know what thoughts you want to voice, sub-personal mechanisms select and assemble the vehicles that express those thoughts.

To argue for this, Hornsby appeals to what speakers seem to tacitly *know without observation*, as opposed to what they come to know by listening to themselves speak.[16] In fact, her argument is a *modus tollens* on the following conditional:

> If the speaker did exercise procedural knowledge of how to voice her thoughts, then, even if the procedure were something of which she was not explicitly aware, she would be in a position to know of it "without observation" as she spoke.
>
> (Hornsby 2005, 121)

According to Hornsby, the phenomenology of everyday speech shows that the consequent of the conditional does not hold. So, she denies the antecedent and concludes that voicing thoughts in a sentence is basic. For her, a speaker who becomes aware of the vehicles of her utterances does not do it attending to something already known by her without observation (Hornsby 2005, 122–123). If she does it at all, she does it in the capacity of listener of her own speech, for example, noticing ex post that she has said the wrong word.

There is evidence, however, that Hornsby is mistaken about this—at least, the phenomenological evidence she invokes might be misleading. One recurrent observation in naturalistic and controlled settings is that slips can actually be corrected before they happen, even before there is a motor manifestation of them. In the literature, these are known as micro-slips: slips corrected early enough not to count as mistakes (Reed and Schoenherr 1992). Sometimes speakers report micro-slips: you catch yourself about to say the wrong word, or about to make a spoonerism. But they too can be inferred from overt corrections that seem to be anticipations of the slip.

Micro-slips are evidence of non-perceptual editing exercises that operate prior to processes involved in phonation and voicing (Motley et al. 1981; MacKay 1987; Wheeldon and Levelt 1995; Postma 2000). Further, the existence of such exercises is evidence that, contrary to what Hornsby argues, voicing your thoughts in a sentence is not basic, at least not according to the criterion proposed here. Awareness of the micro-slip normally involves an exercise of procedural knowledge at the level of the vehicles of those thoughts. Or, in the terms used earlier, it involves a plan in which the agent, *not merely* some sub-personal mechanism, represents the utterings of elements of the sentence.

Obviously, Hornsby's is not the only knowledge-based criterion for basic actions. But her proposal illustrates a problem that such accounts are likely to face. If one wishes to sort out which actions are basic by appealing to the criterion that one's knowledge of how to do them is basic, one had better have a way of determining when knowledge is basic. Here, as in other cognitive domains, appeals to phenomenology can be highly misleading.

In contrast, because the detection of slips can be operationalized, the present proposal has the resources to accommodate the problem. Slips mark the joints of plans, signaling the lower bounds of the agent's planning. Thus, to the extent that basic knowledge and basic planning coincide, the empirical criterion offered here helps also to identify instances of basic knowledge. In short, basic knowledge is procedural knowledge of how to perform slip-proof actions.

9. Bodily Movements

Most action theorists agree that basic actions (except for pure omissions and mental actions) are restricted to bodily movements. Some theorists think that, insofar as exercises of agency involve a person causing an event, actions are those events caused (Davidson 1971; Smith 2010). Others insist that actions, rather than being events, are the causings of those events (Hornsby 1980; Dretske 1988). Accordingly, some claim that basic actions are bodily motions; others claim that they are causings of bodily motions.

Regardless, the essential contrast is between moving one's body and the effects the resulting motions have beyond the surface of one's skin. And the point of the contrast is to isolate the contribution of the agent from the contribution that the world, so to speak, makes. Isolating the former is a way of picking out basic actions. The latter are supposed to occur within the limits of one's skin.

Given what has been said so far, I would seem to agree. Taking speech as a paradigmatic case of agency invites one to locate the search for basic actions within the bounds of the skin. For one thing, speaking is done with your body without contributions from the world other than those that allow for bodily control. Further, uttering phonemes in one's language seems to be a good example of what would be a pure moving of one's body.

It would be a mistake, however, to restrict *all* slip-proof actions to mere bodily movements. Simply, the generalization is not empirically supported. Whereas the restriction holds in the verbal domain, in some nonverbal instances the evidence seems to indicate that what is basic for an agent (that is, slip-proof) goes beyond the limits of her body.

To see this, consider skilled typists. As has been observed, typing slips among experts (typos) always involve the wrong key being struck by a finger appropriate to that key. Most often they involve hitting the key horizontally or vertically adjacent to the intended one: typing mi*a*take or mi*x*take, instead of mistake. Less often, they consist of homologous finger substitutions: typing *d* instead of *k* (Lessenberry 1928; Rosenbaum 2010, ch. 9).

One kind of mistake not observed, however, is finger-to-key confusions. That is, if you are an expert (roughly, if you do not hunt-and-peck) and make a typo by hitting *w* instead of *e*, you would *not* hit the *w* with your middle finger. You would do it rather with your ring finger. Due to years of practice each key becomes associated with one particular finger. So, having figured out which key to hit, you do not face the additional problem of deciding which finger to use.[17]

These observations show, first, that typos do not occur as common sense has it. It is not as though the finger is moved too far to the left or the right, too high or low. Rather, the mistake occurs at the level of key selection. More important, the observations provide evidence that keystrokes are for skilled typists what phonemic segments are for competent speakers. One substitutes, transposes, or omits them, but one doesn't normally err in producing them, say, by confusing which key goes with which finger. The evidence indicates, in other words, that what is basic for skilled typists are not finger movements but keystrokes.

It would be surprising if typing were the exception. It is well known that with repeated use tools become "transparent" to their users—extended cognition theorists have provided various examples.[18] And it might turn out that using the tool, not moving one's body in such and such a way, is what's basic in some of these cases. Think about moving the cursor on your screen. Or consider people who regularly use prosthetic limbs or other extensions of their body. After a period of adaptation, for instance, leg amputees report having to consciously think about the position of their limbs before getting up, yet once in motion being able to "just walk."[19]

10. Conclusion

In this chapter, I've put forward a criterion for identifying basic actions. The criterion generalizes results from psycholinguistics in the light of a philosophical theory of action. In addition, it helps defuse skepticism about basic actions, illuminates discussion of basic knowledge, and shows why it might be a mistake to restrict the search for basic actions within the limits of the agent's body. Overall, the proposal illustrates how much can be gained by thinking systematically about certain everyday mistakes and by taking seriously the idea that speaking is a paradigmatic way of acting.[20]

Notes

1. Classic explanations for this structural parallelism can be found in Sellars (1956), Davidson (1975), and Fodor (1975).
2. To be fair, Aristotle recognized that skill drives out deliberation. And in some cases acquiring a skill renders some complex actions basic. But this is not always the case. The acquisition of a "high-level" skill might make it easier for an agent to cobble together basic actions, without changing what's basic for her.
3. Thus, I remain neutral in the discussion between unifiers (Anscombe 1963; Davidson 1971) and multipliers (Goldman 1970) regarding the interpretation of the accordion effect. The present proposal can be rephrased in terms of actions being slip-proof under a description.
4. Slips have been documented in various languages: Arabic (Abd-El-Jawad et al. 1987), Dutch (Cohen 1966), French (Rossi et al. 1998), German (Meringer 1908), Mandarin (Chen 1993), and Spanish (Viso et al. 1991); in signed languages (Klima and Bellugi 1979; Hohenberger et al. 2002); and among children and bilinguals (Poulisse 1999; Jaeger 2005).
5. Rulon Wells III was thus inspired to formulate his *First Law of slips*: "A slip of the tongue is practically always a phonetically possible noise" (1951, 86). See also Shattuck-Hufnagel (1983), Dell (1986), and Meyer (1992). Dissenting voices are Mowrey and MacKay (1990) and Pouplier (2004).
6. Exceptions include Searle (1983) and Hornsby (2005).
7. For discussion of slips, see Norman (1981), Reason and Mycielska (1982), Sellen (1990), Jónsdóttir et al. (2007), and Amaya (2013).
8. The notion of plans adopted here comes from Lashley (1951), Miller et al. (1963), and Brand (1984). It differs somewhat from Bratman's (1987) view of plans as structures for coordinating *future-directed* intentions.
9. For accounts of slips along these lines, see Norman and Shallice (1986), Reason (1990), Baars (1992), and Sellen and Norman (1992).
10. An action can be part of a plan without requiring from the agent *further* planning. This is the case if the action is a constituent of a larger whole, but its execution does not require further subordinate actions.
11. Normative principles, such as the principle of charity, might provide some limits. But the limitations apply to patterns of actions or belief systems, not to individual actions or beliefs.
12. In lexical slips, for instance, more frequent elements are less likely to be affected by error, and less frequent elements tend to be replaced by more frequent ones (Dell 1990; Stemberger 1991a).
13. Of course, walking for a given distance is not basic; it is a complex action made up of tokens of a basic action unit, the alternating stride. Experiments with split-belt treadmills provide further evidence that the stride pattern is a

non-breakable unit (Reisman et al. 2005; Choi and Bastian 2007). In them, human subjects have been shown to adapt to different patterns of locomotion and exhibit adaptation aftereffects. However, they always maintain the pattern of alternating steps with no airborne periods.

14. As a competent speaker, one might also break down the uttering of phonemes into smaller actions, say, if one is trying to teach someone else the correct pronunciation. This does not show that uttering phonemes is not basic for the speaker. It just shows that there are alternative (non-basic) ways for her to utter phonemes.

15. It is hard to say how exactly the proposed criterion would work out in Thompson's example, given how schematic the latter is. But the idea would be that if moving from A to C were basic, one would not observe the mover slip, say, by moving from A to B* instead of A to B. Of course, the agent could move through B* out of clumsiness, but that's a different kind of failure.

16. For the notion of knowledge without observation, I follow Hornsby:

    if she (the agent) were to attend to, or to reflect upon, what she is doing, then it is something she could find herself doing, and in finding herself doing it, she would not need to make observations of the sort that a spectator might make. (Hornsby 2005, 121)

17. Grudin (1983) was the first to report this observation, which played a major role in validating computational models of typing (e.g., Rumelhart and Norman 1982) that assumed finger movements to be circumscribed by the keys associated with each finger.

18. The terminology of tools being transparent comes from Norman (1999). For discussion in relation to extended cognition, see Clark (2008).

19. For discussions of adaptations to prosthetic limbs, see Fraser (1984) and Murray (2004, 968). For adaptation in laparoscopic interventions by expert surgeons, see Verwey et al. (2005) and Wilson et al. (2010).

20. Versions of this chapter were presented at the Time and Agency conference at George Washington University in November 2011 and at the 2013 Pacific Division Meeting of the American Philosophical Association. I would like to thank Roman Altshuler, Andrei Buckareff, Valentina Cuccio, Luca Ferrero, Kim Frost, Steven Gross, John Heil, Kirk Ludwig, Michael Pauen, Michael Sigrist, Roy Sorensen, David Velleman, Eric Wiland, and Jeff Zacks for insightful comments and helpful discussion.

References

Abd-El-Jawad, H., & Abu-Salim, I. (1987) Slips of the tongue in Arabic and their theoretical implications. *Language Sciences*, 9(2), 145–171.

Amaya, S. (2013) Slips. *Noûs*, 47, 559–576.

Anscombe, G.E.M. (1963) *Intention*. Ithaca, NY: Cornell University Press.

Baars, B. (1992) The many uses of error: Twelve steps to a unified framework. In B. Baars (Ed.), *Experimental slips and human error* (pp. 3–37). New York: Plenum Press.

Baier, A. (1971) The search for basic actions. *American Philosophical Quarterly*, 8(2), 161–170.

Brand, M. (1968) Danto on basic actions. *Nous*, 2, 187–190.

—— (1984) *Intending and acting: Toward a naturalized action theory*. Cambridge, MA: MIT Press.

Bratman, M. (1987) *Intention, plans, and practical reason*. Cambridge, MA: Harvard University Press.

———— (2010) Agency, time, and sociality. *Proceedings of the American Philosophical Association*, 84(2), 7–26.

Butterfill, S., & Sinigaglia, C. (2014) Intention and motor representation in purposive action. *Philosophy and Phenomenological Research*, 88, 119–145.

Chen, J. (1993) A corpus of speech errors in Mandarin Chinese and their classification. *World of Chinese Language*, 69, 26–41.

Choi, J., & Bastian, A. (2007) Adaptation reveals independent control networks for walking. *Nature Neuroscience*, 10, 1055–1062.

Clark, A. (2008) *Supersizing the mind: embodiment, action, and cognitive extension*. Oxford: Oxford University Press.

Cohen, A. (1966) Errors of speech and their implication for understanding the strategy of language users. *Zeitschrift für Phonetik*, 21, 177–181.

Danto, A. C. (1965) Basic actions. *American Philosophical Quarterly*, 2(2), 141–148.

Davidson, D. (1971) Agency. In *Essays on actions and events* (pp. 43–61). Oxford: Clarendon Press.

———— (1975) Thought and talk. In S. Guttenplan (Ed.), *Mind and language* (pp. 7–23). Oxford: Oxford University Press.

De Boysson-Bardies, B., Sagart, L., & Durand, C. (1984) Discernible differences in the babbling of infants according to target language. *Journal of Child Language*, 11(1), 1–15.

Dell, G. S. (1986) A spreading-activation theory of retrieval in sentence production. *Psychological Review*, 93(3), 283–321.

———— (1990) Effects on frequency and vocabulary type on phonological speech errors. *Language and Cognitive Processes*, 5, 313–349

Dretske, F. (1988) *Explaining behavior: Reasons in a world of causes*. Cambridge, MA: MIT Press.

Enc, B. (2003) *How we act: Causes, reasons, and intentions*. Oxford: Clarendon Press.

Ferrero, L. (2009) What good is a diachronic will? *Philosophical Studies*, 144(3), 403–430.

Fodor, J. (1975) *The language of thought*. Cambridge, MA: Harvard University Press.

Fraser, C. (1984) Does an artificial limb become part of the user? *British Journal of Occupational Therapy*, 47, 43–45.

Fromkin, V. (1971) The non-anomalous nature of anomalous utterance language. *Language*, 47, 27–52.

———— (1973) Introduction. In V. Fromkin (Ed.), *Speech errors as linguistic evidence* (pp. 11–45). The Hague: Mouton.

Goldman, A. (1970) *A theory of human action*. Englewood Cliffs, NJ: Prentice-Hall.

Grudin, J. (1983) Error patterns in novels and skilled typists. In W. E. Cooper (Ed.), *Cognitive aspects of skilled typewriting* (pp. 121–143). New York: Springer-Verlag.

Hohenberger, A., Happ, D., & Leuninger, D. (2002) Modality-dependent aspects of sign language production: Evidence from slips of the hands and their repairs in German sign language. In R. Meier, K. Cormier & D. Quinto-Pozos (Eds.), *Modality and structure in signed and spoken languages* (pp. 112–142). Cambridge, MA: Cambridge University Press.

Hornsby, J. (1980) *Actions*. London: Routledge and Kegan Paul.

———— (2005) Semantic knowledge and practical knowledge. *Proceedings of the Aristotelian Society*, 79, 107–130.

Jaeger, J. (2005) *Kid's slips: What young children's slips of the tongue reveal about language development*. Mahwah, NJ: Lawrence Erlbaum.

Jónsdóttir, M., Adólfsdóttir, S., Cortez, R. D., Gunnarsdóttir, M., & Gústafsdóttir, H. (2007) A diary study of action slips in healthy individuals. *Clinical Neuropsychologist*, 21(6), 875–883.

Klima, E., & Bellugi, U. (1979) *The signs of language*. Cambridge, MA: Harvard University Press.

Lashley, K. S. (1951) The problem of serial order in behavior. In. L. Jeffress (Ed.) *Cerebral mechanisms in behavior*. New York: John Wiley & Sons.

Lessenberry, D. (1928) *Analysis of errors*. Syracuse, NY: L.C. Smith and Corona Typewriters.

MacKay, D. G. (1987) *The organization of perception and action: A theory for language and other cognitive skills*. New York: Springer-Verlag.

Mele, A. (1992) *Springs of action*. New York: Oxford University Press.

Mele, A. R., & Moser, P. K. (1994) Intentional action. *Nous*, 28(1), 39–68.

Meringer, R. (1908) *Aus dem Leben der Sprache*. Berlin: B. Behr's Verlag.

Meyer, A. S. (1992) Investigation of phonological encoding through speech error analyses: Achievements, limitations, and alternatives. *Cognition*, 42(1–3), 1–3.

Miller, G., Galanter, E. & Pribram, K. (1963) *Plans and the structure of behavior*. New York: Holt, Reinhart & Winston.

Motley, M., Baars, B., & Camden, T. (1981) Syntactic criteria in prearticulatory editing: Evidence from laboratory induced slips of the tongue. *Journal of Psycholinguistic Research*, 10, 503–522.

Mowrey, R., & MacKay, I. (1990) Phonological primitives: Electromyographic speech error evidence. *Journal of the Acoustical Society of America*, 88(3), 1299–312.

Murray, C. (2004) An interpretative phenomenological analysis of the embodiment of artificial limbs. *Disability and Rehabilitation*, 26(16), 963–973.

Norman, D. (1981) Categorization of action slips. *Psychological Review*, 88, 1–15.

——— (1999) *The invisible computer*. Cambridge, MA: MIT Press.

Norman, D., & Shallice, T. (1986) Attention to action. In R. Davidson, G. Schwartz & D. Shapiro (Eds.), *Consciousness and self-regulation* (pp. 1–18). New York: Plenum Press.

Pacherie, E. (2006) Towards a dynamic theory of intentions. In S. Pockett, W. P. Banks & S. Gallagher (Eds.) *Does consciousness cause behavior?* (pp. 145–167). Cambridge, MA: MIT Press.

Postma, A. (2000) Detection of errors during speech production: A review of speech monitoring models. *Cognition*, 77, 97–131.

Poulisse, N. (1999) *Slips of the tongue: Speech errors in first and second language production*. Amsterdam: Benjamins.

Pouplier, M. (2003) Units of phonological encoding empirical evidence. Thesis (Ph.D.), Yale University.

Reason, J. (1990) *Human error*. Cambridge: Cambridge University Press.

——— (1992) Cognitive underspecification: Its variety and consequences. In B. Baars (Ed.), *Experimental slips and human error* (pp. 71–91). New York: Plenum Press.

Reason, J., & Mycielska, K. (1982) *Absent-minded? The psychology of mental lapses and everyday errors*. Englewood Cliffs, NJ: Prentice-Hall.

Reed, E., & Schoenherr, D. (1992) The neuropathology of everyday life: On the nature and significance of microslips in everyday life. Unpublished manuscript.

Reisman, D., Block, H., & Bastian, A. (2005) Interlimb coordination during locomotion: What can be adapted and stored? *Journal of Neurophysiology*, 94, 2403–2415.

Rosenbaum, D. A. (2010) *Human motor control*. Amsterdam: Elsevier.

Rossi, M., & Peter-Defare, E. (1998) *Les lapsus: Ou comment notre fourche a langué.* Paris: Presses Universitaires de France.

Ruben, D. H. (2003) *Action and its explanation.* Oxford: Oxford University Press.

Rumelhart, D., & Norman, D. (1982) Simulating a skilled typist: A study of skilled cognitive-motor performance. *Cognitive Science*, 6, 1–36.

Sandis, C. (2010) Basic actions and individuation. In T. O'Connor & C. Sandis (Eds.), *A companion to the philosophy of action* (pp. 10–17). Oxford: Wiley-Blackwell.

Searle, J. (1983) *Intentionality: An essay in the philosophy of mind.* Cambridge: Cambridge University Press.

Sellars, W. (1956) Empiricism and the philosophy of mind. *Minnesota Studies in the Philosophy of Science*, 1, 253–329.

Sellen, A. J. (1990) Mechanisms of human error and human error detection. Thesis (Ph.D.), University of California San Diego.

Sellen, A., & Norman, D. (1992) The psychology of slips. In B. Baars (Ed.), *Experimental slips and human error* (pp. 317–339). New York: Plenum Press.

Shattuck-Hufnagel, S. (1983) Sublexical units and suprasegmental structure in speech production planning. In P. MacNeilage (Ed.), *The production of speech* (pp. 109–136). New York: Springer.

Smith, M. (2010) The standard story of action: An exchange. In J. Aguilar & A. Buckareff (Eds.), *New perspectives on the causal theory of action* (pp. 45–55). Cambridge, MA: MIT Press.

Stemberger, J. (1991a) Apparent anti-frequency effects in language production: The addition bias and phonological underspecification. *Journal of Memory and Language*, 30(2), 161–185.——— (1991b) Radical underspecification in language production. *Phonology*, 8(1), 73–112.

Stoel-Gammon C. (1985) Phonetic inventories, 15–24 months: A longitudinal study. *Journal of Speech and Hearing Research*, 28(4), 505–12.

Thompson, M. (2008) *Life and action.* Cambridge, MA: Harvard University Press.

Verwey, W., Stroomer, S., Lammens, R., Schulz, S., & Ehrenstein, W. (2005) Comparing endoscopic systems on two simulated tasks. *Ergonomics*, 48(3), 270–287.

Viso, S., Igoa, J., & García-Albea, J. (1991) On the autonomy of phonological encoding: Evidence from slips of the tongue in Spanish. *Journal of Psycholinguistic Research*, 20(3), 161–185.

Wells, R. (1951) Predicting slips of the tongue. *Yale Scientific Magazine*, 26, 9–30.

Wheeldon, L., & Levelt, W. (1995) Monitoring the time course of phonological encoding. *Journal of Memory and Language*, 34, 311–334.

Wilson, M., Vine, S., McGrath, J., Brewer, J., Defriend, D., & Masters, R. (2010) Psychomotor control in a virtual laparoscopic surgery training environment: Gaze control parameters differentiate novices from experts. *Surgical Endoscopy and Other Interventional Techniques*, 24(10), 2458–2464.

3 The Antinomy of Basic Action

Kim Frost

1. Basic Action and Practical Atomism

At what point does the will get involved in changing the world? If one takes the wording of the question seriously, and believes that the will gets involved in changing the world *immediately*, at some *point*, then one believes in basic actions.[1]

Basic action theories take many forms, but they all agree that there are actions (or for Davidsonians, descriptions of actions) that are distinctively *immediate*. Theories are individuated by the kind of mediation in question. Some say the mediation is causal (basic actions *cause* non-basic actions, and are not caused by other actions); others say it is instrumental (basic actions are *means* to non-basic actions, and are not performed by other means); yet others say it is psychological (basic actions are, for example, *thought of* as means to non-basic actions, and are not themselves thought of as requiring further means). Theories also differentiate themselves in non-essential ways with claims about how to recognize basic actions in the wild, independently of their definition: it is variously claimed that they are volitions, or neural events, or bodily movements, or (graceful skilful) things done "all in one go."[2]

Basic action theorists think that the will *must* get involved in changing the world through something immediate: non-basic actions cannot come to be except via the mediation of some basic action(s). The reason for this necessity is often left opaque but is usually associated with a vicious regress suited to the kind of mediation in question. As Danto claims, it couldn't be that *every* action is caused by another action of the same agent. Or as Davidson claims, it couldn't be that *every* true description of an action under which it is intentional makes reference to a contingently achieved effect of the action. Or as Hornsby claims, it couldn't be that an agent has an *infinite* number of beliefs about her means to her ends, or has an *infinite* number of "items of knowledge" about how to achieve her end. To the extent that the viciousness of the regresses is left implicit, as it usually is, the question of exactly why causes, or thoughts of contingent effects, or beliefs about means, or "items of knowledge," can't go on forever is left to the reader's imagination.

Belief in basic action is usually accompanied by commitment to *practical atomism*. Practical atomists believe that all non-basic actions are caused by, or constituted by, or composed of (or otherwise essentially mediated by) basic actions.[3] This promises a simplified account of agency. For a practical atomist, practical *thought* of non-basic action mirrors the metaphysics of the case, whereby non-basic "molecules" are grounded in relations with, or between, basic "atoms." When an agent does A by doing B (intentionally), and B is a basic action, the agent thinks that, for example, B *causes* A (mirroring causation), or that to B *is* to A (mirroring constitution), or that B *is a part of* A (mirroring composition). Under the supposition that all basic actions are intentional, once we have an account of the metaphysically fundamental relationship between an agent and her basic actions, then the non-basic actions will fall into place with some minimal theory-jiggling, exploiting the kind of thoughts and relations described (and perhaps some extra conditions of knowledge, skill, or reliability) to extend the status of intentional action from the basic to the non-basic case.

Belief in basic action, and the kind of practical atomism that usually accompanies it, is an orthodox position in contemporary philosophy of action. Given this, it would be remarkable if a simple argument could show that belief in basic action is deeply mistaken, and so that the metaphysical foundations of much contemporary philosophy of action are rotten. Later I shall present an argument that purports to do just that.

For ease of exposition, I will assume that an *instrumental* conception of basic action is the most promising, and will mean *instrumentally* basic action by the term "basic action." Basic actions are things an agent does intentionally without doing them *by* doing anything else intentionally as means to that end. I will make no further assumptions about how to recognize basic actions in the wild.

2. Thompson's Argument

Thompson (2008) has a simple argument against the very idea of basic action that relies on the continuity of time. Consider an intentional movement from A to C as an example of a supposedly basic action. (We choose an action that takes time, so as to address Thompson on his strongest ground.) The action takes time, so it has parts that take time. These parts could intelligibly be *rationalized* by the whole. One could intelligibly ask of someone moving from A to C, "Why are you moving from A to B?", where B is a point that is halfway to C, provoking a rationalization such as "because I am moving from A to C." If so, then one could ask the same kind of question of the movement halfway to B, provoking a similar rationalization in terms of moving to B, C, or some point beyond C. Thompson notes that in the minimal cases such questions would be rather strange "conversationally speaking," but could be seen as less strange under the supposition that the tiny movement is all of the action that the questioner can see. Thompson's conclusion is that we have no reason to deny that the parts discussed are intentional actions. After all, they are distinct from the action of which they are a part, yet they serve the whole as compositional means, and they are rationalized in the same way that more obvious "macro-cases" of compositional means are rationalized, and they too can serve as rational ground for still smaller parts. So there are further rational grounds, all the way down, and the supposedly "basic" action is not basic after all. Whenever one performs an intentional action one performs an *infinite* number of component intentional actions. That is: there are no basic actions, only *more basic* ones.

3. The Antinomy of Basic Action

We seem to have wandered into an antinomy: what I call the *antinomy of basic action*. On the one hand we have the basic action theorists with their (often inchoate)[4] horror of the infinite, and on the other we have Thompson with his (barely repressed)[5] disdain for the immediate. Yet if there is something absurd in the idea that the will engages with the world at an immediate point, there is also something absurd in the idea that *any* change that is formally subsumed by a change the agent has in mind to make is *thereby* a bona fide intentional action in its own right. Let me reflect on these apparent absurdities by way of motivating subsequent discussion.

Recall that the practical atomist requires basic actions to be *intentional*, so that they can form the foundation of a practically atomistic account of intentional agency. Intentional action carries with it the implication of *success*—doing something one has in mind to do. Accordingly, basic action is a *perfect* case of the will getting to grips with the world, even if only in an action that is small and limited in scope. If there is an *imperfect* case of the will getting to grips with the world, due to false beliefs, lack of grace or skill, or contingent interference, then either the thing that is imperfect is a non-basic (unintentional) action, or else *there is no action at all* (that is, something may happen, but the will does not express itself, whether perfectly or imperfectly, in action).[6]

The requirement that basic actions be intentional, and so successful, puts pressure on the practical atomist to locate basic action in some supposedly safe realm free from the possibility of contingent perversion, interruption or interference *during* the action. This puts pressure on the practical atomist to say that basic actions are somehow *temporally immediate*.[7] If basic actions took time to resolve into successes or failures, then they could be diverted from their intended course, and there would be "basic actions" that were *not* intentional, in that they would not be what the agent had in mind to do. But if the practical atomist submits to these pressures, then basic actions won't be recognizable as *actions*. For Thompson is surely right that actions, at least in the central cases, take time to resolve. We need different words to talk about things that don't take time to resolve (such as instantaneous ingressions, achievements, and accomplishments). This pole of the antinomy seems absurd because there is pressure to say that basic actions are, and are not, actions (that they do, and do not, take time to resolve). (Note that the pressures are a product of the theoretical goals of practical atomism. Later we will consider an alternative that gives up on these theoretical goals.)

What seems absurd about Thompson's pole of the antinomy is that it mistakes *formal play* for our *actual practice* of rationalization, where we offer *particular* reasons for what we do. It is surely not *merely* conversationally speaking that Thompson's rationalizations concerning tiny geometrically identified movements are odd: they seem to bridge the gap that exists between our actual practice of rationalization and (somewhat nerdy) geometrical banter. (Why is the chicken moving half a trillionth of the way across the road? To get a trillionth of the way across the road. Supposing the chicken sapient, this is a *paradigm* of action-rationalization for Thompson.)

If our aim is to avoid both absurdities, we should start with Thompson. Thompson's position is dialectically stronger because his regress serves to make his point *explicitly*. The support that the basic action theorist's regress offers for her conclusion is only as strong as the case made for the *viciousness* of the regress, and this viciousness is usually left implicit, so it is unclear whether the regress constitutes a *reductio* or is simply a feature of the case that we ought to accept. The next two sections respond to Thompson's argument. In responding to Thompson, a positive view emerges that explains what is right, and wrong, about Thompson's view, and what is right, and wrong, about traditional basic action theory.

4. Thompson's Argument Is Inconclusive

The obvious claim to reject in Thompson's regress is that the relevant parts are intentional actions just as the encompassing "basic" action is supposed to be. In what follows I assume that the appropriateness of Anscombe's special question "Why?", understood as a demand for reasons for action, delimits the domain of intentional action. If the question is not applicable to something the agent is doing then that thing is not an intentional action. If Thompson's argument is sound, then the question "Why are you moving from A to B?", understood as a demand for reasons for action, must have application to the movement from A to B, no matter how small the part (and movement) described happens to be.

Anscombe says that her special question "Why?" is denied application when the agent is unaware of whatever is asked about. When we ask for reasons for action we ask after something the agent has in mind to do: if she is completely unaware of the aspect in question, then obviously she doesn't have that in mind at all. But Thompson's agent probably is aware of going halfway (quarter-way, etc.) in some sense of "aware." Every minimally educated subject knows that one must go halfway if one goes all the way. Even those who are not *au fait* with the concept of division might see or imagine that one goes halfway (*this* far) when one goes all the way (*that* far).

That said, it makes a difference *how* the agent is aware of moving to the halfway point. Sometimes one may observe or infer that one is centrally involved in some change, whilst also truly claiming that one does not intend to do *that*, and does not do *that* intentionally, and does not have any reasons for doing *that*. Anscombe thinks the kind of *practical* awareness of what one is doing that is relevant to demands for reasons for action is so different from other ways of knowing what one is doing that her special question "Why?" is denied application when the agent must rely on observation or inference in order to answer it.[8] This claim is tied to Anscombe's conception of *practical knowledge* as distinct from theoretical knowledge. There is a puzzle about whether practical knowledge is knowledge of what one is actually doing or only knowledge of what one has in mind to do (whether or not one is actually doing *that*). We can dodge the puzzle, for we only need the weaker claim. We need not decipher Anscombe's views about practical knowledge in order to appreciate that it is not by observing or inferring that an agent knows what she has in mind to do in the way that is relevant to demands for reasons for action. Even if I need a moment to become alert and find the words to express what I have in mind to do as I drive (quite habitually, with a minimum of care and attention) to work, I do not *look* at my current movements, nor *infer* from some premise ("Well it's 8.50 a.m. on a Monday morning . . .") in order to work out that what I have in mind to do is to drive to work. The same is true of all those actions undertaken as compositional means to the end of driving to work: taking the shortcut, speeding through the stop sign, and so forth.

To capture our subject matter by stipulation, let us say that if the agent has it in mind to do something, so that demands for reasons for action apply to what she has in mind to do, then the agent *intends* to do that. The agent intentionally driving to work intends to drive to work, intends to take the shortcut, and so forth, and has a special practical awareness of what she has in mind to do. We may ask the driver many questions about her driving, but it is inappropriate (or anyway, missing her point) to ask her why she is dribbling snot out of one nostril (unless she intends to be gross).[9]

The stipulation is sufficient to show that Thompson's argument is inconclusive. Although the agent in question intends to move from *A* to *C* and does so intentionally, and although she probably knows there is some halfway point *B* on the way to *C*, we do *not* know *how* the agent knows this, or whether she *intends* to move from *A* to *B*. If the agent knows about the halfway point *only* by means of observation or inference, then (following Anscombe) she is not *practically* aware of the halfway point. Thompson's argument is inconclusive because he has not shown that the agent knows about the halfway point otherwise than by observation or inference.

One might object that the agent who intends to move from *A* to *C must* intend to move from *A* to *B*. After all, (halfway rational) agents must intend the *necessary means* to their ends, and going halfway is a necessary compositional means to going all the way. So it would be incredible to suppose that an agent who intends to move from A to C might not intend to move from A to B.

The objection fails because it needs to be shown that the movement from A to B is a necessary *means* rather than a mere necessary *aspect* (part, presupposition, consequence) of what is done. Even granting that agents must intend the necessary means to their ends, agents need not intend all the necessary aspects of something they intend to do.

Suppose I am pounding a nail into a board to fix it to a post, and that there's no way in the circumstances to hammer without making noise, and that I know this, and that a would-be questioner cannot see *any* of the movement I am making, but can hear the godawful racket. Just as Thompson's question "Why are you moving from A to B?" makes sense when the questioner cannot see all of what the agent is doing, it would be perfectly natural, "conversationally speaking," for the questioner in the hammering case to ask "Why are you making that godawful racket?", as if that were something that I really meant to be doing. But I don't conceive of making a godawful racket as promoting any of my ends, nor do I conceive of it as desirable in any particular way (as manifesting justice, etc.). I don't have a *reason* for making a godawful racket (not even the null reason: "I just thought I would"). Making a godawful racket is accidental to my intent, although it is something I am doing, and it is something I *must* do if I am to hammer. The fact that it is a necessary aspect of what I am doing (perhaps even one I *must* know about) does not show that it is something I do *intentionally*, or that it is a necessary means to some end of mine, or that it guides or delimits my activity in any interesting way.

The hammering case exhibits systematic similarities with the relevant tiny movements from A to B. When I respond to the question "Why are you making a godawful racket?" by saying "because I am hammering in a nail," I do not rationalize what was asked about. Such a response does not use the "because" of rationalization, but rather the "because" of (something like) mere efficient causal explanation. My response provides a *starting point* for a conversation about reasons for action, rather than providing a reason for action, because the questioner has latched onto an aspect that is accidental to the *particular* instrumental order I have in mind to pursue. Similarly, the agent who responds to the question "Why are you moving from A to B?" with the answer "because I'm moving from A to C" need not rationalize what was asked about. Such a response could use the "because" of (something like) mere *formal* causal explanation, instead of the "because" of rationalization, exploiting the ratio 2:1 to explain why she goes halfway to C.

What seems to go missing in Thompson's argument is the connection between intentional action and having *particular reasons* for what one does intentionally. The hammerer is *practically indifferent* to making a godawful racket—he neither intends to make one, nor intends not to—because he has no particular reason to make a godawful racket. (Should he realize he is pissing off the neighbors thereby, he might then *intend* to do so, for the reason that they deserve it.) To the extent that it is up to the hammerer whether or not he intends to make a godawful racket, so too might an agent moving from here to there think that it is up to *her* whether or not moving to any particular point short of her final destination is an intentional movement or not. And in some cases, particularly for short or arbitrarily chosen trajectories, she may not see any particular reason to think that a movement from A to B is an *intentional* movement of hers. She can say (truthfully) that she is practically indifferent to the movement, because it is open to her to treat the movement as one of the many aspects of the case—such as neurons firing, muscle contractions, the precise placement of her foot, accompanying noises, and movement of air that is in the way—that in some sense take care of themselves.

So: Thompson's argument is inconclusive. We have no obvious reason to think that particular practical thoughts (i.e., the intention to A) *must* reach down to further particular practical thoughts (i.e., the intention to B, where B is a proper part of A).

5. Thompson's Conclusion Is False

Intentional actions need not be *perfectly* articulated in their parts. Someone who intentionally crossed the Sahara may have unintentionally fallen and rolled down a sand dune during the action, without thereby ruining the intentionality of the action as a whole. It does not matter when or where the accidents occur so long as they aren't overwhelming: people get off on the wrong foot, stumble, and fall exhausted across finish lines yet still manage to win races. A counterexample to Thompson's regress then takes the following form. What is an overwhelming accident—the kind that destroys intentionality—with regard to one movement need not be overwhelming with regard to another. So imagine a case in which an accident occurs at the beginning of a movement, so that the agent moves from A to B but does not move from A to B intentionally, and yet *does* move from A to C intentionally (she gets off on the wrong foot and then rescues the movement overall from disaster). Such a case seems to block Thompson's regress at point B.

One might object that this counterexample turns on a bad analogy. Although we can make sense of the movement across the Sahara as encompassing embarrassing unintentional mistakes, pratfalls, and so forth, we cannot make sense of the *tiny* movements relevant to Thompson's argument as mistakes or failures in their own right. Who cares if you move one micron *forward*, say, rather than one micron backward at the beginning of your movement from A to C? The objection helps our case more than it hinders it. If we cannot make sense of the tiny movements as failures or mistakes, then neither can we make sense of them as *successes*—as the agent having done something *in particular* that the agent intended to do (according to our stipulated use of intention).

One may still object that we only credit the agent who moved across the Sahara with an intentional movement because of the things she did *right* in the course of that deed. Had her progress been an endless series of pratfalls, we could not credit her with an intentional movement overall, but only a lucky (probably hilarious) accidental one. All Thompson needs for his style of argument is the claim that for any completed intentional action, there is some *proper part* of that action that is an intentional action in its own right. The proper part need not encompass any particular geometrical point (like halfway), but there has to be *something* the agent did right.

As far as it goes, this latter objection is a good one. We cannot credit an agent with an intentional action if she doesn't do *anything* right. But the objection does not show that in cases where no pratfalls and so forth occur *what* the agent must do right is something that is a *proper part* of the completed deed.

When the agent encounters certain kinds of obstacles, we have good reason to require *more* of her in the way of practical thought if we are to credit her with an intentional deed overall. The agent prior to falling had no thought about *getting up*; after she has fallen we require such a thought of her, dividing her project into parts and addressing intentions to *particular* means that can make amends for the *particular* setback. But where there are no unexpected obstacles or difficulties, we have no good reason to posit further particular practical thoughts, and so no reason to think that the agent must intend to do something that is a proper part of a larger project. To think otherwise would be to treat life as if it were one *continual* overcoming of error, misfortune and particular (infinitely small) practical problems (a tempting but perhaps overly neurotic thought).

Given this, we might question the setup of Thompson's central case. Thompson says that the halfway point B is a *particular* place along some particular path to C, where the path is given to the agent in sense or imagination, so that it would be as much true to say of the agent that she is heading to B as to say that she is heading to C when she sets out from A. If so, then moving to B is probably not a *necessary* means to moving to C. Most of the time one's movement from one place to another is not constrained by tunnel walls so that there is only *one* way to get to one's goal. In most cases, for any *particular* B in sight, one could miss B and still move from A to C successfully.

An argument against Thompson's conclusion then takes the following form. Imagine a case in which an agent moves from A to C intentionally but does so *without* an intention to follow some determinate particular path from A to C that is clearly distinguished from other possible paths to the goal (such as a road, or some other narrow, complete, determinate trajectory). Such an agent intends to move to C, but leaves her practical thought at that level of relative determinacy. The explanation of why she takes the *particular* path she does take will then appeal to her skills and habitual ways of moving rather than to her particular intentions. (We might note in passing that the explanation of why the agent starts *when* she does could similarly appeal to her skills and habitual ways of moving.) For such an agent, at each moment of the movement there is some segment of an ultimately successful trajectory from A to C that she has completed, but given the absence of an intention to move to those places in particular, or to follow a particular path that included them, we may say that prior to reaching them she was *practically indifferent* to them. If she was practically indifferent then, she is practically indifferent now: by the time a particular intention might be relevant, the places are already taken care of, just as the starting point was. Although it is true of such an agent that during the whole time of the deed she was moving to C intentionally, at no time was there some particular B halfway along a particular path to C such that she was moving *there* intentionally (and similarly for any other place one might fix upon short of C itself). The agent lacks an intention to do something that is a proper part of moving from A to C intentionally, so Thompson's conclusion is false.

One might object that such an agent must choose a direction to move in as a means to her end, even if she does not thereby choose a determinate path from point of origin to goal. If her employment of her skills and habitual ways of moving is to count as a manifestation of *practical thought*, rather than a blind response to some psychological occurrence, then she must determine her own activity *in* thought so as to make it instrumentally relevant to her end: she must intend to go *towards* C (or *forwards*, or *that way*).

If this is right (and I will argue later that this is only partly right), then at each moment of the movement the agent will have completed a stretch of activity of moving *towards* C (or *forwards*, etc.). When completed, the stretch of activity constitutes a successfully completed intentional action of moving from A to C. Dividing up this mass of activity, we can isolate an infinite number of component stretches of activity in thought, by isolating each stretch in time. Supposing such division is legitimate in the realm of *practical* thought, each component stretch of activity looks to be a successfully completed intentional action, and a variation on Thompson's conclusion—a *strong* variation, which grants intentionality to *all* the parts of this kind of action, rather than just some—seems to go through.

The division in question is not legitimate in the realm of practical thought. Consider the content of the proposed infinite series of intentions in Thompson's original example. The contents are all different: "I intend to go to B", "I intend to go to AA", "I intend to go to AAA", and so forth. If one accepts the case, one has good reason to think that *distinct* intentions address these distinct endeavors, because the agent has chosen a path to follow, the places on the path are distinct, the content of each intention varies with the place on the path identified, and intentions are individuated by their content. Now consider the content of the proposed infinite series of intentions in the new example of an agent who intends to move towards C but does not settle on some determinate path. These intentions all have the *same* content:

"I intend to move towards C (forwards, etc.)." We cannot distinguish these intentions on the basis of their content, so we have no reason to posit a multitude of intentions. It won't help to say they are differentiated by implicit temporal indexes: we have no reason to suppose that temporal indexes must be part of their content. We would do better to say that the agent has *one* practical thought here that unites her continuous activity during this period of time into *one* continuous motion to C (and perhaps beyond). Were there an unexpected obstacle, we would require more of her in the way of practical thought, for merely moving forwards, say, will not carry her all the way to her goal. But in the ordinary case there are no obstacles, and her practical thought, in its relative simplicity, is sufficient to sustain the relevant intentional action all the way to completion.

So: if it is possible to move intentionally to C without settling on a determinate path then basic actions are possible.

6. The Limitations of Practical Thought

We have not established that every intentional action *must* be resolvable into an ordered structure of basic and non-basic actions. Let me explain why.

Intentional actions are partly individuated by criteria of success, much as beliefs are partly individuated by truth conditions. We *insist* on criteria of success when there is *particular reason* to do so: the agent has a particular reason to move to C, which is why it counts as a success, but she needn't have a particular reason to move to any particular B, which is why moving to the particular B she does move to need not count as a success, and so need not count as an intentional action. (It could, had she *planned* on moving to B.)

Now suppose I intend to perform a metaphysical feat to prove that there need not be basic actions in every case: I will straighten my finger by means of an infinite number of compositional stages, and the stages will themselves be performed by means of an infinite number of further compositional stages, and so on *ad infinitum*. I will perform an infinite number of intentional actions, none of them basic, in under a second (or so the circus advertisement reads).[10]

This is a silly philosophical example. But what reason do we have for denying that I can think of and perform the infinity of tasks required? Basic action theorists grant that agents can exploit anything they know in the service of their ends. I know about infinity, and like the mathematician proving something (intentionally!) by mathematical induction, I exploit this knowledge in the service of my peculiar end. The individual means are very *easy* to perform, certainly. But I could fail at some of them. I could fail at all of them if I do not move my finger when the time comes to act. Apart from the relationship to reasons for action, the criterion of success is what matters for delimiting something the agent has in mind to do.

What this example shows is that even if in most cases there are basic actions at the "bottom" of an instrumental order, this is not a *general* thesis about intentional action, because we can have infinite numbers of practical thoughts if we have reason to have them (whether those reasons are silly and philosophical, or serious and mathematical, or some other kind). The example is important because basic action theorists implicitly limit our capacity to think practically without providing a good argument for the limit (i.e., they do not *say* why their regress is *vicious*). Thompson, for his part, is right that there is an infinite *potential* for rationalization in any case of intentional action, but his mistake is to treat the potentiality as an actuality. The problem is not that we *can't* encompass the required infinite series in thought, but just that we usually *don't*.

7. Activities, Actions, and Instrumental Orders

The last argument against Thompson turned on practical thought of an *activity*—intending to move *forwards*—and here the argument against Thompson meets up with problems faced by the other pole of the antinomy. Where they meet, we learn something about how time figures in the form and content of practical thought.

First, let me stipulate terms.

By an "activity" I mean something an agent does, so that if the agent *is doing* that then they *have already done* that (even if only for a short time). *Moving* and *moving forwards* count as activities in this sense—for each moment that one is moving one has already moved (even if only a little bit). Activities are open-ended and analogous to things described by mass nouns. There is no natural duration or limit included in the concept of an activity, so we may call them temporally *infinite*.

By an "action" I mean something that an agent does, so that if the agent *is (still) doing* that then they have *not yet done* it. Crossing the road counts as an action in this sense: whilst one is crossing the road one has not yet crossed it. Actions have bounds and are analogous to things described by count nouns. There is a natural duration or limit included in the concept of an action, so we may call them temporally *finite*.

Activities figure in the specification of actions, as when one intends to move forwards *until one gets to C*. But the converse only seems to hold for cyclical activities. Moving as such does not *essentially* depend on a component specification of a kind of action, but walking (*qua* cyclical activity) may well do.[11] It is plausible that there are intentional *activities*, in the sense defined, as well as intentional *actions*, and that the term "intention to A" covers both of these.

Consider how actions and activities figure in the last argument against Thompson. I claimed that the agent must intend to move *towards C* (*forwards*, etc.) in order to make her practical thought instrumentally relevant to her end. But this is only partly right. "If I move towards C I will reach C" is not the specification of a sufficient means to the end. C might be a meter away, and it is not true that if I move towards C I will move a meter. That thought does not represent a premise of sound instrumental reasoning, and cannot be required of a rational agent. What would represent a premise of sound instrumental reasoning is something like this: "if I move towards C *until I get to* C I will reach C." To make activities serve finite ends, we have to *limit* them, in thought and deed. When they are limited this way, we end up with a specification of an *action*—moving towards C until one gets to C—rather than a mere activity.[12]

The intention to move towards C is not *by itself* sufficient to sustain an intentional movement to C. But the intention to move towards C until one gets to C *is* sufficient, and that is enough to scupper Thompson's argument.[13] For here too, there is but *one* intention that unites the agent's continuous activity into *one* continuous motion to C, rather than an infinite series of intentions that divide her activity into an infinite series of tinier and tinier accomplishments.[14]

By way of concluding, let me apply this point about activities and actions to a little-remarked problem found in the literature on basic action.

When faced with the nasty contingencies that plague non-basic actions, basic action theorists often *retreat* in order to identify the basic thing done in some case. Davidson (2001, 60), for instance, says that when I fail to turn on the light at least I flick the switch. Flicking the switch is certainly more basic than turning on the light; but is it basic as such? Davidson, who is famous for saying "All we ever do is move our bodies," thinks not. He says that sometimes I may even fail to flick the switch, in which case at the very least *I move my hand*. There is a shift here from talk of an *action* to talk of an *activity* at the basic level. Davidson did not say "At the very least I move my hand *from A to B*," which would specify an *action*. That wouldn't help him find something basic, because one might *miss* one's target (*B*), so one would have to look further back in the chain of mediation for something *really* basic (i.e., something immune to perversion or interference). Davidson's talk of "moving one's hand" is better understood as talk of an *activity* that has no inherent bounds and no inherent target, and which is "always already perfected", so that it can play the role of being an original, perfect expression of the will, in which the further imperfections (failures to do this and that) are grounded.

Moving one's hand, as such, is not a *means* to the end of flicking the switch. (Compare "if I move my hand, I will flick the switch" to "if I move my hand from here to there, I will flick the switch, because it is between here and there.") So the movement of my hand is not an intentional *action* in the way that the flicking of the switch might have been, and it need not be an intentional activity either, unless I have some reason for moving *as such* (never mind where to).[15]

I have argued that the relation of instrumental mediation fails at the basic level if one retreats to basic activities: such activities have the wrong temporal form to be means to ends when those ends consist in the performance of *actions*. Instrumental mediation is only one kind of mediation. But the same kind of point applies to any practical atomist looking for a safe haven in which to locate basic action, no matter what kind of mediation they appeal to. The points made here apply straightforwardly to causation, for instance. If I am pushing a rock then I am causing the rock to be pushed. It's nonsense to say that if I am pushing a rock then I am causing the rock to be pushed *a meter*. Causation, conceived as without an inherent limit, should not be mistaken for causation conceived as possessing an inherent limit.

We are now (finally) in a position to say what is right and what is wrong with the very idea of basic action. Basic action theorists are right to think that there are some things that agents must be able to do without the mediation of further distinct intentional action, else they could not perform any intentional actions at all. Agents must be able to engage in *activities*, as these form part of what it is to be an action. But the word "do" is ambiguous between activities and actions, and activities need not be *intentional* activities in order to ground the possibility of intentional action. They could instead be thought of as exercises of skills or capacities.

One might ask how stretches of mere non-intentional activity are supposed to "add up" to an intentional basic *action*. This is a bad question. Any intentional *action*—such as moving from A to C—must be *constituted* by a stretch of activity that is the exercise of a skill or capacity—such as moving (as such). The source of limits on that activity, in thought and deed, is the agent herself, and the instrumental rationality of what's done depends on her pursuing an instrumental order represented in those limits. But it would be a mistake to think of those limits and that order as something extra, added to an activity in which the subject is already *non-intentionally* engaged. Rather, moving as such is one mode or aspect of exercise of the general capacity to do what one has in mind to do. When one has in mind to do something *limited* and *finite* and *instrumentally rational*, one's "mere activity" *already* has the character that would make it constitute an intentional *action*, were the activity to be brought to a successful completion. That does not imply that one's "mere activity" is *intentional activity*; that would require an independent reason to, for example, *move as such*. We might instead say that it is *thoughtful* activity: motion permeated by (in this case, limited, finite) thought.

Notes

1. Throughout this chapter my focus is on intentional actions of thinking subjects, rather than actions of pistons, slugs, and acids. I often omit the qualifier "intentional."
2. Basic action theories include Danto (1965), Goldman (1970), Davidson (2001), Hornsby (1980), and Ginet (1990). For more references see O'Connor and Sandis (2010, 16–17).

3. For Davidsonians: non-basic action *descriptions* are essentially mediated by basic action descriptions.

4. For example, Danto (1965) suggests that it is absurd to think of the agent as *always* having to do an infinite number of things first before they do the thing they have in mind to do, but that is all he has to say about the matter. One wonders how he would deal with Zeno's paradoxes.

5. As Thompson (2008) puts it, he is tempted to adopt the manner of Quine and declare himself deep amongst the "don't cares" when considering the limit that supposedly marks off basic action from non-basic action (110).

6. I don't claim that there couldn't be some other form of expression of the will on such occasions. In Frost (2012) I argue that basic action theorists ought to countenance "processual" expression of the will in addition to the expression of the will in action.

7. This pressure is felt in different ways by different theorists. Davidson (2001) was content to locate basic action in short, but presumably not instantaneous, bodily movements. Hornsby (1980), following a line of thought from Davidson, pursues basic actions inside the body and identifies them with neural events that are "tryings." (Hornsby now thinks that this line of thought constitutes a *reductio*, rather than a development, of Davidson's view.) Volitional theories of action such as McCann (1974) locate basic actions in mental acts that do not have the temporality characteristic of ordinary bodily movements, but which are not clearly means to an end. It is open to a basic action theorist to resist the temptation to make basic actions temporally immediate, but I don't know of any theories that loudly publicize the claim that a basic action could take, say, a couple of hours.

8. Anscombe (2000, 13–14 and 49–51).

9. Nothing is claimed about whether intentions are beliefs, or desires, or more nebulous states (or habits, or frames) of mind: the stipulation is supposed to be as ecumenical as possible. Nothing is claimed about whether the adoption of intentions requires a feeling, sign, or specially attentive act of consciousness. Nothing is claimed about whether one must adopt an intention *prior* to executing it. Nothing is (yet) claimed about whether the adoption of one intention must be accompanied by the adoption of some others: perhaps if I intend to move from A to C I must intend to go halfway, but then again, perhaps not. It *is* claimed that if one intends to Φ one knows that one intends to Φ without recourse to observation or inference.

10. I won't pause between the stages, because physiology prevents me, but that is beside the point. The feat is not that impressive when you actually see it performed. There are no refunds at the philosophical circus.

11. Mourelatos (1978) suggests that actions, in the sense defined earlier, can always figure in the specification of *activities*, as when one thinks of running a mile as a qualitatively different kind of *activity* from running a 100-meter dash. What Mourelatos contrasts is better described as a contrast between sprinting and long-distance running. Someone could, for instance, run a 100-meter dash by exercising their skill at long-distance running until they have run 100 meters (they probably wouldn't win), or, if they were very fit, they could run 100 meters for a mile without a break in this *single* stretch of activity.

12. I don't claim that activities can't figure in sound instrumental reasoning; it is just that when they do they are *constitutive* means to ends, and the ends they serve have a similarly open-ended temporal form. See Rödl (2007, 34–38).

13. Similarly, the intention to *move to C* is by itself sufficient to sustain intentional movement to C, but only because it is (usually) equivalent to the intention to move towards C until one gets to C.

14. Activities have their own "success conditions": I am not moving towards *C* if I am not moving. But we needn't worry that such success conditions lead to another version of Thompson's argument. Were one to ask "Why are you moving towards *B*?" and receive the answer "because I'm moving to *C*," this "answer" would not rationalize what was asked about. The response would instead provide a *starting point* for a conversation about reasons for action, in the sense rehearsed in the hammering example.
15. This move to activities at the basic level is not peculiar to Davidson. Goldman (1970), for example, describes all cases of basic action in terms of continuous activities (e.g., *raising* my arm). He never describes a basic action in terms of, for example, raising my arm *from here to there*. A similar point applies to volitional theories of basic action (Prichard [1949]; McCann [1974]) that treat basic actions as willings: whilst one *is willing* that *p* one has already *willed* that *p*.

References

Anscombe, G.E.M. (2000). *Intention*. Cambridge, MA: Harvard University Press.

Danto, A. (1965). Basic Actions. *American Philosophical Quarterly*, 2(2), 141–148.

Davidson, D. (2001). *Essays on Actions and Events*.Oxford: Oxford University Press.

Frost, K. (2012). *Mental Capacities and Their Imperfect Exercises* (Doctoral dissertation), University of Pittsburgh. Available at http://d-scholarship.pitt.edu/13226/1/KF_ETD_23072012rev.pdf (UMI No. 3532831).

Ginet, C. (1990). *On Action*. Cambridge: Cambridge University Press.

Goldman, A. (1970). *A Theory of Human Action*. Englewood Cliffs, NJ: Prentice-Hall.

Hornsby, J. (1980). *Actions*. London: Routledge and Kegan Paul.

McCann, H. (1974). Volition and Basic Action. *Philosophical Review*, 83(4), 457–473.

Mourelatos, A. (1978). Events, Processes and States. *Linguistics and Philosophy*, 2(3), 415–434.

O'Connor, T. & Sandis, C. (Eds.). (2010). *A Companion to the Philosophy of Action*. Oxford: Blackwell.

Prichard, H. (1949). *Moral Obligation*. Oxford: Clarendon Press.

Rödl, S. (2007). *Self-Consciousness*. Cambridge, MA: Harvard University Press.

Thompson, M. (2008). *Life and Action*. Cambridge, MA: Harvard University Press.

4 Second Nature and Basic Action

Ben Wolfson

1. A Puzzle about Basic Actions

The line of thought that leads to the postulation of basic actions is straightforward. Most of our actions are indisputably complex, composed of subordinate actions, means to or specifications of our ends, actions with their own structure of sub-actions. But, so the line goes, it can't be turtles *all* the way down; it must bottom out in an atomic action somewhere. For however unafraid of invoking infinities we may be in general, the analysis of an action should reveal its intentional structure as it might have been entertained by the agent, and how could the agent entertain an infinitely long practical syllogism? Or how, if every action decomposes into more, could the agent get started at all? Only an action that is *not* complex, not carried out by performing subordinate actions for its sake, could stop the regress. These are the teleologically basic actions.[1]

Everyday experience supports this line. When I type on a keyboard, my fingers move around and press the keys, my muscles contract, and much else goes on besides; nevertheless, it seems that I type *straight off*, and that the proper description of my action is simply that I'm typing, be it the letter *a* or the word "action"—one can quibble about the level of granularity here; the point is that putative means such as "by moving my pinkie to the left and depressing the key" or "by contracting such and such muscle" *escape* the description under which I acted. My pinkie does move to the left, my muscle does contract, but these are mere movements, not actions I perform.

Compare a complex act, such as making a pie. Not only does it involve several subordinate further actions, they themselves must be undertaken with an understanding of their relation to, and with the intention of, pie-making—it's not enough, for the agent to be making a pie, that he perform a series of actions and *end up* in the presence of a pie. The ability to perform a complex action is partly predicated on being able to give an account of subordinate steps that might be taken to the end, and engaging in one requires taking such steps with that aim.[2]

Basic actions, then, should strike us as puzzling. A basic action is done "immediately" in that there are no mediating actions, but it's not done *at once*; it can be interrupted. What are we to say about a basic action interrupted in its development? It was begun but was not completed; has the agent *done* anything? If a complex action is interrupted prior to completion, we may still identify in its development actions the agent has performed: he did not make a pie, but he did roll out the dough. The complex action has instrumentally related parts, whose relations are reflected in the agent's understanding, which afford a partial answer to the question why we identify the series of performances as *an* act of pie-making. The basic action lacks a like instrumental structure, and the agent may not understand the movements going into it: base movement, one might worry, is transmuted into golden action only at the moment of completion. Even if we locate a reason for starting the basic action, the agent does not stand to it as an *actor* any longer, merely someone who started things off and now observes.[3]

Such worries have led Thompson (2008) and Lavin (2012) to argue that nothing basic in the intended sense is truly an action, and concomitantly that we ought not recognize, in theorizing about action, the basic in addition to the complex. Part of the concern has to do with the temporal and physical complexity of the (as we might put it) agentially simple basic actions—motion through time and space remain, leading to the question what is done in the case of interruption.[4] But there is a wider-ranging concern as well, based on the relation of the agent to development of the action. Lavin, for instance, explicitly regards the agent of a basic action as alienated from its progress—it's merely kicked off somehow, and develops of its own, without a role for the agent's understanding.

Lavin sees basic actions so understood as "vital to the intelligibility of the *causal theory of action*, according to which physical action consists of a mere event and a condition of mind joined (in the right way) by the bond of causality" (Lavin 2012, 274); his attack on basic actions is thus meant to be an attack on causal theories as well. A basic action plays the role of the correctly occasioned "event"; since it's a mere event, caused by the agent's cast of mind, the manner of its unfolding need not be understood by the agent. Basic actions and the causal theory fit together neatly.[5]

The core of this wider-ranging concern can be put as follows. First, "rational agency is the capacity to apply thought to action: it's the capacity to do things for reasons or on the basis of considerations," and in instrumental agency "the thought about *A* is always that it has some relation to *B*, something else the agent aims to do" (Lavin 2012, 275), though that relation can be rather various in different cases (the means might be sufficient for the end, or necessary for it, or neither though they still promote it). Now, "a basic action is, by definition, not mediated by any such thought" (Lavin 2012, 275), but it still involves *accomplishing* something—it involves something's becoming so that was not so before. But the agent has no "thought about how to bridge the gap" between the initial state and the end state.

If, then, the thought that rational agency applies to action *is always only instrumental thought*, rational agency is not found in basic actions. And the typical description of basic actions as things that the agent "just does" can, indeed, make that conclusion seem apt: I *just* do it—it isn't mediated by an episode of thought, and the capacity for basic action is arational. A causal theorist might find this congenial; an opponent of causal theories of action is likely to find it suspect, and may propose in response that *every* action has an internal structure like that of the baking project with which I began. Suspicion is only called for, however, if rational capacities are *always* linked with instrumental thought: if all rational capacities are cut from the same cloth as the capacity to make pies. Lavin reveals his allegiance to this pre-supposition early on, saying that "it is open to us to consider an alterna-tive [to basic actions], one on which action is not, in the fundamental case, barren of means-end structure" (Lavin 2012, 274). The choice is stark: *either* basic action is an arational kicking-off of a process of whose further progress the agent is merely a spectator, *or* its instrumental thought reigns supreme: rationality cannot be exhibited in something the agent "just does." This is the source of the conviction that a basic action could only be "the arithmetic sum of *what merely happens* and something else" (Lavin 2012, 278): that is, that what goes on in a basic action is constitutively independ-ent of an agent's thought about what is going on. A basic action could not be constitutively bound up with its agent's rationality, because by definition it lacks an instrumental structure, and only an instrumental structure will do for rational capacities.

Lavin's attack thus depends on a particular construal of basic actions, and the assumption that basic actions could *only* be so construed—so that we can identify an attack on basic actions with one on causal theories— goes hand in glove with the contention that something is an action only if it has intentional parts and is brought about by an understanding of its relation to those parts.[6] I will argue that this represents a too-limited conception of what rational abilities come to, that we need middle ground between the sort of capacity exhibited by bakers in baking and the *merely* natural capacities exhibited in, for instance, sugar when it dissolves in water. I'll call the sort of capacity answering Lavin's characterization *tech-nological*, because in it a technique is exploited; the next section will give a (compressed) constructive argument that rational capacities must include more than the technological. The fourth section will address the arguments directed specifically at basic action and its physical characteristics. The missing category of capacity is that of second nature: we should not expect it to have all the articulation of capacities that exploit a technique, but they are not, for all that, merely arational, and I'll conclude with some discus-sion of it. Basic actions will then be exercises of such capacities—to, say, raise one's arm. I'll primarily be concerned, however, with the need for such a category, especially in the basic action case. The rest will have to remain programmatic.

2. Fluency and Technological Capacities

In laying out a familiar line of thought supporting basic actions, I asked how, absent them, an agent could get started at all. This suggests an equally familiar regress argument: the agent couldn't get started, because she would have to *consider* each prospective action before undertaking it. She could never start acting because she would never stop considering. Here there is a quick rejoinder: we are reminded that skill drives out deliberation (Thompson 2008, 108),[7] and that an action can have a form accurately recapitulated syllogistically without the agent having actually deliberated.[8] Andrea Kern puts the point well: we "can admit acts as actualizations of a rational capacity that are not themselves the result of a decision, as long as the subject of these acts *can* comport himself decisively [sc. with deliberation] to them" (Kern 2006, 236; my translation, emphasis in original). What is important is the activity of an understanding of what is to be done, not explicit deliberation. Thus the legitimacy of speaking of an "order which is there whenever actions are done with intentions" (Anscombe 1963, 80).

It's tempting to identify acts done without deliberation—call them *fluent* actions—with basic actions. And then, arguing as earlier that even fluent actions have a deliberative structure, we can, on the basis of the identification, argue that there are no basic actions, only actions done without explicit deliberation.

The fluent/non-fluent distinction is useful, and the observation that there are fluent actions that have a deliberative structure is important. But subjects cannot comport themselves decisively to *all* fluent actions. There is an additional distinction within the class of fluent actions between those done without occurrent thought about how to do them, and those that do not rely on a linguistically expressible understanding of how to do them at all.[9] Only the latter are basic in the intended sense.

Kenny (1976, 53) provides a good example: when I hear spoken English, I understand it. The capacity to understand spoken language is surely rational, and, surely, fluent: I just do hear the words—not mere sounds—and hear them as significant; the sound is transparent to the meaning. There appears to be nothing else I *do* in order to understand the words.

This appearance, I claim, is real. Hearing a language as significant is not something to which one can comport oneself decisively, and is not technological. It's not that skill at understanding English has rendered it unnecessary for me to deliberate afresh on each occasion; attributing a deliberational structure to occasions of my understanding English mischaracterizes them. The possibility of deliberating is denied: one cannot choose not, or incorrectly, to understand another's speech.[10] In what sense, then, can I be said to comport myself decisively to such acts?

We have to engage in the organized stringing together of means to accomplish an end because our powers do not extend immediately to *those* ends. But this is not to say that there are *no* ends to which our powers extend immediately, and when we do what we have come to be able to do straight off our doing will not have the complexity characteristic of taking means. The capacity to understand language is indubitably rational—possessable only by creatures possessed of reason, and something that exhibits reason in its exercises—but is not possessed of all the features had by technological capacities. The temptation to identify rational with technological capacities is strong, partly because the latter wear their rationality on their sleeves: we can see an order of means subservient to ends in each episode of their exercise. But that cannot be the only form of rational capacity.

3. Basic Action

If we already have to recognize non-technological rational capacities in the *one* case, then there can be no *general* objection to recognizing them when basic actions are at issue. But, as I mentioned, Thompson and Lavin have *specific* arguments as well, turning on the temporal and physical aspects of action:

The first point came up already in Section 1: any movement will be in time, hence interruptible; if the basic action is done without means, what will the agent have done if interrupted? Manifestly something has *happened*. It would be perverse to claim what happened would only have been an action had it not been interrupted, that the agent did not *do* anything yet.

Second, any movement will be along a path, and one can ask, of some point *p* before the end point *e*, "Why were you moving to *p*?", and get the answer, "because I'm moving to *e*." (See, e.g., Thompson [2008, 107–108].) This supposedly reveals the partial movement to *p* as an intentional action in its own right: movement to *p*, for the sake of getting to *e*.

The first suggests that we need to find an intentional action that is "smaller" than the basic action, and the second embodies a suggestion for what it might be. In pseudo-Eleatic fashion we find in each putative basic action a shorter action *also* done intentionally as such for the sake of the greater: it *is* turtles all the way down.

Consider the second suggestion first. The opponent of basic actions can acknowledge that one does not think about moving first halfway, then the rest of the way: skill drives out deliberation. The opponent can *also* acknowledge that one neither knows nor deliberates about how, for example, one reaches for cups. The claim is rather that, while one does not know how *in general* one moves one's arm, one does know particular things: that to reach *that* cup one must move one's arm from *here* to *there*, and to do so must first move it halfway there, and halfway to the halfway point, and so forth. Thus Lavin:

> The decisive question is whether the individual actions in which an agent exercises practical knowledge must themselves lack the structure of means-end rationality . . . It might be that I do not know any *general*

procedure for walking, but that nevertheless, when I walk on particular occasions, I perform these actions by knowingly taking certain *particular* means . . . Suppose I exercise my knowledge how to walk in walking from this rock here to that tree over there. I can know that I have intentionally walked to this point here which is about half-way in order to walk to that tree; or again, I can know that part of the way I walked to the tree was by walking to this point . . . Do I not intentionally walk from here to there by intentionally walking every inch of the way?

(Lavin 2012, 286–287; emphasis added)

"Walking from here to there" is not a paradigm basic action, but, as Lavin points out, it's not clear in general where it's plausible to draw the line, especially at the level of action-types. (It's not, however, necessary for the defender of basic actions to furnish a catalog of basic action-types.) Nevertheless, the thrust should be clear: in *particular* episodes of something apparently basic we can find what we may as well call "steps" of their particular executions: namely, the *stages* of their execution. These stages are (following Lavin and Thompson) formally similar enough to the more explicitly step-like steps of actions that have an obviously means-end structure that they can be assimilated: they share, for instance, the formal similarity that one can say "I was φ-ing as part of ψ-ing" or "I was ψ-ing by φ-ing."

We know the objection already; it's the same one aired at the beginning of Section 2. It's not that skill has rendered deliberation unnecessary,[11] but that the imputation of a deliberative structure incorrectly characterizes the action. We can distinguish two senses in which this is the case, depending on whether we take the intermediate points to be specified independently of the path taken, or to depend on it.

In the former case, the agent may have what we might call general particular knowledge—the knowledge that in general the way to get from the particular point A to the particular point B is via the particular point C, as in the imagined report Lavin gives: "I am walking from Athens to Delphi: I walked from Athens to Thebes, now I'm walking from Thebes to Delphi" (Lavin 2012, 291). Such an agent is able, apart from any particular journey from Athens to the oracle, to say that one gets there via Thebes. The case therefore does not resemble putatively basic actions very well, and also does not serve as the best illustration of particular knowledge. Let us try to do better.

One option is to remove from the agent's thought about his journey any reference to the points between his origin and destination. There will then be no guarantee that the agent is aware that he is going to the middle points (under some relevant descriptions) at all, vitiating the claim that he was going to them as such intentionally. But however we restore reference to the intermediate points, we will still lack reason to believe that the agent goes to them *intentionally as such*, as a means to getting to the end, even though he will be able to answer the question "Why were you going to C?" with "because I was going from A to B"—just not in a rationalizing sense.

Suppose that you and a companion are driving along Highway 101 from Los Angeles to San Francisco. You have consulted a map but not investigated the intermediate parts of your route in much detail. All you know is that one gets to San Francisco by getting to and staying on 101: the freeway will take care of the specifics. Your companion may ask you, on catching sight of a sign reading "Now entering Atascadero," why you're going *there*, that is, to Atascadero. The response "because we're going to San Francisco" is perfectly in order. But you weren't going to Atascadero, *qua* Atascadero, *as a means of* going to San Francisco. You may never have heard of Atascadero before. You were going there as a consequence of the route you chose.[12] It's still possible to say that you're going to Atascadero intentionally, in that you're going intentionally to San Francisco and this going is also to Atascadero. The fact that Atascadero has a *name* given independently makes the fact that you may not have been going to it as such somewhat obvious. But this is inessential; after all, when I move my arm to my glass, the halfway point *there* doesn't have an independently given name. In general if we pick out a halfway point independently of the path actually taken, we may doubt whether we go to that point as such in order to go all the way. Knowing in advance that the path one has chosen will take one through a particular point doesn't make it the case that one is intentionally going to that point as such in order to get to one's destination.

Indeed, we may doubt whether we go to a point picked out in that manner *at all*. Looking at my hand and at my glass you may correctly identify some region as halfway between the two—meaning, perhaps, halfway on the shortest path between them. When I move my hand, though, there are several possibilities. I might not move through there at all; I might move through there but not halfway on my path, and I might move through there and have it be halfway along the traveled path. Even in the last case, however, I needn't move through the region *in order* first to get halfway to the glass, as if it stood like a checkpoint between me and the glass that must first be got to, then passed.

Suppose now that we're thinking of an intermediate point on the path actually traveled. It seems to be a theoretical attainment, which self-aware self-movers could in principle do without, to know that movement to a point involves movement halfway to that point. So we should be suspicious of the suggestion that it's part of each intentional movement that one intentionally move to each halfway point along the path taken. And it would be difficult to do that intentionally as such, since where the halfway point on the actual path *is* is not determined until the movement is *complete*. (Observe the play of tense in the long quotation from Lavin mentioned earlier: "I can know that part of the way I walked to this tree *was* by walking to this point"—but often I only know that after I *have walked*. If "this point" is nothing more than some point on the path that I happen to have taken, then often I will *not* prospectively know that I *am walking to it*—and rarely will I be walking to whatever points I'm walking to *in order to* reach the tree.)

But the most telling objection to this way of taking things is that there is just no place for practical thought here: one does not deliberate about what cannot be otherwise, and it's impossible, in tracing a path between two points, not thereby to trace all the partial paths. Agents need not concern themselves with this.

There is, then, a trilemma: if in "intentionally walking every inch of the way" between two points "the way" involves distinguished partial paths specified independently of the actual taking of some path, then either (1) it's possible that the agent takes the partial paths as independent steps in their own right,[13] explicitly seeing to it that they are accomplished. But it's acknowledged that putatively basic actions are not like this. Or, the agent does not explicitly see to it, in which case he may not trace the partial paths he thinks he is, and (2) he goes to midpoint *p* in consequence of going, not in order to go, to end point *e*. If, finally, (3) the way is simply whatever way the agent happened to take, then we have again lost particular knowledge in favor of the most general knowledge about movement one might have, the unavoidable fact that movement from one point to another involves movement between points. As there is no way for the agent *not* to go every inch of *some* way between the start and the end, and as the way in question is whatever way the agent is taking, there is no place for deliberation to the end of going every inch—not because of skill, but because deliberation is out of place. And if one is walking to the end intentionally, then one is *ipso facto* walking every inch intentionally. Intentionally walking every inch of the way isn't like intentionally visiting every bar on the block: the intentional walking covers all inches, but there is not, for each inch, an intentional walking for it.[14]

The first point still remains: if an action is interrupted, we must find an action that the agent had intentionally done at the time he was interrupted. Suppose I'm taking hold of a glass—assuming this is a basic action—when I suddenly die, not yet having even touched the glass. I did not grasp the glass, but I did do *something*—what? If we take it that the only kinds of capacities are the arational and the technological, either I was the subject of an arational process, or I was following a procedure of some sort, assembling means. "Consider the time . . . when *X* is doing *A* [a basic action] . . . Other things have already happened (*X* did *A***), and still others are underway (*X* is doing *A**) . . . The action in progress (*X* is doing *A*) is at once an ever increasing stack of *have done's* and an ever shrinking list of *still to do's*" (Lavin 2012, 292; emphasis in original). One of the "have done's" here would be: extended my arm some distance. But if the action is not technological, how can the question be answered?

We can treat this in the manner of Moline (1975). For Moline, the attribution of a capacity already implicitly recognizes that it depends on the obtaining of certain circumstances: "it is not the case that sugar has the property of being-soluble-under-all-circumstances-provided-nothing-external-interferes. Rather it *has* under all circumstances the property of being-soluble-under-certain-special-circumstances, namely, the ones under which it will invariably dissolve" (Moline 1975, 253)—so that a sugar cube simply does not have the capacity to dissolve-when-encased-in-wax or in saturated liquid.[15] It's not that in such cases something has interfered with the capacity to dissolve, which it still did have in those circumstances; it had no such capacity. (Or rather, it still did have in those circumstances the capacity to dissolve in *other* circumstances.)

Not only sugar but also our capacities are like that. (I have the capacity to swim and to comprehendingly hear English—but not amid crashing, ten-foot-high waves.) We are dealing with an external interference, but one that does not begin until the action interfered with is already underway. So, while I do not have the capacity to continue reaching when I am dead, when I began I was in a position to exercise my capacity to reach—death interrupts the exercise of this very capacity, not a stringing together of the exercises of other subsidiary capacities. What I did was to exercise my capacity in circumstances that *later* became unsatisfactory—something that is possible precisely because of the temporal extent of the action.[16] More prosaically, what I did was *try* to reach the glass; I failed, not because I did anything incorrectly by way of reaching (I took no means at all), but because as I was underway the circumstances became (highly) unsatisfactory; I could not *continue* to exercise the ability.[17] For the exercise's being begun is no guarantee it will finish.[18] If φ-ing is a basic action that one has begun, and one is interrupted before completing it, then one did perform an intentional action: one tried to φ.[19]

This conclusion may seem to leap right over Lavin's argument—what of A* and A**? In a way it does: for while it follows from the temporal structure of an action that other *things* have *happened*, and still others are *underway*, it does not follow from that temporal structure that those things are themselves actions. Lavin's parenthetical identifications are unjustified.

Both Lavin and Thompson agree that some of the A* that happen in the course of the performance of an action are not themselves actions, and that agents are ignorant of them. Thompson acknowledges that "a description in terms of muscular contractions" would be "alien to the agent's mind in the intentionalness-destroying way" (Thompson 2008, 111); Lavin admits the same of "the activation of efferent motor neurons, the contractions of certain muscles" (Lavin 2012, 294). But what is special about such descriptions that an agent can be ignorant of *them* without prejudice to the status of the goings-on of which they are a part as rational actions? Lavin rejoins:

> It is one thing to be unaware of *something* "implicated in the production" of an action; it is another to be unaware of *everything* going into this. On our present supposition, the (putative) agent of a basic action lacks awareness not only of, say, underlying physiological processes . . . but also of the intrinsic articulation in the development from *doing* to *did*.
>
> (Lavin 2012, 294; emphasis in original)

This however does not answer the question. It's equally unclear why, when the agent is moving from A to B, the happening-to-happened developments

involving moving to first C do, and the happening-to-happened developments involving muscular contractions do not, count toward the "intrinsic
articulation." Muscular contractions are certainly an important part of getting places (and are more important than movements to such actions as
clenching an already shut jaw).[20]

Lavin is correct that it's "another [thing] to be unaware of *everything*"
implicated in our action. But the agent of a basic action is not like that. Such
an agent knows (at least) that she is trying to φ. The presumption that something more is needed is a relic of precisely the conception of action that Lavin
and Thompson, in other moods, seek to overcome. For they seem to assume
that because we can locate a physical movement in the course of an action,
that movement must itself be an action, because the movement *qua* movement is part of the action *qua* action: they are implicitly thinking of action as
just a certain kind of physical movement, such that the divisibility of the one
implies a corresponding divisibility of the other. Thus is derived the requirement that the agent know about the phases of the motion: since those phases
are themselves actions, the agents must have the knowledge of them that is
characteristic of action. Having started from an identification of movements
of a certain sort with actions, we need to ensure that the agents really do know
about them in the right way. But if we *start* from the position that "it is the
agent's knowledge of what he is doing that gives the descriptions under which
what is going on is the execution of an intention" (Anscombe 1963, 87)—if,
as Anscombe goes on to assert, such knowledge is a formal feature of anything
that can be called an intentional action—then we are in a position simply to
deny that the bodily motions that occur in the course of a basic action are
themselves actions. The only (or the least) *action* on the scene—the only (or
the least) thing of which the agent has practical knowledge—is the φ-ing itself.

What of Lavin's concern about the "articulation in the development" of
the action? We reply that there *is* no articulation in the *action*. That does
not render the fact that the agent performs the action over time a mystery
to her, nor should it to philosophers: while the agent of a basic action may
not know—at least not practically—what happenings enable her action, she
does know that she can perform it, should she call on herself to do so.[21]

4. Conclusions

The contrast between rational and merely natural capacities is clearest in the
case of technological capacities.[22] But technological capacities' assembling of
means cannot be all there is to rational capacities. Here we may turn again
to Kern, who, when introducing rational capacities, refers to second nature,
which deserves to be called "nature" precisely because it has an "immediate
efficacy," but is "second" because it's "intrinsically bound up with a specific
process of acquisition" (Kern 2006, 195), a process of learning and habituation seen in the linguistic competence at play in hearing and the bodily
self-mastery of basic actions.[23] "Second nature" is not *just* a set of rational
capacities to φ, ψ, and so forth without needing to deliberate; it's a set of

rational capacities to φ, and so forth *directly*: deliberation and decision are no longer at issue. That ought not strike us as paradoxical or surprising, especially when we consider deliberation itself: the capacity to deliberate is a capacity to do something directly, in several senses. Undertaking to deliberate when faced with a problem cannot in general be the upshot of deliberation, and when, in the course of deliberation, one thinks of *p* or *q*, or transitions from *p* to *q*, that too cannot be the result of its own deliberation.[24]

Like technological capacities, second-natural capacities are products of our rational natures, but in a different way; second-natural capacities are the result of an education.[25] Also like technological capacities, the exercise of a second-natural capacity entrains our rational faculties:[26] my movement would not be a basic *action* if there were not a place for the application of the questions of what I'm doing and why. And as with all actions, it's my answers to these questions that reveal what, among the many things going on with me, *are* my actions.

One can take the claim in the theoretical case that the capacity to comprehendingly hear is second-natural as an instance of the general claim, familiar from McDowell (1994), that our understanding is active in reception; we do not take means to understand some not already understood item. This could not be so if the capacity's exercises were not immediate. In the practical case immediacy means that an agent is an originator, not merely an assembler, of actions.[27] Granted, even actions immediately done involve many happenings. But these lack practical significance for the agent, and may vary substantially across different executions of the same action-type. (Basic actions are multiply realizable.)[28] The *action* first appears on the scene with the general, practically significant thing to be done, which would appear in the agent's self-knowledge in the form "I am doing . . .": I'm reaching for the cup, balancing the pole, and so forth. As Husserl says, in acting "we have understanding of what happens practically rather than of the process in terms of its physical causality" or of "a merely mechanical motion" (Husserl 1989, 272–273).[29] Certainly "my hand is a thing *too*, and if *I* execute a subjective 'I move' there is also executed a physical process" (Husserl 1989, 272–273; emphasis in original). But if we take seriously the idea that action is *not* the arithmetic sum of a mental nexus and a physical event whose independently assessable characteristics determine the final sum, the fact that my hand does something when I φ may be at most causally relevant to the carrying out of the action.[30] Recall that Lavin argued against basic actions because he opposes causal theories of action and the metaphysical picture he thinks accompanies them. But the non-causal theorist only has to oppose basic actions if, like Lavin, she holds that the only alternative to a bare event is something structured by means-end rationality, and such a move can be supported neither in the general case nor when restricted to actions.

We seem to have wound up at some distance from the event-causal theories that the common acceptance of "causal theory of action" encompasses: basic actions understood as the exercise of capacities to do something straight off not only involve no commitment to event-causal theories, they are *prima facie* inimical to them. To say that "an event that merits the title 'action' is *a person's* intentionally doing something" (Hornsby 2004, 19; emphasis in original) is to set oneself against the "arithmetical" conception that Lavin associates with event-causal theories, which reduces actions to the sum of addends to which agential concepts are alien.[31] For here we say that actions bottom out in a person's exercising her capacities. Such a saying needs filling in: it's at best programmatic. But it's a program that an anticausal theorist can pursue without falling into the problems that vex Lavin and Thompson.

Notes

1. For other senses in which an action might be thought "basic," see Baier (1971).
2. I argue for this in detail in Wolfson (2012).
3. Frankfurt (1978) objected to the causal theory along similar lines, but his proposal simply adds epicycles in the form of periodic interventions. His agent who guides an action simply performs *more* actions of essentially the same type as the one who engages in *one* initiation. Cf. Lavin (2012, 301n44).
4. I will for the most part ignore putatively semelfactive actions (such as reaching in the sense of attaining, as in Vendler [1967, 102]) or quasi- or fully mental actions such as recognizing or realizing (as in Rothstein [2004, 36]). The issues afflicting basic actions cannot be resolved by attributing semelfactivity or mentalness to them *all*.
5. Mightn't the causal basic action theorist object that the "right way" (or content of the condition of mind) will ensure that such knowledge is had? Perhaps: but this is unlikely in itself and unlikely to meet Lavin's concerns. First, because part of the motivation for most supporters of basic actions is the commonsensical claim that in many cases we do things and we *don't* have the kind of knowledge about its proceeding or inner structure that we do in other cases. Second, because as long as the knowledge is just *there*, while the event unfolds *of its own*—as long as we do not have practical knowledge active in the unfolding of the process—the agent will still be estranged from the *actual* unfolding of *this* process.
6. The assumption that basic actions can be construed only in one way may explain why Lavin, having argued that a causal theory requires basic actions, does not address whether basic actions require a causal theory.
7. Lavin, who endorses this move, thus cannot mean an explicit application when he writes of "the capacity to *apply* thought to action" (Lavin 2012, 275; emphasis added).
8. Cf. McDowell (1998, 107) and Heidegger (1978, 98).
9. "Linguistically expressible" here is meant to capture the possibility of decisive comportment, since what is envisioned is an explicit course of deliberation, which I take to be linguistically mediated. An understanding not so expressible (something like what Noë [2013, 192] seems to have in mind) would not perform that role.
10. Perhaps trained linguists can attend *just* to the sounds of languages they speak, as one hears that trained painters can attend *just* to the play of colors before them. This would be a *new* ability, not evidence that we *all* attend first of all just to sounds.
11. We might also ask what the relevant skill is. The action in question is not generically moving my arm, or walking, but moving my arm, or walking, from this particular place to that. Odd skill!

12. I am drawing on a similar example from Haugeland (1998).
13. An example in the case of bodily movement might come from lifting a weight over one's head: first—oof!—to the shoulder, then—heave!—over the head.
14. One should compare this with Aristotle's solution to Zeno's Achillean paradox. Aristotle argues that "in the act of dividing the continuous distance into two halves one point is treated as two, since we make it a beginning and an end" (Aristotle 1984, I:263a23–24), and if one does this one "will get not a continuous but an intermittent motion" (Aristotle 1984, I:263a29–30), in which (so to speak) one must visit each point *as such*. If the intermittency of the motion is forced all the way down, motion will indeed be impossible (*Physics* 263b5). But the motion need not be intermittent; if it's continuous it will move through potential, not actual, halves—it won't move to, and then from, each point as such.
15. Cf. § 5 of the commentary on Θ 5 in Makin (2006).
16. The same sort of thing could happen with a sugar cube that began to dissolve but didn't dissolve all the way because it was destroyed by a bomb.
17. Cf. Wilson (1989, 161–162).
18. Note that trying and being interrupted does not mean that I have exercised the capacity *incorrectly*, or even that I could have succeeded. Capacities for action are fallible and always require the world's cooperation to some extent. Indeed, the idea that one might exercise a capacity for a basic action incorrectly threatens regress, as it carries with it the suggestion that one exercises the capacity according to some method. Moreover, since there are no parts to basic actions, when one tries and fails to perform one, one cannot be said to have done something *wrong*. One merely failed to do what one tried to do. (Of course one can have done associated things wrong—for instance, one might have incorrectly checked the state of one's environment before getting started.)
19. This does not mean that a successful φ-ing involves two actions, the φ-ing and the trying. A successful attempt to φ *is* the φ-ing one has carried out, not an additional action alongside it. The point here is that an *unsuccessful* attempt to φ does not baffle our action explanations: trial-talk is as it were *for* explaining what one was about when one did not succeed.
20. Hornsby (2011) mentions a class of "non-mediated" actions some of whose members are similarly interesting: *carrying*, for instance, or *holding*. Too, muscular contractions, as the name suggests, are movements!
21. The phrase is David Hills's (in conversation). This admittedly leaves unanswered the question how she knows that she is calling herself to φ, when she is; however, this same question arises in the case of complex and putatively basic actions—how do I know that it's a cherry pie I'm bending my efforts toward?—and cannot be addressed adequately here. The point of the foregoing remark is that an agent engaged in a basic action needn't be concerned with a stepwise articulation of the action.
22. Note that Aristotle's leading examples of rational capacities, the "arts" and "productive forms of knowledge" (*Met.* θ, 1046b3) are technological.
23. Second nature as an acquired immediacy is appropriate in areas beyond those discussed here. Consider the conception operative in McDowell's ethical thought: We observe a habituation into virtue and an immediate efficacy in the virtuous person. The immediacy manifest in ethical virtue goes beyond the mere lack of explicit deliberation by the virtuous person: the continent person can also, sometimes, do without deliberation. Virtue is manifest as well in what things are not even *potential* deliberanda. While the virtuous person *can* offer a reason for why it wouldn't have been appropriate to do some vicious thing ("because it's wrong" being the most anemic), it's not off the initial deliberative agenda because it appeared on, and was dismissed as wrong from, a prior agenda. A reconstruction

of the absence of the vicious option in terms of a prior deliberation would make the virtuous agent a very quick-witted continent agent.

24. See Müller (1992, 165–167) for a subtle exploration of this thought. The point about transitions can be seen already in Carroll (1895).

25. "Education" in the sense of an upbringing with components of habituation, an *Erziehung*—clearly technological capacities can be acquired by teaching.

26. "Entrains" because the involvement of our capacities must differ widely in the cases of action and perception I've discussed: we cannot help hearing comprehendingly, but act for reasons.

27. One salient difference between the practical and theoretical cases here discussed is that while reason may be involved in reception, one does not need *a* reason to receive, but one does need a reason—at least a degenerate reason—to act. The argument for the existence of rational but non-technological capacities did not, however, hinge on the fact that reception does not require *a* reason to be kicked off; it hinged on the fact that there is no place *in the receiving* for the instrumental rationality characteristic of technological capacities.

28. One may wish to think here of the discussion of substitutes in Merleau-Ponty (1983).

29. Cf. McDowell (2007, 367).

30. Basic actions are often supposed to have an intimate connection with the body, but the preceding reflections suggest that anatomical connectedness is not (except causally) directly relevant: the practical significance of my arm's being attached by tendons to my trunk is nil. Indeed, O'Shaughnessy's perceptive discussion of the "immediate object" of the will effectively identifies the body-with-which-one-acts as that which has an immediate practical significance for one (so that "this is how a man relates to his legs [or] an amputee to his prosthetic workable hand; but it is not how these people relate to their kidneys or brain or hair or clothes or furniture" [O'Shaughnessy 1980, I:147]). (See also Gallagher [2005, 29].) We are left with a picture of agency as founded capacities to do certain things with a "body" characterized precisely by its potential for being deployed toward various ends.

31. Talk of what *a person* does can only ever set a *problem* for causal theories, which must attempt to reconstrue person-talk in terms of the mutual relations of impersonal events and states.

References

Anscombe, Elizabeth. 1963. *Intention*. Oxford: Basil Blackwell.

Aristotle. 1984. *Complete Works*. Edited by Jonathan Barnes. Vol. I. Princeton, NJ: Princeton University Press.

Baier, Annette. 1971. "The Search for Basic Actions." *American Philosophical Quarterly* 8 (2): 161–70.

Carroll, Lewis. 1895. "What the Tortoise Said to Achilles." *Mind* 4 (14): 278–80.

Frankfurt, Harry. 1978. "The Problem of Action." *American Philosophical Quarterly* 15 (2): 157–62.

Gallagher, Shaun. 2005. *How the Body Shapes the Mind*. Oxford: Oxford University Press.

Haugeland, John. 1998. "Mind Embodied and Mind Embedded." In *Having Thought*, edited by John Haugeland, 207–37. Cambridge, MA: Harvard University Press.

Heidegger, Martin. 1978. *Being and Time*. Translated by John Macquarrie and Edward Robinson. Oxford: Blackwell.

Hornsby, Jennifer. 2004. "Agency and Actions." In *Agency and Action*, edited by Helen Steward and John Hyman, 55:1–23. Royal Institute of Philosophy Supplement. Cambridge: Cambridge University Press.

———. 2011. "Actions in Their Circumstances." In *Essays on Anscombe's Intention*, edited by Anton Ford, Jennifer Hornsby, and Frederick Stoutland, 105–27. Cambridge, MA: Harvard University Press.

Husserl, Edmund. 1989. *Ideas Pertaining to a Pure Phenomenology and to a Phenomenological Philosophy, Second Book*. Translated by R. Rojcewicz and A. Schuwer. Dordrecht: Kluwer Academic.

Kenny, Anthony. 1976. *Will, Freedom and Power*. Oxford: Blackwell.

Kern, Andrea. 2006. *Quellen des Wissens: Zum Begriff vernünftiger Erkenntisfähigkeiten*. Frankfurt: Suhrkamp.

Lavin, Doug. 2012. "Must There Be Basic Action?" *Noûs* 47 (2): 273–301.

Makin, Stephen. 2006. *Aristotle's Metaphysics* θ. Translated by Stephen Makin. Oxford: Clarendon Press.

McDowell, John. 1994. *Mind and World*. Cambridge, MA: Harvard University Press.

———. 1998. "Some Issues in Aristotle's Moral Psychology." In *Ethics*, edited by Steven Everson, 107–28. Companions to Ancient Thought 4. Cambridge: Cambridge University Press.

———. 2007. "Response to Dreyfus." *Inquiry* 50 (4): 366–70.

Merleau-Ponty, Maurice. 1983. *The Structure of Behavior*. Translated by Alden Fisher. Pittsburgh, PA: Duquesne University Press.

Moline, Jon. 1975. "Provided Nothing External Interferes." *Mind* 84 (334): 244–54.

Müller, Anselm. 1992. "Mental Teleology." *Proceedings of the Aristotelian Society* 92: 161–83.

Noë, Alva. 2013. "On Overintellectualizing the Intellect." In *Mind, Reason, and Being-in-the-World: The McDowell-Dreyfus Debate*, edited by Joseph Schear, 178–93. London: Routledge.

O'Shaughnessy, Brian. 1980. *The Will: A Dual Aspect Theory*. Vol. I. Cambridge: Cambridge University Press.

Rothstein, Susan. 2004. *Structuring Events: A Study in the Semantics of Lexical Aspect*. Oxford: Blackwell.

Thompson, Michael. 2008. *Life and Action*. Cambridge, MA: Harvard University Press.

Vendler, Zeno. 1967. "Verbs and Times." In *Linguistics in Philosophy*, edited by Zeno Vendler, 97–121. Ithaca, NY: Cornell University Press.

Wilson, George. 1989. *The Intentionality of Human Action*. Stanford, CA: Stanford University Press.

Wolfson, Ben. 2012. "Agential Knowledge, Action and Process." *Theoria* 78 (4): 326–57.

5 Making the Agent Reappear
How Processes Might Help

Helen Steward

A recurring theme in critiques of what has come to be called the "standard story of human action"[1] is that the standard story cannot provide an account of the causal genesis and ontological structure of action which gives the *agent* of the action an appropriate role in the account.[2] Actions are said by the standard story to be events (generally speaking, bodily movements) which are caused by some suitable conjunction of preceding mental events and states—things such as desires, beliefs, changes in desires and/or beliefs, episodes of intention formation, and so on. But none of these antecedent events and states, self-evidently enough, is itself an agent. The question whether the standard story can survive this particular worry is therefore the question whether the insistence that agents *ought* somehow to figure in any adequate representation of the initiation and control of action is a foolish demand that mistakes the whole purport of the standard story—that to demand that the agent show up in the account would be like demanding, as Velleman (1992, 462) puts it, that a cake appear in its own recipe—or whether, on the contrary, it is a fundamental objection that no theory of the usual event-causal sort will ever be able to meet.

I have, on the whole, been inclined to subscribe to a qualified version of the second view.[3] However, I think the issue needs more ontological resources for its full resolution than some others who have taken issue with the standard story have supposed. It is quite commonly held, amongst some of those who oppose the standard story, that more careful attention to the logic of action explanation would, in effect, suffice entirely to undermine the claims of any given set of particular mental events and states, of the kind which the standard story postulates, to be the causes of our actions, because such attention would disclose that action explanations (particularly those which invoke *reasons*) rarely, if ever, advert to such particular event/state-causes in the first place.[4] These philosophers have pointed out, for example, that the questions by means of which we most naturally demand causal elucidation of agents' doings are generally questions of the form "Why (was it the case that) *p*?", where *p* is a *fact*[5] (e.g. "Why did he eat the banana?"), not questions of the form "What caused A?", where A purports to be a singular term for a particular action (e.g. "What caused *his eating of the banana*?").[6]

The relevant proposition p, moreover, may be negative (e.g. "Why didn't he turn up to the meeting?"); stative (e.g. "Why is he just standing there like that?"); or general (e.g. "Why does she always knock so loudly?")—and it is not in the least obvious that any of these sorts of questions is asking about the cause of the occurrence of any particular event, even while it remains patently obvious that the kind of elucidation sought in these negative/stative/general cases is of the same nature as that which is sought in cases where what is to be explained *does* seem to involve the occurrence of a token action. And even if we think that in asking such questions, we *are* (at least sometimes) asking why a ø-ing event occurred (as for example with a question such as "Why did he press the button?"), to ask why a ø-ing event occurred is still not to ask, of any individual ø-ing, why *it* occurred. Such questions, as pointed out by Davidson (1967a, 1967b, 1969) are of the general form "Why (was it the case that) p?", where p is normally a true proposition which may *quantify* over events but does not usually *refer* to any individual occurrence.[7] To ask "Why did she raise her arm?", for example, is not to ask of any given, token arm-raising of hers why *it* occurred—it is rather (at best) to ask why there was *an* event of a given kind (e.g. why there was *an* arm-raising).[8] Much of what is usually called "action explanation" in philosophy is thus argued by these philosophers really to be a matter of explaining *truths*, or *facts* (generally facts about why agents have done/not done the things they have done/failed to do), rather than pointing out the causal antecedents of token events. Moreover, these truths or facts are often explained by things which are themselves ostensibly truths or facts (e.g. "because *he wanted to blow up the bridge*", "because *she thought it was the right thing to do*")—and so it is highly debatable whether we can readily derive from folk psychology a commitment to the existence of a realm of particular, causally efficacious events and states of the sort that might compete with agents in a full account of the causal provenance of actions. Later in this chapter, I shall need to return to this point, which is a source of constant confusion in the philosophy of action. But it remains questionable, I think, how much work such observations can do *on their own* to dispel the worry that it is hard to make sense of the relationships between the agent of an action, the action itself, and the action's causes (whatever these are conceived of as being, ontologically speaking). Even if one thinks of action explanation (as I and many others do) as a matter mainly to be understood in terms of relationships amongst facts, still, there is no denying that there are such things as particular actions, which take place at particular times, in particular places, even if much of the philosophy of action asks questions which are not (or should not be) primarily concerned either with these particulars, or with their causal provenance. Moreover, these particular actions are exceedingly interesting occurrences, metaphysically speaking, even if ordinary thoughts and talk mostly does not mention them very much or attempt causally to explain them (as opposed to explaining why agents have done things of certain *sorts*). And so, even granted the correctness and importance of the claim that it is really *truths or facts* that are generally the focus of our action explanations, the question remains to be asked, what sorts of things these particular actions are—and how best to give an account of them which properly permits us to position the agent in the account in her rightful place, not indeed as the action's *cause*—but as the doer of the doing in which the action consists. In this chapter, I want to consider the question whether the category of *process* might help with this difficult task.

In the first section of the chapter, I shall make a distinction between two versions of the worry about disappearing agents, so as to sharpen up the description of the worry which I believe the category of process might be able to allay. In the second section, I shall attempt to give a brief summary of an account of processes (and their distinctness from events) which I have developed elsewhere. Then in the third section, I shall explain some respects in which I believe a processual account of actions is better placed than one which regards actions as events to respect the central role of the agent in the phenomenon of action.

1. The Problem of the Disappearing Agent

For many, the so-called Problem of the Disappearing Agent is probably most closely associated with the name of David Velleman. Velleman (1992, 461) argues that the standard story fails to "cast the agent in his proper role." He continues as follows:

> In this story, reasons cause an intention, and an intention causes bodily movements, but nobody—that is, no person—*does* anything. Psychological and physiological events take place inside a person, but the person serves merely as the arena for these events: he takes no active part.
> (Velleman 1992, 461)

Thus expressed, I think many of those who have worried about disappearing agents would accept Velleman's description as a good statement of their own disquiet about the standard story. But as Velleman's paper progresses, it becomes clear, I think, that his worry is actually much more specific than might at first be suggested by the aforementioned quotation. Velleman's concern, it becomes evident, is not about the capacity of the standard story to serve as an account of action *in general*, but rather its potential to deliver an accurate account of a more narrowly delimited class of actions that he terms "full-blooded actions." Velleman seems happy to accept that there are many actions which are not full-blooded in the relevant sense, and in which the agent does not "tak(e) . . . an active part" (462)—and suggests that the actions of non-human animals are amongst them; he also appears to believe that there are many *human* actions which fail to be "full-blooded." But he does not express any doubts about the capacity of the standard story to account for these *non*-full-blooded actions. His anxiety about the disappearance of the agent is an anxiety that what he calls "the distinctively human feature" (462) is missing from the standard story. What this distinctive feature is, exactly, is not entirely easy to pin down; it is, at any rate, something which is lacking in cases in which a person acts "half-heartedly, or unwittingly, or in some equally defective way" (462). For present purposes, fortunately, it is not really necessary to characterize precisely what it is for an action to be "full-blooded," in Velleman's sense. We need only to note that Velleman's concern is not with the inability of the standard story to deliver an adequate account of what happens when someone acts in *all* instances of action, but only in a certain privileged subset of cases.

By contrast, though, it seems possible to have a much more basic version of the disappearing agent worry, one which questions the capacity of the standard story to deliver an account even of actions that clearly fail to be "full-blooded" in anything like Velleman's sense, including: tedious actions which are part of one's daily routine, say, carried out with a minimum of deliberation, thought and attention, such as making a sandwich for lunch, or cleaning one's teeth; actions which are done because of desires one would prefer not to have, such as taking a drug one wishes one were not addicted to; and even what might be called "absent-minded" actions or activities, such as walking up and down aimlessly whilst waiting for someone, or scratching one's chin. This was first pointed out (to the best of my knowledge) by François Schroeter (2004); and the point is developed in admirable clarity in an unpublished paper by Matthew McAdam. Schroeter's overall concern is to oppose what he calls "endorsement models" of autonomous agency, of the sort which he claims are offered by Frankfurt and Velleman. According to such models, autonomy is secured with respect to a given action only when the agent in some way or another endorses the desire which leads to it. One of Schroeter's central objections is that on such models of what it is to act autonomously, no agential involvement in the action appears to take place at what Schroeter calls "the executive stage" of the action—that is to say, during the period in which the action is actually taking place and throughout which the agent must put in place the various elements of his or her action plan, adjusting the plan in response to new information, making changes if necessary, and so on. Rather, truly agential involvement all occurs "upstream" of the action, as the agent selects from amongst the whole panoply of his desires those with which he wishes to identify, or which he wishes to endorse. But Schroeter points out that the difference between an autonomous action and one which is non-autonomous seems to be a distinction *within* a class of things which are still intuitively actions (rather than, say, a class which also contains mere reflex responses or events such as seizures which simply befall one), and whose active nature is apparent at the executive stage, no matter whether the action accords or not with some specially endorsed desire. He notes, for example, that even an unwilling addict of the sort that both Frankfurt and Velleman would wish to characterize as unfree in important respects is scarcely a "passive bystander" to "forces that move him," as Frankfurt seems at one point to suggest (1988, 22). In taking the drug, the addict will normally be perfectly well aware of what he is doing, and crucially, he must usually formulate an action plan in order to achieve his desired end—that is to say, he must work with some sort of representation of the task which is to be performed, breaking it down into its essential components, settling on the most efficient means to ends, and so on. He may need to call his dealer, for instance, to obtain a supply; and even if the drug is on hand, there will still be choices to make—choices about how to take the drug (smoke? inject?), how much to take, whether to do so alone, or call another addict to participate. There may be equipment to gather and prepare and so on. It can hardly be acceptable to represent this executive stage as something from which the agent's role as guide and decision-maker is entirely absent, even if the addict is in some important sense alienated from what he is doing. Still, what he is doing is something he is *doing*, not something which is merely happening to him, and any acceptable account of action will have to have something to say about what makes it the case that this is an instance of action rather than of something which merely befalls the agent.

It might be responded, perhaps, that endorsement models of autonomous agency are trying to capture the conditions not of *agency* in general, but rather of *autonomous* agency, and that it is hardly fair to criticize these models for not succeeding in a task for which they were never intended. There is some justice in this complaint. Yet it cannot be denied that on occasion, defenders of the endorsement view speak as though it really is the distinction between active undertakings and merely passive befallings which the presence/absence of endorsement is supposed to characterize. The previous quotation from Velleman, for instance, complains that on the standard story, "no person *does* anything" (1992, 461), which certainly makes it sound as though the essence of the complaint is that *activity per se* fails to be represented properly by that story.[9] And Frankfurt's talk of the unwilling addict being a "passive bystander to the forces that move him" is similarly suggestive of the idea that it is the distinction between activity and passivity that we will be helped to understand by the endorsement model. But it is highly doubtful that this is really the case, given that activity still seems to be characteristic even of the alienated engagements with the world that endorsement models seek to separate from the "full-blooded" actions in which our will as agents is truly manifest. McAdam accordingly distinguishes helpfully between two versions of the Problem of the Disappearing Agent, which he calls "the rationalist version" and the "basic version." The rationalist version, the one which Velleman seeks to address, is characterized by McAdam as "a moral psychological problem about accounting for the role that our reflective capacities play in human action in terms of CTA" (3). The basic version, on the other hand is "a metaphysical problem about how the activity of parts relates to the activities of the wholes they comprise" (3), a question about how bodily movements which are caused by what McAdam refers to as "sub-agential items" like mental states can be permitted to count as movements caused by the whole agent of whom they are the states. McAdam argues that Velleman accuses the standard story of suffering from the basic version of the Disappearing Agent worry, but of responding only to the rationalist version. What is essential for present purposes is to understand that there are, then, (at least) two distinct versions of the Problem of the Disappearing Agent, and that it is essential to be clear which is at issue. The problem for which I am proposing that processes might form part of the solution is what McAdam calls the *basic* version—it is a problem about whether the standard story can account properly for the agent's role in *any* case which is truly characterizable as an instance of action, and not merely about a special category of actions which are truly expressive of the agent's will, or which constitute peculiarly human forms of engagement with the world.

It might, of course, be doubted whether there *is* any such problem as the basic version of the Problem of the Disappearing Agent. Velleman himself seems to doubt it: his thought seems to be that the standard story might do perfectly well for half-hearted, unwitting, or similarly defective sorts of action; and that the mere fact that the standard story represents agency as a series of interactions amongst states and events is, in and of itself, probably unobjectionable. Indeed, he appears to go somewhat further than this when he asserts that "any explanation of human action will speak in terms of some such occurrences, because occurrences are the basic elements of explanation in general" (1992, 468). But for reasons I have already mentioned, it is in fact highly questionable whether this is true. If we take the term "occurrences" strictly and literally, the claim is false even by Velleman's own lights, since states (which are *not* occurrences) are clearly elements of explanation on the model of agency he favors (in particular, the state of being motivated by reason). But even if (ill-advisedly, in my view)[10] we loosen up and regard both states and events as "occurrences" in some more general sense, it remains deeply controversial whether these things constitute the "basic elements of explanation in general." As I have already suggested, an alternative view, which has been widely endorsed by many of those who have reflected on the phenomenon of explanation, is that its basic elements are *facts*: that is to say, that they are true propositions.[11] As already noted, these things are our most usual explananda ("Why is it the case that *p*?" "Why is it the case that *not p*?" "How did it come about that *q*?") and they are also our commonest explanantes ("because *r*").[12] Some of these facts may of course be facts which *imply the existence* of occurrences. For example, *that Doris capsized the canoe* implies at least one particular occasion of capsizing.[13] But when we undertake to explain the capsizing (in answer to the question "Why did Doris capsize?"), we undertake to explain it *qua* capsizing, and not, say *qua* wetting of Doris; or *qua* the most terrifying event of Doris's life; or *qua* effect of insufficient training on the part of Doris; even if the capsizing *is* (identical with) a wetting, or the most terrifying event of Doris's life, or the effect of insufficient training on the part of Doris. Such explanations as we might give of the fact that Doris capsized (e.g. that Doris was practicing to improve her abilities to re-right the canoe) are therefore really better seen as explanations of why an event of a certain *kind* (the capsizing-by-Doris kind) took place.[14] They are not well thought of as explanations of *particular*, multiply redescribable occurrences. Moreover, the *explanantes* in this case are likewise facts or propositions ("because Doris was practicing to improve her abilities"—or, converting to the belief plus desire form favored by many philosophers, "because Doris *wanted* to improve her abilities and *believed* that practice would help her to do so"). So Velleman's premise actually seems to me eminently contestable.[15] It is not in the least evident that occurrences are the basic elements of explanation in general—it seems much more plausible, indeed, that facts play this role.

Someone might think that Velleman's conclusion that occurrences must figure in the correct story of human *action* remains plausible nonetheless. For even if occurrences cannot be said to be "the basic elements of explanation in general," actions surely *are* occurrences, and when one explains why an occurrence of a certain kind took place, it might be suggested that we always need to advert somehow to further occurrences. In fact, of course, even this is not *strictly* true. I can explain why a bridge collapsed by adverting to a structural weakness, rather than to its being traversed by a heavy lorry. But still, it might be said, the "complete explanation" of the bridge's collapse will have to mention the lorry's crossing of the bridge as well as the structural weakness. Even this might be questioned for any of a variety of reasons—one might be skeptical (as I am) about the idea that anything has a "complete explanation," or one might note that some events appear to take place when they do for no discernible reason (e.g. radioactive emissions) and that there are many others for which, though we assume a prior event-cause, we have no actual *knowledge* of any (e.g. when a book slides off a shelf on which it has been happily propped for years, all of a sudden). But even if we allow Velleman the suggestion that the causal explanation of an event in the inanimate world must always include some sort of reference to the occurrence of another, prior event, it is not clear that we are entitled to insist without further argument that this same model of event causation must apply also throughout the *animate* world. The question at issue, indeed, between event-causalists and many agent-causalists in the philosophy of action might be said precisely to be the question whether this is so.

To see why an agent-causalist might demur, it is essential to realize that there is an awkward ambiguity in that phrase of Velleman's, "explanation of human action." For the standard story purports to offer not an *explanation* of human action, exactly, but an *analysis* of the ontological structure that must underlie any instance of the phenomenon. The whole causal chain described is supposed, *in toto*, to *constitute* an agent's active contribution to the world, in much the same way as a certain succession of causally related events can count as a ball's breaking of a window (the ball's colliding with the window, the glass shattering as a result). The contribution of the mentioned events and states to the causal order, it might be said, simply *amounts to* the agent's contribution, when an appropriate bodily movement ensues and is caused in "the right kind of way."[16] But in the view of many agent-causalists, there is a crucial difference between mere inanimate substance causation and any instance of true agency. The reason is that it is an utterly central part of the way we tell the folk psychological story of any actively produced result that such things as believing, desiring and intending alone are never enough to get limbs into bodily motion—for that to occur, the agent also has to *act*. I can sit here wanting to write the next section of this chapter, and intending to do so, for hours on end, but I may still procrastinate and none of my intending will be in the least effective in seeing to it that my fingers actually move upon the keyboard if I do not actually *execute the action*. Execution is needed to get any of our intentions realized: so much is a truism, but an important one. We always need to *do* something if our intentions are to be translated into reality. And it is this important truism that seems misrepresented by the standard story. In the standard story, once an intention is in place, and certain relevant beliefs are also held by the agent (perhaps, e.g., the belief that *now* is the time to act on the intention in order to realize it most effectively), the bodily movement which is an essential component of the action simply follows causally as a result. But we all know, I submit, that this is not how it is.[17] The events which are actions are absolutely required to move the world on from intentions, on the one hand, to realizations, on the other. It seems deeply to misrepresent the phenomenology of action to suggest that the appropriate bodily movements simply follow inexorably upon the presence of the relevant intention. But in that case, it seems we need an action to appear *in the middle* of the causal chain offered us by the standard story, the causal chain which was supposed to tell us what it was for an action to occur in the first place. We seem to need an action to appear between the intentions and beliefs on the one hand, and the bodily movement on the other, to make the bodily movement happen, as it were, pursuant to the relevant intention. Our involvement is necessary here, at this crucial point. It is the disappearance of this executive stage from the standard story that makes it seem essentially incapable of representing the crucial agential role of *executor*—and which makes it possible to argue that the causal chains involved in the translation of thought into reality will not reduce to chains of events in which the agent's involvement does not appear—that is to say, in which *actions* do not appear. And it is this, I suggest, that is the basic issue for the standard story. Actions cannot be constituted by the causal chains described in the standard story if they are needed, crucially, to serve as links within those chains.

What, then, should replace the standard story, so that it can surmount the Problem of the Disappearing Agent? If particular actions are occurrences (as I have conceded), what is to be our model of their causal generation, if it is not to identify actions with bodily movements caused by a certain type of prior structured series of mental events and states? I have already distanced myself from those "agent-causal" accounts which insist that the agent herself must be the cause of an event which is an action. But if event-causal accounts such as the standard story are no good, and if agent-causal accounts in which the agent herself appears as the cause of an action will not serve to replace them, what is the alternative? My suggestion will be that we need to look to the category of *process* for help. In the next section of this chapter, I shall try to explain what I mean by a process, before going on, in the final section, to explain how the view that actions are processes might help with the Problem of the Disappearing Agent.

2. Actions as Processes

What are processes, and how do they differ from events? For reasons of space, I cannot do more here than provide a brief summary of an account of processes that I have developed and argued for in more detail elsewhere,[18] but I hope it will be possible to say enough to make it appear that the category is a hopeful resource for the philosopher of action. To begin with, it is essential to note that processes, as I understand them, are not merely chains of events, nor are they simply relatively *long* events. Chains of events simply compose longer events, and long events are still events. If the event/process distinction is to do any serious metaphysical work, processes have to be characterized in such a way that they are more fundamentally distinct from events than this.

How, then, is the distinction to be characterized? I begin from the idea that the most usual way in which we form expressions which purport to refer to such things as events or processes is by a certain sort of *nominalization*. From a sentence such as "Shirley ate a biscuit," for example, one can obtain the nominalization "Shirley's eating of a biscuit," an expression which most philosophers would regard as referring to an *event*. But in my view, there are reasons for thinking that although one *can* use this nominal to refer to an event, the most usual referent of the phrase is not an event, but something I call a *process*. To help us to disambiguate, let us speak of eatings$_p$ (or, more generally, of ø-ings$_p$), which are the processes, and eatings$_E$ (or more generally, of ø-ings$_E$), which are the events. What exactly is the difference, and why should anyone believe that there might be ambiguity here in the first place?

The first reason for thinking that "Shirley's eating of a biscuit" may not always refer to an event derives from the appealing thought that events do not change (at any rate, in respect of their intrinsic properties). There are strong arguments for this view in the literature, which I cannot rehearse in detail here,[19] but the basic thought is this. Change requires the persistence of an entity over time, during which the persisting entity first has one property and then loses it (generally to gain another from the same quality space). But events do not persist through time in the requisite way to undergo change. It is true that someone's eating of a biscuit, for example, might begin by being noisy, say, and finish by being much quieter, as the eater realizes that her noisy eating might be considered bad manners, and attempts to eat more quietly as she finishes the biscuit. But the whole event is not noisy at any time—and the whole event is never quiet either. Rather, there is a certain kind of succession of phases within the event itself—a noisy phase, followed by a quieter one. This is not change in the *event*; it is mere succession. The event which is constituted by the whole eating does not change—it has the auditory profile it does, and that is that. Analogies with the spatial distribution of certain sorts of properties are often used to make the point. Just as a curtain which has faded towards its top edge—so that it is dark green at the bottom but pale green at the top—does not change, so an event which is noisy at the beginning and quiet towards the end does not change either. The event is constituted by changes in *other* things—it does not itself undergo any change at all. It simply *is* a change.

What, then, are the things that may be changing when events occur? Evidently, there will often be changing *substances* involved—for example, Shirley's mouth will be moving as she eats the biscuit, the biscuit will be disintegrating, and so on. But we also wanted to say, in addition, that there was a change in Shirley's *eating*—that it *became quieter*. This seems to imply that there was something that began noisily and got quieter *as it went on*. Must we insist, in view of the earlier argument concerning the impossibility of change for events, that this cannot be strictly and literally true? Surely there is something very implausible about this thought. And yet the argument for the view that events do not change seems sound. The solution, in my view, lies in realizing that there *is* something occurrent that changes in a case such as this, but it is not an event. Rather, it is a *process*—something which *goes on* through time and can change as it does so.

Others who have developed similar views have drawn comparisons with the realm of continuants to help soften the initial surprisingness of the idea that we might need to recognize more than one class of occurrent entities.[20] Just as some have suggested that we need to recognize not only individual horses, say, but also the portions of matter of which they are composed, existing in the same place and at the same time as the individual horses, so, they say, we need to recognize processes alongside events. On the whole, though, those who have argued in this way have wanted to regard the processes as the things comparable to the portions of matter and the events as the things comparable to individual substances such as horses.[21] But for various reasons, I think this gets things rather the wrong way about. It is, in my view, the processes which are comparable to the horses, and the events which should be relegated to a role comparable to that of mere portions of matter. The reason for thinking so is that it is the processes, and not the events, which are mereologically and modally robust, just as a horse is mereologically and modally robust. By this, I mean that a process can acquire extra parts and still be the same process—it can *grow* (just as a horse can grow). As Shirley eats her apple, for instance, temporal parts are constantly being added to the process of eating in which she is engaged—but it is intuitively the same individual process of eating that is going on throughout. But the *event* of her eating the apple just consists of the parts it consists of, and that is that; it cannot grow. Moreover, it is plausible to think that there is a certain robustness in the cross-world identity conditions of processes, parallel to that which exists for individual substances.[22] This horse, for instance, might have had other properties of various kinds—it might have lost its leg as a foal, or had a longer mane. Similarly, Shirley's eating$_p$ of the biscuit might have been quieter all round if she'd only realized earlier that she was crunching unforgivably, right at the beginning. But her eating$_E$ of the biscuit has all its properties essentially. Changes to any of its parts would plausibly have constituted a different event. Of course, it is possible to deploy terminology differently. Others have undoubtedly utilized the term "event" to pick out the items I am here calling "processes." But those who have done so have generally failed to recognize that there are two distinct categories of entity available to be recognized here, not one. I believe clarity can be immeasurably aided by making the distinction between events and processes—and that having made it, we will be in a position to make progress on a number of knotty problems in a range of different areas of philosophy, including, especially, the philosophy of action.

3. Processes and the Disappearing Agent

Why should one think that the idea that actions are processes might help us with the Problem of the Disappearing Agent? After all, it might not be readily apparent how replacing one kind of occurrent with another could possibly help reveal the involvement in action of an entity which is *not* an occurrent at all, but a continuant. However, processes differ from events in a number of important ways, all of which help, in my view, to bring the agent of the action into the picture.

The first important difference between processes and events relates to the ways in which we are inclined causally to explain them. Events, we tend to think, should have causes—and by and large, as noted earlier, the causes of events are likely to include other events. Even if the event is relatively long—like the First World War, say—we can ask what caused it, and what we mean by that is something that caused its *start*, that initiated it. When we ask for the cause, or causes, of an event, even if it is a long and complex event, intuitively composed of many others, we are implicitly declaring ourselves essentially blind to, and uninterested in, any *internal* structure the event might have, even if we concede that a different token event would have occurred had things gone differently during its course. For instance, if someone wants to know the cause of the First World War, it would not be apposite to mention the refusal of the Allied Powers to accept the German peace offering of 1916, even though, had peace negotiations succeeded at that point, the actual First World War—that actual event, as it were, in its entirety—would not have occurred. That does not make the refusal of the Allied Powers to accept the offering into a cause of the First World War. So far as its causation *qua event* is concerned, the primary question is: What was the *trigger* for this event? What set it off? And this, in and of itself, tends to exclude consideration of the role of the agent from the metaphysics of action. Agents, whatever else they are, do not look like the right sorts of entities to be triggers.[23]

But I submit that we do not think about the causation of processes in the same way. The question "What caused this process?" is indeed rather strange. We might possibly want to know what started a process off, but with processes, a range of other causal questions jostle for attention. Why, for example, did the process unfold in this way rather than that? Why has it not stopped? What is sustaining it and keeping it going? What is responsible for any regular patterns we may observe in its progression? And so on. And to answer these questions, we seem to need to advert not to triggers, but to what might be called *sustainers*—things which perhaps explain the maintenance of a pattern, or a movement towards a goal, over the course of time. Once again, it is important to note that the appropriateness of questions about how a process is maintained and kept on track is not restricted to cases in which a process takes what is intuitively a fairly long period of time. On the contrary, the course of time involved may be very short. Many processes are over in a few milliseconds; think of those that go on in a computer, for example. But the same sorts of causal questions still arise, even for these microprocesses—namely, how is it ensured that they will be carried through to completion, that their different parts will succeed one another in a way resistant to interference and obstruction? Their *initiation* is only one of the aspects we may seek causally to explain.

The reason this seems important for philosophy of action is that agents seem much better suited to function as sustainers than as triggers. Things which can sustain or maintain other things have to be things which themselves *last*. This is not to say that they *must* be agents, or even that they must be substances of any kind—processes can themselves serve as sustaining causes for other processes, and states are another kind of lasting entity to which we might make appeal when looking for answers to questions about how processes are sustained. It is not a sufficient condition of an account of agency that makes proper space for agents themselves that it conceptualize actions as processes, rather than events, and therefore as things in need of sustenance. But it may well be a *necessary* condition. Even if agents are not the only sorts of sustaining entities there are, they at least have the potential to function as sustainers of processes, whereas it seems evident that they are not the right sorts of entity at all to function as triggering causes. If we conceptualize actions as events of bodily movement, and then ask questions about the causation of these events, we have arguably already adopted a metaphysics of action which precludes finding agents to be of central interest, since we have adopted a view which makes the triggering event the matter of central and crucial relevance.

A second advantage of the move to processual thinking relates to the fact that whereas events are changes, occurrences suitable to serve as the causes of other events, processes are *changings*, suitable not so much to be causes as to be *causings*. In this half-baked noun, "changing," in which, to quote Vendler, there remains a verb "alive and kicking" (Vendler 1967, 131), we have a ready grammatical reminder of the thing which is doing the changing—the doing or the undergoing in which the process consists. And whereas it is perfectly respectable to ask what caused a given event, it is not obvious that it is always perfectly respectable to ask what caused something which is itself a causing. If it were, then we would be able to ask, when an *event* causes another, what caused the causing of the second event by the first—which seems to be a question that is out of place. Why then should we think it must be in place when the causing is by an agent? Causings, that is, perhaps are not the sorts of things that themselves need to be caused—and so the chain of questions about actions which proceeds back beyond the action to ask "Yes, but what caused *that*?" might legitimately be refused. This need not mean that there is no causal explanation of why the agent has done what she has done, no explanation of *the fact that p*, for relevant p (e.g. that S has robbed the poor box or gone to University or danced a tango), a causal explanation in terms of further facts about S and/or her circumstances. It is only that we are not obliged to search for a *particular* cause for a causing.

A third and final benefit of the move to processual thinking—one connected to the second—relates to how we are inclined to ask about the initiation of processes. I have said that many causal questions arise concerning processes that are not visible when we ask merely for the cause of some event (e.g. questions about how they are maintained over time). But still, anything that is an individual process must begin at some time, if it has any kind of well-defined identity as an individual at all. And we might want to know why a process began when it did. But we are likely, I think, to ask the question in a rather different way from the way we will be inclined to ask it when an event-causal model dominates our thinking. In asking for the cause of an event, we have already implicitly set the stage for an answer that mentions another event. But as noted earlier, the question "What caused this process?" is odd—as are its more specific variants, for example, "What caused this running/dancing/writing . . . ?" Much more natural, if we are interested in the initiation of such things, is "Why did S start to run/dance/write?" and so forth. And this is quite clearly a question about S, the agent of the action, and invites an answer in terms of S—perhaps, for example, just "because S suddenly felt like running/dancing/writing." Here it is quite evident that our explananda are *facts*, not particulars—and so will be our explanantes. (The explanandum in this particular case is *the fact that* S started to run/dance/write; and the explanans is *the fact that* she suddenly felt like running/dancing/writing.) We are, I think, much less likely to lose sight of this important fact if the causal structures we envisage as realizing the changes in which we are interested are processual ones. Because processes are not all "there," as it were, until they are completed, it is much less tempting to think of them as potential usurpers of a capacity to produce and initiate that really belongs not to occurrences of any kinds, but to agents and other substances. Their dependence upon the substances which sustain them is clear, whereas events masquerade much more successfully as purveyors of their own distinctive form of causal influence.

In conclusion, then, what would those all-important diagrams look like if we were to replace events with processes in our thinking about agency? Something like this, perhaps:

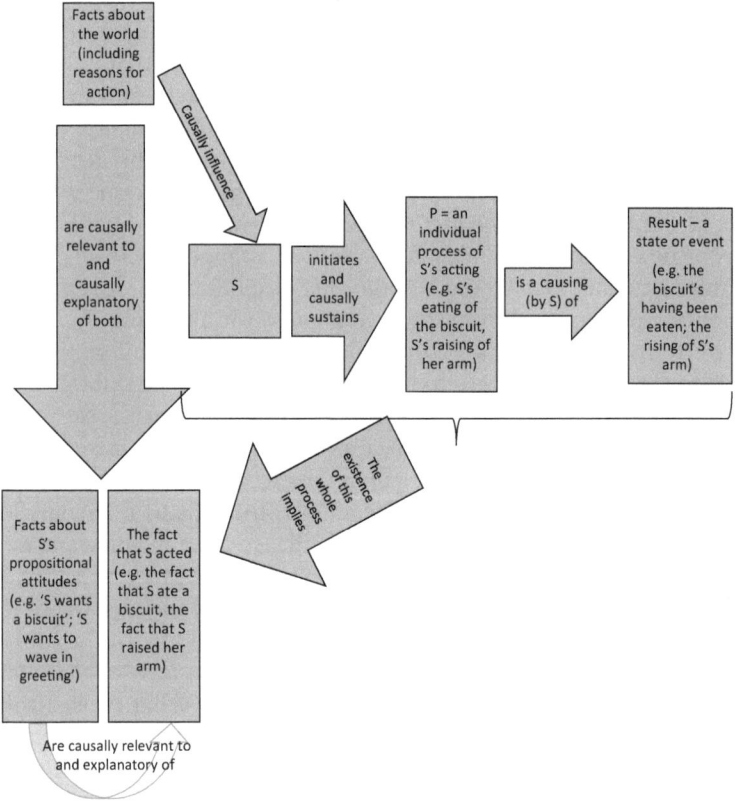

Figure 5.1 Facts, processes, and events in the causation and explanation of action

In this figure, I have represented both causal-explanatory relationships between facts and extensional causal relations between particulars. It is acknowledged by the line of arrows which moves across the diagram horizontally, that particulars of a special kind are implied by action sentences, but these are conceived of as individual processes, and not events. And unlike events, these processes are causings already—it is therefore inappropriate to insist that they must themselves have particular causes of an event-like sort, though their initiation may admit, of course, of a reason-giving explanation, which indicates the reasons which influenced S to act. Actions appear unreduced in the middle of the diagram, their existence causally explained, perhaps, by certain facts about the agent, but not triggered by particular causes, since causings do not require to be caused by other particular causes. And though in its resistance to a reductive event-causal model, this figure might be deemed to endorse a variety of agent causation, it does so without making the fatal assertion that S is the cause of any action. S is rather a *causer* of results; and her causing those results is a process, something which goes on through time, changing and growing as it does so, sustained throughout by S, who cannot be deleted from the picture and replaced by a remorseless chain of extensional causal relationships amongst entirely other varieties of entity.

Notes

1. As far as I know, it is Velleman who first coined the phrase "the standard story of human action"; see his (1992, 461).
2. Various versions of this worry can be found, for example, in Velleman (1992), Searle (2001), Hornsby (2004), and Steward (2012b).
3. For a full explanation and defence of the extent to which I endorse a form of agent causation, see my (2012b, ch. 8).
4. See e.g. my (1997), and also Dancy (1993), Stout (1996), and Alvarez (2010).
5. By a "fact" I mean merely a true proposition.
6. See my (1997) for a detailed explanation of the importance for philosophy of causation of understanding the difference between *facts*, which are expressed by sentences and have a propositional structure, and token *events*, which are particular entities suitable for picking out by means of singular terms, such as definite descriptions.
7. This point is explained by Davidson in "The Individuation of Events" as follows:

   We recognize that there is no singular term referring to a mosquito in "There is a mosquito in here" when we realize that the truth of this sentence is not impugned if there are two mosquitos in the room. It would not be appropriate if, noticing that there are two mosquitos in the room, I were to ask the person who says "There is a mosquito in the room," "Which one are you referring to?" On the present analysis, ordinary sentences about events, like "Doris capsized the canoe yesterday" are related to particular events in just the same way that "There is a mosquito in here" is related to particular mosquitos. It is no less true that Doris capsized the canoe yesterday if she capsized it a dozen times than if she capsized it once; nor, if she capsized if a dozen times, does it make sense to ask "Which time are you referring to?" as if this were needed to *clarify* "Doris capsized the canoe yesterday." (Davidson 1969, 167)

8. Even this may be too much to claim. As Hornsby notes, "asking why a ø-ed, we hope to learn something about a, the person, but if we asked why a's ø-ing occurred, a might not be a subject of concern at all." (2004, 134)
9. Perhaps Velleman intends the word "person" to be significant here—indeed, the context ("nobody—that is, no person—*does* anything") suggests as much. Does

Velleman intend to leave open the possibility that somebody that is *not* a person (an animal, perhaps?) might be doing something, even though the person is not? The answer is not clear; but I believe myself that the metaphysics that would be required to make sense of the idea that an animal might be doing something that the spatiotemporally coincident person was not, is in the end untenable (see Snowdon 1990). Insofar as the idea encodes important intuitions, there are surely better ways to encapsulate them.

10. See my (1997) for an extended argument against the idea that it is safe to regard states—even so-called "token" states—as events in a broad sense of the term "event."
11. See e.g. Van Fraassen (1980) and Mellor (1995).
12. And one must also remember "Why is it the case that not-p" (e.g. "Why didn't you put the cat out?"). Attempts to insist that answers to such questions must be explanations of occurrences are likely to involve us in spurious and quite unnecessary difficulties with "negative events," which there is simply no need to countenance once the essentially propositional nature of explanation is understood.
13. This example is Davidson's—see his (1969).
14. As before: at best. See note 8.
15. One can see this fairly readily by reflecting upon one of the examples which is central to Velleman's overall case for thinking that the agent is importantly missing from the standard story in some sorts of cases. The case he envisages is an occasion on which I have arranged a meeting with a friend for the purposes of resolving a disagreement. As the conversation proceeds, the friend's "offhand comments provoke me to raise my voice in progressively sharper replies until we part in anger" (Velleman 1992, 464). The object of explanation here is not a particular occurrence. It is a set of *fact*s of a fairly complex sort—the fact that Velleman's voice rose progressively, that his replies increased in sharpness, and that the replies became so sharp, eventually, as to precipitate an angry parting.
16. This qualification (or something similar) is generally added to the standard story in order to get around the notorious problem of deviant causal chains.
17. Sometimes, Velleman seems inclined to make the same point: "the agent . . . moves his limbs in execution of his intention; his intention doesn't move his limbs by itself" (1992, 462). But this is true no less of the non-full-blooded cases than it is of the "full-blooded" ones for which he demands an improved account.
18. See especially my (2012a), (2013), and (2015).
19. See Dretske (1967), Mellor (1981), Hacker (1982), Simons (1987), and Galton and Mizoguchi (2009) for arguments of this kind.
20. See especially Mourelatos (1978).
21. There is a further debate, which I hope to avoid here, about whether one needs to recognize only processual *activity* (corresponding, as it were, to *stuff*) or whether there is also a case for the recognition of individual processes, alongside events. See Hornsby (2012) for the former view, and my (2013) for the latter.
22. See my (2013) for a development of the idea of what I call "modal robustness in virtue of form," a property common to individual substances and individual processes.
23. This point is famously made, in effect, by C.D. Broad in his objection to the notion of substance causation—see his (1952, 215).

References

Alvarez, M. (2010). *Kinds of Reasons*. (Oxford: Oxford University Press).
Broad, C.D. (1952). 'Determinism, Indeterminism and Libertarianism', in *Ethics and the History of Philosophy* (London: Routledge and Kegan Paul): 195–217.
Dancy, J. (1993). *Moral Reasons* (Oxford: Blackwell).

Davidson, D. (1967a). 'Causal Relations', *Journal of Philosophy* 64 (1967); repr. in *Essays on Actions and Events* (Oxford: Oxford University Press, 1980): 149–192.

—— (1967b). 'The Logical Form of Action Sentences', in N. Rescher (ed.), *The Logic of Decision and Action* (Pittsburgh: University of Pittsburgh Press); repr. in his *Essays on Actions and Events* (Oxford: Oxford University Press, 1980): 105–148.

—— (1969). 'The Individuation of Events', in N. Rescher (ed.), *Essays in Honor of Carl G. Hempel* (Dordrecht: Riedel); repr. in his *Essays on Actions and Events* (Oxford: Oxford University Press, 1980): 163–80, to which page references refer.

Dretske, F. (1967). 'Can Events Move?', *Mind* 76: 479–492.

Frankfurt, H. (1988). 'Freedom of the Will and the Concept of a Person', in *The Importance of What We Care About* (Cambridge: Cambridge University Press): 11–25.

Galton, A. and Mizoguchi, R. (2009). 'The Water Falls but the Waterfall Does Not Fall: New Perspectives on Objects, Processes and Events', *Applied Ontology* 4: 71–107.

Hacker, P.M.S. (1982). 'Events and Objects in Space and Time', *Mind* 91: 1–19.

Hornsby, J. (2004). 'Agency and Actions', in J. Hyman and H. Steward (eds.), *Agency and Action* (Cambridge: Cambridge University Press): 1–23.

—— (2012). 'Actions and Activity', *Philosophical Issues* 22: 233–245.

Mellor, D.H. (1981). *Real Time* (Cambridge: Cambridge University Press).

—— (1995). *The Facts of Causation* (London: Routledge).

McAdam, M. 'Velleman on What Happens when Someone Acts', unpublished manuscript.

Mourelatos, A. (1978). 'Events, Processes and States', *Linguistics and Philosophy* 2: 415–434.

Schroeter, F. (2004). 'Endorsement and Autonomous Agency', *Philosophy and Phenomenological Research* 69(3): 633–659.

Searle, J. (2001). *Rationality in Action* (Cambridge, MA: MIT Press).

Simons, P. (1987). *Parts: A Study in Ontology* (Oxford: Oxford University Press).

Snowdon, P. (1990). 'Persons, Animals and Ourselves', in C. Gill (ed.), *The Person and the Human Mind: Issues in Ancient and Modern Philosophy* (Oxford: Oxford University Press): 83–107.

Steward, H. (1997). *The Ontology of Mind: Events, States and Processes* (Oxford: Oxford University Press).

—— (2012a). 'Actions as Processes', *Philosophical Perspectives* 26(1): 373–388.

—— (2012b). *A Metaphysics for Freedom* (Oxford: Oxford University Press).

—— (2013). 'Processes, Continuants and Individuals', *Mind* 122: 781–812.

—— (2015). 'What is a Continuant?', *Proceedings of the Aristotelian Society, Supplementary* 89: 109–123.

Stout, R. (1996). *Things that Happen Because They Should* (Oxford: Oxford University Press).

Van Fraassen, B. (1980). *The Scientific Image* (Oxford: Oxford University Press).

Velleman, D. (1992). 'What Happens When Someone Acts?', *Mind* 101: 461–481.

Vendler, Z. (1967). *Linguistics in Philosophy* (Ithaca, NY: Cornell University Press).

Part II

Diachronic Practical Rationality

6 "What on Earth Was I Thinking?"

How Anticipating Plan's End Places an Intention in Time

Edward S. Hinchman

How must you think about time in order to form an intention? When you intend to φ at some future time *t*, you must of course think about *t*; perhaps you must in some related way think about the stretch of time between now and *t*. But must you place your endeavor in any broader prospect or retrospect? In what follows I argue that you must: in forming an intention, you commit yourself to a specific prospect of a future retrospect—a retrospect, indeed, on that very prospect. I argue that this broader temporal attitude articulates the species of self-accountability necessary for diachronic practical commitment. In forming an intention you project a future from which you will not ask regretfully, referring back to your follow-through on that intention, "What on earth was I thinking?"

In thus thematizing "plan's end" I build on Michael Bratman's approach to the stability of intention, on which the commitment at the core of intention puts you into a complex species of rapport not only with the later *you* that will execute the plan but with the still later *you* that will look back on that relation of self-influence.[1] Bratman plausibly argues that the key to undertaking a plan that will rationally resist subsequent shifts in your preferences lies in how you anticipate the retrospective stance that your future self will, if rational, take toward your having brought the plan to completion. You make your practical commitment stable by looking thus forward to how you'll look back. If you anticipate looking back with fair regret on your having executed an intention—that is, on your having done what you intended to do—then you simply cannot form the intention, because you cannot institute the species of rational stability that defines an intention. As I develop the approach, the problem lies not in any disappointment targeting the results of follow-through—we can hypothesize that you wouldn't be disappointed—but in regret targeting the self-relation that you would institute in following through.

What exactly is wrong with intending in a way that you expect to regret? Note, first, that you may aim to avoid regret in individual acts of intending without aiming to live a life altogether free from regret—just as you may aim at truth in individual acts of belief without aiming to live a life altogether free from false belief. We're talking about an aim to avoid regret in the particular case, not in the general. Expecting that you'll regret a particular relation of self-influence posits three temporally distinct perspectives. Because you now expect that your plan's-end self will prove unable to make retrospective sense of the self-trust relation whereby you performed the action—"What on earth was I thinking? How could I have trusted that intention?"—you cannot now make prospective sense of how you could aim to institute such self-trust. When you claim the intrapersonal authority characteristic of intention, you expect that the relation whereby you execute the intention will continue to strike you—all the way out to *plan's end*—as in this way "speaking for" you. I aim to explain the stability of intention in terms of that self-trust relation, with its cognate species of self-accountability.[2]

In focusing specifically on the aim to avoid such regret, which I'll call "trust-regret," my approach differs crucially from Bratman's. Though Bratman earlier emphasized a pragmatic need for diachronic stability in intention, he later came to ground the no-regret condition in a broader account of *agential authority*.[3] The "problem of agential authority" poses this question: what kind of logical functioning is sufficient for you to count as engaging in full-blown agency?[4] The problem of agential authority thus poses a question of attributability: what does it take for the action to count as straightforwardly attributable to its agent? I aim to provide an alternative to both Bratman's later emphasis on attributability, via agential authority, and his earlier emphasis on the pragmatic need for stability. Beyond a merely pragmatic emphasis, we need to see how the stability of intention rests on a norm of intrapersonal self-constitution. But we cannot explain this species of intrapersonal self-constitution in terms of attributability.

My approach differs from Bratman's not merely because I'll argue that there can be stable intentions without attributable agency—for example, an unwilling but planning addict. My approach more fundamentally differs from his because I doubt one of Bratman's central claims: that there is a "metaphysical imperative" to maintain your identity through time.[5] I'll argue that we must distinguish two aspects of how personal agency unfolds through time. On the one hand, we think that the person who performs an action must remain the same person through the interval of this performance. On the other hand, we think that the action itself must amount to a single performance—that is, a single instance of agential self-governance—through the interval in which it occurs. The first assumption poses a core facet of the issue of diachronic personal identity, but I do not believe that the issue of personal identity, even in this facet, need directly inform the issue of diachronic self-governance. On my approach, the second assumption poses the most fundamental explanandum for a theory of diachronic self-governance. The core question is not "How does the agent's settled identity establish the identity conditions for this instance of agential self-governance?" but "How does the agent's *claim* to a settled identity establish identity conditions for this instance of agential self-governance?" My approach emphasizes how the claim of attributability informs an aim to avoid the reactive-attitudinal sanction of trust-regret, rather than any aim at attributability as such.

1. Toxin and Temptation

Why should we regard intention as sensitive to future regret? I agree with Bratman that appeal to the perspective from which you might look back with regret provides the basis of a particularly compelling resolution of Gregory Kavka's toxin puzzle.[6]

To get Kavka's puzzle case, imagine that an eccentric billionaire with a reliable intention detector reliably promises to give you a million dollars if at midnight tonight you form the intention to drink a certain toxin tomorrow at noon. You must do so, he stipulates, without ignorance, manifest irrationality, or such external mechanisms as a toxin-administering machine or a side bet. If you do thus form the intention at midnight, the billionaire will deposit the money in your account tomorrow morning as soon as the bank opens. He doesn't care whether you drink the toxin, which you know will make you quite ill for a day or two but leave you thereafter unharmed. To get the money, you need merely form the intention to drink it. Kavka's puzzle is that forming this intention seems impossible under the circumstances.

Let's assume that Kavka's treatment of his original case is correct: you cannot get the million dollars merely for forming the intention because you foresee that it would be irrational to follow through on that intention. How exactly does a toxin case differ from a run-of-the-mill temptation case, in which you expect to undergo a transient shift in your preferences as the time to act grows near? And why should this difference make such a difference, preventing you from even forming the intention in a toxin case? I agree with Bratman that our explanation should appeal to a perspective that comes after the time of action, but why should that perspective have such authority?

Had the billionaire offered to reward you for drinking, rather than for forming the intention to drink, it would have been perfectly possible for you to form the intention to drink. In forming and retaining that intention, you'd have had to expect that you'd be tempted to reconsider when presented with the noxious liquid. How might you have countered those temptations? One natural strategy is by reminding yourself that you'd not regret following through on this intention. One salient change introduced by the billionaire's offer to reward you for forming the intention to drink, rather than for drinking, is that this strategy no longer works. You expect that, with the million dollars in your bank account, you *will* regret drinking the toxin. This shift in your expectations—with a reward for drinking you would expect not to regret, but with a reward for merely forming an intention to drink you expect to regret—helps explain why you are unable to form the intention in Kavka's puzzle case.

Bratman more generally distinguishes ordinary "temptation" cases from "toxin" cases modeled on Kavka's in terms of expected regret (1999c, 79ff.):

(TOXIN) As you form or retain the intention, you anticipate that you will regret following through on it and that you will not regret not following through.

(TEMPTATION) As you form or retain the intention, you anticipate that you will not regret following through on it and that you will regret not following through.

Bratman treats the anticipation as a matter of expectation. In a temptation case, you expect you will eventually regret having given in to the temptation to act contrary to your intention, but in a toxin case you expect you will not regret. Moreover, in a temptation case you expect you will not regret having followed through on your intention, but in a toxin case you expect you will regret. Bratman argues that this "no-regret condition" explains why you can form the intention in the temptation case but not in the toxin case.

The no-regret condition imposes a key distinction between what we may call the "once-future" perspective of action and the "twice-future" perspective from which you might look back on that action with regret. That distinction is especially clear in a toxin case, in which forming the intention would require expecting both that you will regret following through on it and that you will not regret not following through, expectations that in turn prevent you from forming or retaining the intention. Bratman's moral from the toxin puzzle is that in order to form an intention you must expect both that, given what you expect about the circumstances in which you will act, you will not later regret following through on the intention and that you will later regret not following through. This expectation targets the once-future circumstances of your future action *as they will be regarded* from a still later, *twice*-future perspective.

Let me acknowledge straight off two worries about the no-regret condition. First, one might worry that the no-regret condition does not formulate a necessary condition on present-directed intentions, since we all commonly perform impulsive actions while thinking "I'm going to regret this." Though in other work I argue that an analogue of the no-regret condition does apply to present-directed intentions (Hinchman 2016b), I lack space to pursue that issue here. For present purposes, I claim only that the no-regret condition serves as a necessary condition on future-directed intentions. Still, one might worry that we can form future-directed intentions that are akratic, if not exactly impulsive, and that we might express the akratic element in these intentions by thinking "I'm going to regret this." I agree that this is possible, but the question is what it shows. In the next two sections, I'll argue that the possibility cuts against Bratman's approach, with his emphasis on expectations and attributability, and vindicates my alternative.

2. Why Plan's End Must Be Projected, Not Expected

Why should this difference in what you expect to happen long after the intended action—at what we're following Bratman in calling *plan's end*: the twice-future point beyond which you will not change your attitude toward your once-future action[7]—register in your ability to form the intention? Why should it be a necessary condition on forming an intention that you give any regard, even dispositionally,[8] to how you will view the action from plan's end? Now that we see the need for a no-regret condition, we can see the need for what I'll call a "projection" out to plan's end. The need for such projection arises from a problem that I'll now develop for Bratman's simpler view on which the no-regret condition appeals to mere expectations about the future.[9]

Consider any case in which you expect both that your sensibility will change after you've acted on the intention and that after the change your regret will have no bearing on whether you should have acted on the intention. Say you're planning to join a cult that will, among other things, "reprogram" you into hating your parents. This isn't the reason you're joining the cult, but you can foresee that it will have this effect. Still, you do now love your parents and want to spend some time with them to make vivid your love for them, expecting that you'll soon thereafter come to hate them and that the context of your changed attitude will help them to see that the change is not their fault. So you form an intention to spend a week with them in which you will make your love for them as clear as you can, expecting that you'll soon thereafter adopt a stable attitude of regretting having followed through on that intention. The fact that you expect that you will have this regret at plan's end obviously has no bearing either on your ability to claim agential authority when you form the intention or on the stability of your intention once you form it.[10]

If Bratman's interpretation of the no-regret condition were correct, then the cult case would be a toxin case, wherein you cannot even form the intention because you expect you'll regret following through on the intention at plan's end. But the cult case is clearly not a toxin case. Clearly, it is a temptation case, wherein you may naturally expect a preference shift as the time of action grows near and you begin to appreciate how much easier your life would be if you acted as if you'd already undergone the "reprograming" that you expect you'll undergo after visiting your parents. Expecting you'll come to hate your parents, you may be tempted to act as if you already do, since you may prefer the simplicity of a consistent attitude to the anguished complexity projected by your intention. But, no: you're resolved to resist that preference shift, not because it is merely transient—by hypothesis, you believe it isn't merely transient—but because plan's end lies in a projected future that happens not to coincide with your expected actual future. And we may say the same of the less cartoonish cases that share this structure.

It obviously won't help simply to prohibit applications of the no-regret condition that depend on a shift in sensibility. Sometimes looking ahead to a shift in sensibility is entirely appropriate and not to do so irrational. Think of any case wherein a shift in sensibility marks the stereotypical effects of moving forward in your life. You expect you'll grow more "conservative" as you get older. Or you expect you'll have a different take on your life post-parenthood, no longer taking for granted, as you now do, that suburbs are for sell-outs. You do not expect that this change of mind will reflect better values, merely values better attuned to your needs as you expect them to become as you move into a later stage of your life. Everyone undergoes such shifts in sensibility over the course of their lives. You might even expect that such a shift of sensibility will emerge as a causal byproduct of some intention that you have formed and now retain, without regarding the shift as determining how the no-regret condition should apply. The irrelevance of such expectations may become especially clear in hindsight. Perhaps you find your sensibility shaped by your decision to have children but don't think your earlier bouts of diachronic agency should have been shaped by expectations of a parenting reality that you were then barely able to imagine. Or perhaps you've always expected that you'd wind up depressed and destitute, just as your parents did, with an abiding sense of the pointlessness of all your efforts to avoid this fate. Why should these expectations of where you'll actually wind up have any normative bearing on your capacity to intend, plan, and resolve well before the expectations are confirmed (or not)? Some of the expectations may mark a horizon at plan's end in your actual life, since you may actually look back with an altered sensibility and regret your earlier follow-through. We need to understand how the no-regret condition can work within a projection that continues the form of sense-making that informs your intention.[11]

3. What Is It to Project Plan's End? The Key Is Accountability, Not Attributability

In light of these problem cases, plan's end cannot figure quite as Bratman proposed, as the horizon of *expectation* beyond which you will not change your attitude toward the action. But we can nonetheless make use of his idea, by viewing plan's end as a notional perspective internal to the sense-making projection that informs your intention. I'll now argue that the best explanation of the normative role of plan's end treats the projection as grounded in relations of intrapersonal accountability.

The no-regret condition itself motivates the shift to accountability. What exactly do you think you won't regret when your intention is governed by the no-regret condition? It is clearly too strong to require that you project a future in which you'll not regret the action that you intend to perform, since you know that any action may have unforeseen consequences or turn out to violate some principle in a way you hadn't foreseen.[12] The projected regret must target not the action as such—whether in its nature or in its consequences—but how you perform it. What aspect of how you perform the action? I'll now argue that the question of projected regret targets the self-relation that you manifest in this instance of self-governance, given what you now expect about the circumstances of your action. What aspect of the self-relation? I'll argue that you must project a plan's end from which you will not regret the dimension of your self-relation—specifically, the self-trust relation—that would be manifested in your following through on the intention. To form an intention, you must project a future free from what I'll call *trust-regret*, by which I mean regret targeting specifically that relation of self-influence.[13] It is this appeal to self-trust that will get self-accountability in play.

To see how, let's see how this focus on trust-regret generates a deeper difference from Bratman's approach. As I mentioned at the outset, in recent work Bratman aims to ground the no-regret condition in a broader account of *agential authority*.[14] An account of agential authority would explain a specific dimension of your authority over your action, when you have it: it would explain what makes it the case that you are in charge of what you do in a way that would make the action attributable to you—rather than to, say, a force within you. My account also appeals to attributability, but in a completely different way that does not draw any link between attributability as such and authority. On my approach, the question of attributability arises from the agent's own point of view and asks not whether onlookers can coherently attribute the action to this person but whether this person coherently attributes the action to herself in the course of holding herself accountable for it. This approach suggests a different link between attributability and agential authority. If your would-be intention projects a future in which you become, as I'll say in shorthand, *self-attributively unsettled* by the thought that it was you who performed the action, then you cannot coherently claim the authority distinctive of that intention, and you thereby fail in your attempt to form or self-consciously retain the intention. To say that you are "self-attributively unsettled" is to say, in fuller formulation, that within the projection informing your intention you come to regret the self-trust relation whereby you followed through on the intention and thereby implicitly claimed "ownership" of both intention and action. When that happens, you cannot—without confusion or manifest irrationality—claim agential authority, because you see that any such claim would, by your present lights, lack authority for your acting self. Within your projection you see that when the time came to follow through on the intention, you'd be irrational if you did not abandon it. You see you'd be irrational to follow through because you see that you'd regret the relation of self-influence from plan's end. You see you might express the regret with retrospective self-chastisement: "What on earth was I thinking?"

This is why you cannot form the intention in Kavka's toxin case: the most fundamental problem is not that you cannot *be* an authority (from your own point of view) but that you cannot coherently *claim* authority (from your own point of view). For Bratman, you cannot coherently claim authority because you do not satisfy the conditions for being an authority (from your own point of view). I do not, by contrast, aim to explain why you cannot coherently claim authority in terms of any deeper explanation of what constitutes your authority as such (from your own point of view). For present purposes we need not give any deeper explanation of agential authority. We need merely see why you cannot coherently claim such authority when you project that you would be self-attributively unsettled by the self-trust relation that would emerge if you let that claim of authority influence you in the way that it aims to influence you. That's all we need in order to explain the species of agential authority distinctive of an intention.

While I lack space for a full exploration of this difference between my approach and Bratman's, let me offer one argument for separating the core issue of the rational stability of intention from Bratman's conception of the issue posed by agential authority. Assume for the sake of argument that ambivalence of a sufficiently deep or systemic sort can undermine the attributability of an action.[15] Imagine that you are stricken by such ambivalence but also that you engage in what to all appearances looks like planning agency, like the unwilling addict who pursues complex means in order to obtain his next batch of drugs. Given our assumption, the actions that you perform through such planning will not in fact be (fully) attributable to you. But, setting that aside, it seems that we can nonetheless explain how your intentions are rationally stable. How might we vindicate the rationality of your expectation, as you intend, that you will be rational in letting your intention override your deliberative perspective as you act? Here is where my approach diverges most sharply from Bratman's. Again, Bratman explains stability in terms of attributability: it is rational to let the force of your intention override your deliberative perspective as you act just when doing so satisfies the conditions sufficient for making that action attributable to you. So Bratman cannot allow that your intention could be stable in our ambivalence case. My approach, by contrast, appeals not to attributability but to the *claim* of attributability, a claim that you do make, even in such a case, insofar as you aim to avoid your own trust-regret by aiming to perform only those actions that you will not later find self-attributively unsettling. It is part of the self-trust dynamic that you do *regard* as attributable to you the action that you would perform by self-trustfully following through on your intention. If you did not so regard the action, you could not find the self-relation manifested by the performance self-attributively unsettling.[16]

We thus explain the projective form of sense-making that informs your intention. Even if you do not expect to regret trusting your intention in your actual future, if you project plan's-end trust-regret, you cannot claim the species of agential authority that would inform that intention, and so you cannot—without confusion or manifest irrationality—form the intention. But you can claim that species of authority when the course of action to which it would lead is not attributable to you. It is hard to see why we shouldn't allow that an unwilling addict in the grip of his addiction can nonetheless plan how to purchase his drugs and when or how to use them. Even if these actions are not (fully) attributable to him, his planning manifests his agential point of view and as such can prove rationally stable. Gripped by his addiction, he makes a plan to obtain drugs at his dealer's house, but upon arrival our hero finds himself struggling with what seems like an akratic temptation to abandon the plan.

Can we make real sense of such apparent "akrasia within akrasia"? Imagine two possible cases in which our hero might thus pace nervously on the dealer's front step, reluctant to knock. In one case, our hero struggles with a conflict between his akratic intention to buy drugs and the unabandoned resolution to give up drugs that makes that intention akratic, his addicted planning agency thus in conflict with a "better judgment" now functioning as an apparent source of temptations pulling him away from his plan. If he gives in to this temptation, though without thinking of himself as having abandoned his plan to buy drugs, that would count not as "akrasia within akrasia" but as a resolution of the original akrasia into a confused form of ambivalence: he has put himself back into motivational touch with his better judgment, but he is strangely confused by this newfound enkrateia, believing that the intention to buy drugs continues to "speak for" him.

But that is not the only form that "akrasia within akrasia" might take. In a second case, imagine a different temptation pulling our hero away from his plan to buy drugs—the all too familiar temptation to flee that grips him every time he arrives at his dealer's doorstep. He fears his dealer, always has, and coping with this fear is a familiar obstacle on the way to each of his drug purchases. This drug purchase, unlike those others, is akratic, since it is countered by an unabandoned resolution to end his drug habit. But that difference makes no difference to the familiar temptation now gripping him at the dealer's doorstep. If he gave in to this temptation, it wouldn't rescue him from the akratic plan that left him there. Rather, it would make his akratic predicament worse. Giving in to this fear of his dealer would not, on its own, amount to abandonment of his plan to buy drugs. Say he paces back and forth for a few minutes grappling with the fear. That does not amount to grappling with the temptation to buy drugs, given that he intends to end his drug habit. It amounts to grappling with the temptation to give in to fear and thereby fall short in this plan to buy drugs, given that he nonetheless retains the plan. If he does give in, that would amount to akrasia within akrasia—as such, a violation of the rational stability of his intention.

How might we explain rational stability in light of this possibility? The possibility of such cases appears to show that the rational stability of intention cannot be a matter, most fundamentally, of attributability. So how else might we explain it? I have begun to offer an alternative that emphasizes not attributability but the *claim* of attributability: our addicted hero can form an akratic intention to buy drugs because, even though the intention is not actually attributable to him, he can nonetheless claim that it is. A proponent of Bratman's explanatory approach may reply that we cannot make sense of this appeal to a mere claim of attributability without understanding the role of attributability in constituting the intrapersonal relations at the core of rational stability. Is our hero merely claiming that his case is a normal case, in which the intrapersonal relations do suffice for stability? That's how we might have to interpret my talk of a "claim of attributability" if there were no other way to make sense of it. As I've already suggested, however, we can make sense of it within the economy of the agent's self-accountability relations as they project a future out to plan's end.

Our hero can akratically intend to buy drugs, even though he retains his resolution to end his drug habit, because the intention and the resolution project different futures.[17] When he resolves to end his drug habit, he projects a plan's end from which he will not regret having given up drugs and will regret not having done so. But when he intends to buy drugs, he projects a plan's end from which he will not regret continuing his habit ("just one more time") and will regret not having done so ("since anyone who cares about me won't want to see me in this pain"). Notice the parenthetical rationalizations. We needn't imagine our hero blindly lashing out in his addiction, as if he could score some drugs by having a temper tantrum. Though he has resolved not to do so, he gives in to the temptation to project a plan's end that conflicts with the projection at the core of his resolution. We could say that his addicted projection expresses the addiction rather than the judgment with which he identifies; that formulation captures the intuitive force of our hypothesis that this entire course of action, including its planning, is not straightforwardly attributable to him. Even so, there is planning, and the planning occurs in his psychology. From the perspective of this planning, there is rational stability: from this perspective, it would be irrational of him to give in to his fear-based temptation to flee. The projection, with its rationalizations, reveals the influence of the very addiction that he has resolved to overcome. But it enables him to formulate and pursue, through intrapersonal rational commitments, a diachronically robust course of action.

In the next section I'll more fully explain the species of intrapersonal accountability that gives shape to the projection at the core of such diachronic rational commitment. My alternative to explaining rational stability as a matter of attributability begins from my emphasis on how a *claim* of attributability lies nested within intrapersonal accountability from plan's end. But before we move on, let me use the dialectic developed thus far to reply to a worry about how I am using the concept of a projection. I claim that your intention rests on a projection into a future that you need not regard as your actual future. But if you do not expect plan's end to lie in your actual future, how could you could distinguish actual correctness from a mere feeling of correctness? As Wittgenstein framed the problem for a different but structurally parallel aspect of rule-following (1953, sec. 202), how could you distinguish being right from merely seeming right? If you expect that you'll go off the rails in the way of the cult case, how could any projection into a future that you expect to be non-actual serve to keep you in line?

This worry presupposes the emphasis on attributability that I'm rejecting. It is true that if we combined Bratman's appeal to attributability with my appeal to projection, we'd have to say that you constitute yourself as a unified subject of your actions through your projection out to plan's end. But the worry reveals why that would not work in many cases: your projection leaves open the possibility that you will fail to have actual future "partners" in the normative endeavor—that is, actual future selves within the projection. By contrast, it is not hard to see how a species of self-accountability could unfold through a projection. After all, we often feel accountable "in the eyes of others" even when there are no others around, or when we know that no actual others in our community will hold us accountable. The link between accountability to others and accountability to *actual* others is isomorphic to the link between self-accountability and accountability to *expected* actual future selves. In each case, you could not learn how to feel accountable if no one ever held you accountable. (Self-accountability could not be something that you only ever imagine. You learn how to hold yourself accountable in this dimension by coming to feel actual regret!) But in each case, whether interpersonal or intrapersonal, accountability is possible when there is no actual person in position to hold you accountable.

4. Projecting a Retrospect: Regret as Reactive Attitude

We need to understand how this species of self-accountability works in detail. What *are* trust and regret such that they should play this role in diachronic agency? I'm arguing for the importance of a species of agent-regret that I'm calling "trust-regret," since what you regret is that you entered into an unwise trust relation.[18] And I'm arguing for the importance of a species of trust that we could describe as "reasonable but not deliberated confidence in someone's anticipation of what you'll regret," where the "someone" might be your own earlier self. How do these species of trust and regret combine to serve as a backward-looking reactive attitude?

Following an approach pioneered independently by Susan Wolf (1990), Jay Wallace (1994), and Gary Watson (1996), we must understand two things in order to see how trust-regret functions as a reactive attitude: how it involves a sanction, and how application of this sanction involves a norm of fairness. The fairness norm applies as follows. Just as you don't hold someone fully responsible when you don't think she had a reasonable opportunity to avoid this sanction, so you don't regret your intention-to-action self-trust relation when you don't think your earlier selves had a reasonable opportunity to avoid the regret. When I say that you "don't" react or regret in these ways, I mean that we agree that it would be wrong to.

Several different types of case illustrate the intrapersonal fairness norm. First, you don't regret—that is, *trust*-regret[19]—when you don't think your acting self could have foreseen the circumstances in which the self-trust has come to seem foolish. Second, and more complexly, you don't regret when both of these conditions hold: (1) your acting self acted on self-trust—that is, without redeliberation—and (2) you don't think your *intending* self could have foreseen the circumstances that make the self-trust foolish. Third, there is an analogue of the point emphasized by Wolf: you don't regret self-trust informed by *deep* deficiencies of character. This is typically because you can't see them: since the deficiencies inform the perspective from which you deliberate, intend and act, you typically can't recognize yourself in that description. Even if you can recognize the deficiency, your inability to do much of anything to change—which follows from the hypothesis that the deficiencies are "deep," in Wolf's metaphor—makes regret (again, *trust*-regret) the wrong reaction. What you feel is better classified as disappointment.

This role played by fairness is revealed less in the reactive attitude itself than in your efforts to avoid it. Just as there is no constructive deliberative role for the worry that drink or disease will make you regret—apart, of course, from the strategic question of coping with that consequence—so there is no constructive role for the worry that you'll regret unfairly. Though in forming an intention you aim to avoid twice-future regret at the relation that you thereby aim to institute with your once-future self, there is no constructive role for a worry either that you will regret for reasons that you cannot foresee—as opposed to those you can foresee but merely overlook—or that you thereby manifest deep deficiencies in your character, or some other deficiency that it would be unfair for you to regret. Still, such worries can *infect* practical reflection—precisely because they cannot inform it. When reflection is thus infected, you may expect to regret following through on the intention that you're forming, but this expectation does not prevent you from forming that intention, though you may do so with a feeling of despair. It is part of what makes such self-accountable agency possible—in part by preventing such self-despair from undermining your agency—that the norm you're aiming to meet is thus constrained by fairness. Since you aim to avoid *fair* regret, an inability to respond constructively to all the worries that you might feel about taking responsibility does not impede your ability to act. When such anxieties unsettle you, it is the fairness norm that makes intention possible.

We see here another reason to interpret the no-regret condition in terms of a projection that need not coincide with any expectation. In Section 2, we saw how it is possible to expect that you'll regret following through on your intention without being thereby constrained in your ability to form the intention. In those cases, your expectation of regret targets the far side of an expected "shift in your values" (as we might say, in shorthand for this complex range of normative changes). You expect that the regret will be informed by values that you expect you'll adopt for reasons that fail to recommend those values on their merit: in our cartoonish example, that you'll be brainwashed by a cult into hating your family; or, with more realism, that you'll become narrow-mindedly "conservative" with age, or overridingly concerned with the needs of children for whose existence you haven't even begun to plan. But what if you expect not a shift in values but a sheer deficit in your concern with retrospective fairness? Imagine you expect, not the species of "narrow-mindedness" that marks a shift in values, but the species that marks a lack of concern to do justice to any perspective but your own perspective at that very moment. You see both your parents grow embittered by life's frustrations and take out their bitterness on their younger selves, "regretting" nearly everything they did when young, including things that you can see made perfectly good sense at the time. You may come to expect that this will happen to you; but the expectation does not give you pause as you form and carry out the plans that you expect will thus provoke your embittered self at plan's end. You can see that this is not regret but self-recrimination. The expected self-recrimination may trouble you, but it does not disrupt your efforts to plan in the way that projected regret would do. Recrimination is a charge, a species of psychic aggression, and as such it may be entirely arbitrary. It is not a reactive attitude unless it represents itself as responsive to a norm of fairness. Just as mere hostility may not yet amount to a responsibility-imputing reactive attitude on the order of blame or resentment, so mere self-recrimination may not amount to a responsibility-*self*-imputing attitude on the order of trust-regret.

This reflexivity is crucial to the normativity of the self-relation. Interpersonal hostility may or may not represent itself as responsive to a norm of fairness. Even if it does, and even if it thereby counts as a reactive attitude (transforming mere hostility into, say, contempt), that does not yet mean that it *is* fair. If it is not fair, then it may not succeed in establishing any species or degree of actual responsibility.[20] Intrapersonal regret may likewise fail to prove fair, but such a failure does not in the same way compromise its normative role—that is, the normative role that I'm arguing it has in stabilizing intention. Even if the regret proves unfair, when you project the regret you project it *as* fair. And that—the projection—is what ensures that it plays the normative role. You may hold another person in an attitude of contempt that proves unfair. But if you try to project "unfair regret," what you project is mere self-recrimination that, as such, fails to bear on the stability of your intention. In each case, the interpersonal and the intrapersonal, your attitude is governed by a fairness norm: if unfair, it is wrong or unjustified. What makes the difference is that in the intrapersonal case, unlike the interpersonal, you cannot adopt the attitude insincerely. The intrapersonal attitude cannot be insincere because it functions prospectively: you take a prospect toward a future retrospect. Within that prospect, the retrospect is projected as fair, in part because a retrospect that you did not project as fair, within that prospect, could not function to stabilize your intention. The normative force of the prospect thus derives from the retrospective attitude that it projects, and the whole point of projecting is to stabilize your intention. Within that prospect, you project the retrospective attitude as fair, simply because if you didn't you couldn't thereby undertake the normative attitude—the intention, plan or resolution—that the prospect informs.

You can hold someone in just a degree of contempt. Can you likewise project just a degree of trust-regret? There are at least two ways to answer this question: (1) yes, and the role of projected trust-regret in stabilizing intention involves the idea of a threshold beyond which the degree of projected trust-regret is too high; or (2) no, since any degree of trust-regret serves, within the projection, to undermine the stability of your intention and thereby your ability to form it. I believe that the correct answer is (2), but nothing in my argument depends on that answer. I believe that the answer is (2) because I believe that there is at least this much truth in Harry Frankfurt's proposal (1988) that agential authority requires freedom from ambivalence: the *claim* of agential authority requires *projecting* an unambivalent retrospect from plan's end. Even if you are ambivalent, as in our example of an unwilling though planning addict, when you invite self-trust you project an image of yourself as wholehearted in the retrospect that you adopt at plan's end. It is in this respect, I believe, that diachronic agency involves an ideal. Just as you project fair trust-regret, a normative idealization, so you project unambivalent trust-regret, a psychological idealization. Both idealizations may fall short of ensuing reality: at plan's end, you may regret your self-trust unfairly, or you may regret to this or that degree. But *within* the projection you provide yourself with an idealized image of what you are up to in intending, an image whose idealized nature enables you summon the clarity of purpose that you may need to resist temptation.[21] If we prefer answer (1), with its apparently more realistic psychology, we lose this simple account of how clarity of purpose structures the stability of intention. But it is no part of my argument in the present chapter that we cannot provide such an alternative account.[22]

5. The Sanction of Trust-Regret

Talk of "fair" or "unfair" regret makes little sense if the regret does not impose a sanction. What sanction might trust-regret impose? The no-regret condition specifies how the sanction of regret functions in prospect: when you project that your twice-future self will prove unable to make the right sort of sense of how this action could be retrospectively attributable to your once-future self, you cannot now make the right sort of sense of how you could form the intention and thereby prospectively attribute this action to your once-future self. To understand more fully how this works, it helps to develop the analogy between intra- and interpersonal agency suggested by the similarities between intra- and interpersonal trust.

When you accept another's invitation to share an intention, it makes sense to regard you as entering into an interpersonal trust relation premised on that other's expectation that you'll not regret the union. Your co-worker invites you to accompany her on a tour of the premises, but in trusting her (she gently took your arm, and you reciprocated by walking with her) you played right into one of her schemes—what on earth were you thinking? ("A 'tour'?" you ask yourself from plan's end. "How could I have fallen for that scheme?") This species of regret addresses whether the agential union "speaks for" you in the sense at issue in this chapter: are its actions attributable to you as agent (that is, as one of its agents)? Call this an *identification-expressive* species of regret. Since the invitation to share the intention is an invitation to be thus identified with the action, the invitation must represent itself as manifesting the projection that you will thus identify with the action. But then it must represent itself as manifesting the projection that you won't experience fair identification-expressive regret concerning the action. The need for this projection lies in the invitation's claim of an interpersonal analogue of agential authority—not the union between parts of your self that gets a unified you into your behavior, but the union between two agents that gets them both as a unified "we" into an instance of shared agency. Why should you let your co-worker influence you in the way that she proposes? As she takes your arm, she may be explicit: "Let's take a little tour. Trust me. You won't regret it!"[23]

I'm arguing that the same identification-expressive species of regret shapes the intrapersonal case. What you aim to avoid when you intend to perform an action is a species of regret that amounts to the paradoxical—and therefore unsettling—feeling that the action does not speak for you. If it does not speak for you, how could it be attributable to you? But you attributed it to yourself—you staked an implicit claim of attributability—when you entered into the self-trust relation! From this retrospective perspective the sanction registers as *the disorientation you feel when this self-attribution turns out to look wrong*—wrong (in light of the fairness norm) in a way for which it would be fair to hold you accountable. The regret is thus a form of self-accountability: you staked a claim of self-attribution in which, as things now turn out, you cannot recognize yourself. Prospectively, the sanction registers as an inability to presume the intrapersonal authority over your acting self that defines an intention. As we've seen by reflection on toxin cases, you simply cannot form the intention without holding yourself thus accountable. My suggestion, then, is this: just as another cannot coherently claim the authority inherent in inviting you to share an intention while representing herself as projecting fair regret that you entered into the relation, so you cannot claim the authority inherent in forming an intention while projecting fair regret that you followed through on the intention.

An understanding of this sanction explains why you should care about your own future regret. But why should you care about your expectations of future regret? Since in expecting to regret you give evidence of your untrustworthiness in intending, we can rephrase the question: why care about your own status as trustworthy or untrustworthy in intending? Even if you have other reasons to care, can you derive a reason simply from the thought that your twice-future self will regret your untrustworthiness? Here is my proposal: without projecting a relevantly regret-free future, you cannot find your claim of agential authority intelligible. And without finding that claim intelligible you cannot count as making it.[24]

Forming an intention differs on this point from inviting another to share it. You can make agential sense of a presumption of authority over another when you know that you are not relevantly trustworthy. (Your co-worker knows that her self-presentation is insincere.) But you cannot make agential sense of such a presumption of authority over yourself. You can make causal sense of this presumption, if you expect to succeed in deceiving yourself in relevant ways. But, as the problem of "deviant" causation for causal theories of action makes vivid, mere causal intelligibility does not suffice for intelligibility as an action.[25] We can explain this difference by observing that, as many philosophers have emphasized, you cannot form an intention for the "autonomous benefits" of forming it.[26] When you form an intention to φ at t, you must deliberate only from considerations that concern your φ-ing at t. This, again, is why you cannot form an intention to drink the toxin—thereby getting the autonomous benefit of a million dollars—in Kavka's puzzle. By contrast, to count as sincerely inviting someone to share your intention to φ at t, while you must present yourself as deliberating from considerations that manifest an appropriate concern for your invitee's φ-ing at t, you needn't actually feel that concern at all. If an eccentric billionaire offered to reward you for inviting someone to share a putative intention to drink toxin, you could easily do it. The only rub would lie in your confidence that your invitee wouldn't understand your motive in inviting him to drink and would form his intention to drink entirely on the basis of trust in you. If he thought the case through on his own he would be unable to claim authority over his conduct, but simply trusting your claim of authority he can. Knowing this, you can invite him to claim that authority—thereby bagging your reward.

Note well that the restriction on individual intention doesn't derive from your *owing* anything to your twice-future self. This is not like the relation between you and someone who calls you out on wrongful conduct. Nor is it like the relation between your intending and acting selves when the latter fails to trust the former. It is not, in sum, a justificatory or more broadly forensic relation in which you purport to give reasons. In terms of its content, it is a *finding-intelligible* relation.

I'm arguing that this finding-intelligible relation involves the species of accountability also at issue in forensic relations. We can see that this is a form of accountability by understanding how the unsettled feeling at the core of trust-regret functions like other sanctions in keeping you "in line." You view yourself as accountable to your twice-future self insofar as your aim to avoid this disorientation guides your practical reflection from the onset of deliberation to the formation of your intention. Your practical-reflective stance actually extends farther than that, since it is up to you whether to reopen deliberation in the interval between forming the intention and acting on it. The self-accountability at the core of your self-intelligibility thus guides you from deliberation to action, as you project a regret-free future all the way out to plan's end.

The sanction reveals how trust and regret are both acknowledgments of risk. On a natural but mistaken view, trust would manifest a vulnerability to specifically personal influences, over and above your vulnerability to the rest of the world, and regret would register all this vulnerability in retrospect as disappointment that things did not go your way. That misses what's most interesting in both attitudes. Agential self-trust is not mere self-reliance, nor regret mere disappointment, because what each attitude most fundamentally registers is the basis of enkratic rational requirements.[27] Self-trust registers this basis prospectively by positing the absence of what regret registers retrospectively. This basis is an intrapersonal relationship that such vulnerability—while very real—does not define. It would not be the relationship that it is without that vulnerability, but that is because the risks define a genus of broader trust relationships to which self-trust relations belong, not because they specifically define self-trust relations.

Trust amounts to a way of *coping constructively* with these risks, by putting the agency of another into the service of your own—where that "other" may be your own earlier self. Typically there is no other way to proceed. Much work on trust starts from the thought that it's risky to act in a way that relies on others, and much work on regret starts from the thought that it's generally risky to act. Yet if you did not know how to trust yourself, thereby risking the sanction of trust-regret, you would not be capable of acting commissively through time.

Notes

1. Bratman coined the term "plan's end" (1999c, 85ff.). For some qualifications on how I'm going to use the term, see note 7. See note 3 for more on Bratman's approach.
2. This chapter extends the inquiry into how norms of trust structure diachronic agency that I began in Hinchman (2003, 2009).
3. For Bratman's earlier approach, see his (2007f, 56). This restates his justification of the no-regret condition in Bratman (1999c), where he first proposed the condition. Bratman (2007c) emphasizes the difference between the pragmatic orientation of his own earlier treatment and a new orientation toward agential authority, noting how these orientations would make different use of an appeal to regret. Bratman (2014) appeared too late to discuss here; it revises his position in one respect that I criticize in Hinchman (2016d) (see notes 5 and 7).
4. See Bratman (2007e), where he says "necessary and sufficient" (92). Bratman later states that he no longer wants to formulate the problem of agential authority in terms of necessary conditions (2007b, 4–5, 11).
5. For "metaphysical imperative," see Gary Watson's interpretation of Bratman (Watson 2005, 94–95), an interpretation that Bratman has subsequently confirmed, characterizing his own view of agential authority as "a claim about the

metaphysics of agency, not a normative ideal of integrity or the like (though we may, of course, also value some such ideal)" (2007d, 246); Bratman says in a footnote that he is replying to Watson's interpretive claim. Bratman's (2014) formulation may no longer rest on this imperative, though abandoning it raises problems that I discuss in Hinchman (2016d).

6. For the original case, see Kavka (1983).
7. My formulation makes somewhat more precise Bratman's characterization of plan's end as "the conclusion of one's plan" (1999c, 86). It makes sense to prefer my more precise formula over his vaguer phrase because it obviates the conceptually vexing question of when a plan is ever really "concluded," given that one can always return to it and reason from it to new plans. (You rediscover the stamp collection that you abandoned at the age of twelve and resume the hobby—so the plan wasn't concluded after all!) Moreover, some intended actions (e.g., maintaining your health) simply do not actually have an envisaged "conclusion." (Holton [2009, 158] presses this latter observation against Bratman.) Bratman's phrase does appear to differ from my formulation insofar as appeal to the "conclusion" of the plan appears to exclude deathbed conversions and the like. While one might find that an attractive implication in some instances (why should deathbed delirium matter?), the exclusion imports a substantial assumption about diachronic agency that could not be vindicated in all instances (why shouldn't conscientious deathbed reflection matter?). Note well that any such differences between my formulation and Bratman's will not make any difference for my argument against Bratman's understanding of plan's end, since my argument will target not such details but the general idea that plan's end must be expected (rather than projected).
8. The thesis is not that you're actively thinking about your plan's-end self. The thesis is that you're making an assumption about this self: the assumption that (as we'll see) grounds your presumption of agential authority.
9. Note that Bratman uses "projects" alongside "anticipates" to refer to the forward-looking attitude that is governed, in part, by expectations about plan's end (e.g., 2007c, 275). That is, he doesn't draw the principled distinction between expectation and projection on which I am insisting.
10. Why not? It points in the right direction to observe that, unlike a "projection," a mere expectation does not amount to anything worth describing as "telling a story." When you form the present intention, it is plausible to regard you as projecting a plan's-end perspective from which you'll resonate to this "tragic" narrative: "giving my parents their due before I become unable to appreciate what I owe them." That story can count as well-told only to a future self of yours that *is* able to appreciate what you owe your parents—by hypothesis, not the self that you expect that you will become and forever remain. I hypothesize that a mere expectation fails to stabilize your intention because it lacks the element distinctive of "telling a story" about your future. For more on this appeal to narrative, see Hinchman (2015). But my present argument does not require that particular elaboration.
11. I treat this form of sense-making at greater length in Hinchman (2016b).
12. Bratman usually refers to the agent's "follow through," which is consistent with my preferred interpretation of the no-regret condition, but he sometimes explicitly speaks of regretting the "action" (e.g., 1999c, 88; 2007f, 56). In any case, he does not consider the possibility of regretting your relation of self-influence as opposed to your action.
13. By the no-regret condition, you must also anticipate that you would regret it if you failed to realize that relation. But because you typically (albeit not necessarily) expect that you will follow through, this hypothetical anticipation does not play the same role as the categorical one.
14. For references and discussion, see note 5.

15. I'm not sure this really is so, but I lack space to investigate the issue.
16. Note this respect in which accountability, while different from attributability, nonetheless presupposes self-attribution. I pursue further the idea that anticipating regret amounts to a species of self-accountability in Hinchman (2010, 2016c).
17. Both the resolution not to take drugs in general and the intention to obtain drugs now are intentions. I call the first a "resolution" merely for ease of reference, and to remind us of its status in making the other intention akratic. Having both intentions at once is a contradictory state of mind, and this incoherence poses the traditional problem of akrasia: how could the contradiction (i.e., acting against your own intention or resolution) fail to undermine attributable agency? I cannot treat the traditional problem here (I do so in Hinchman [2009, 2013]), beyond noting that taking seriously the problem of the rational stability of intention—how could it be rational to follow through on an intention that you would abandon if you reconsidered?—presupposes that we can solve the traditional problem of akrasia. If it were simply impossible to act (sc. with attributable agency) against your own intention without thereby counting as having changed your mind, then the "rational stability" of intention would amount to a rational pressure not to change your mind. But that is not the species of stability at issue here, and it is not clear that there actually is any such species of rational stability. The species of rational stability that gives the concept of intention its point provides rational pressure not against changing your mind about what to do but against failing to follow through on your intention *without* having changed your mind about what to do. (This point undercuts Bratman's [2014] attempt to broaden rational stability beyond mere non-reconsideration; see my [2016d].)
18. "Agent-regret" is Bernard Williams's term (1981, 27ff.).
19. Henceforth I'll resume my practice of mostly leaving the qualifier implicit.
20. Wolf (1990), Wallace (1994), and Watson (1996) have in different ways argued for this proposition.
21. Here is where my approach calls for something like the elaboration sketched in note 10.
22. Whichever elaboration we prefer, we'll need to explain how the behavior at issue counts as a single and to that extent unified instance of self-governance. I pursue these issues in Hinchman (2016b).
23. I develop such an account of shared intention in Hinchman (2016a).
24. I agree here with David Velleman: the constitutive aim of diachronic agency is a species of self-intelligibility (1989, 94–100 and ch. 4; 2000). But I disagree with Velleman's interpretation of this insight at two key points: (1) my account is not "cognitivist" (in the respect criticized by Bratman (1999b); and (2) I hold that self-intelligibility matters specifically in the claim of agential authority at the core of an intention.
25. For "deviant" causation, see Davidson (1980, 63–81). The older issue of action explanation unfolds from a third-personal perspective, whereas our issue unfolds within the first-personal perspective of the agent. Mapping our issue of agential intelligibility onto that older issue, despite this difference, we might view *mere* causal intelligibility—for example, you expect to forget your actual deliberative basis for deciding to drink the toxin, deceiving yourself into believing that the billionaire requires that you drink—as a form of causal deviancy: your decision-making causes your own behavior without thereby amounting to the uncompromised performance of an action because it does not cause it "in the right way," that is, in a way that engages your deliberative basis for acting.
26. For some early articulations of this general point, see Farrell (1989; 1993, esp. 58–59).
27. I explain how norms of trust inform enkratic rational requirements in Hinchman (2013).

References

Bratman, Michael 1999a *Faces of Intention* (Cambridge: Cambridge University Press)
—— 1999b "Cognitivism about Practical Reason," in *Faces of Intention* (Cambridge: Cambridge University Press): 250–264.
—— 1999c "Toxin, Temptation, and the Stability of Intention," in *Faces of Intention* (Cambridge: Cambridge University Press): 59–90.
—— 2007a *Structures of Agency* (Oxford: Oxford University Press).
—— 2007b "Introduction," in *Structures of Agency* (Oxford: Oxford University Press): 3–18.
—— 2007c "Temptation Revisited," in *Structures of Agency* (Oxford: Oxford University Press): 257–282.
—— 2007d "Three Theories of Self-Governance," in *Structures of Agency* (Oxford: Oxford University Press): 222–253.
—— 2007e "Two Problems about Human Agency," in *Structures of Agency* (Oxford: Oxford University Press): 89–105.
—— 2007f "Valuing and the Will," in *Structures of Agency* (Oxford: Oxford University Press): 47–67.
—— 2014 "Temptation and the Agent's Standpoint," *Inquiry* 57:3: 292–310.
Davidson, Donald 1980 "Freedom to Act," in *Essays on Actions and Events* (Oxford: Oxford University Press): 63–82.
Farrell, Daniel 1989 "Intention, Reason, and Action," *American Philosophical Quarterly* 26:4: 283–295.
—— 1993 "Utility-Maximizing Intentions and the Theory of Rational Choice," *Philosophical Topics* 21:1: 53–78.
Frankfurt, Harry G. 1988 *The Importance of What We Care About* (Cambridge: Cambridge University Press).
Hinchman, Edward S. 2003 "Trust and Diachronic Agency," *Noûs* 37:1: 25–51.
—— 2009 "Receptivity and the Will," *Noûs* 43:3: 395–427.
—— 2010 "Conspiracy, Commitment, and the Self," *Ethics* 120:3: 526–556.
—— 2013 "Rational Requirements and 'Rational' Akrasia," *Philosophical Studies* 166:3: 529–552.
—— 2015 "Narrative and the Stability of Intention," *European Journal of Philosophy* 23:1: 111–140.
—— 2016a "How to Settle on a Shared Intention," unpublished manuscript.
—— 2016b "Intention and Time," unpublished manuscript.
—— 2016c "On the Risks of Resting Assured: An Assurance Theory of Trust," in Paul Faulkner and Tom Simpson (eds.), *New Philosophical Perspectives on Trust* (Oxford: Oxford University Press).
—— 2016d "What is the Rational Stability of Intention?", unpublished manuscript
Holton, Richard 2009 *Willing, Wanting, Waiting* (Oxford: Oxford University Press).
Kavka, Gregory 1983 "The Toxin Puzzle," *Analysis* 43:1: 33–36.
Velleman, J. David 1989 *Practical Reflection* (Princeton, NJ: Princeton University Press): 1–31.
—— 2000 "Introduction," *The Possibility of Practical Reason* (Oxford: Oxford University Press).
Wallace, R. Jay 1994 *Responsibility and the Moral Sentiments* (Cambridge, MA: Harvard University Press)

Watson, Gary 1996 "Two Faces of Responsibility," *Philosophical Topics* 24:2: 227–248.

———— 2005 "Hierarchy and Agential Authority," in John Fischer (ed.), *Free Will: Critical Concepts in Philosophy* (New York: Routledge), volume IV: 90–97.

Williams, Bernard 1981 "Moral Luck," in *Moral Luck* (Cambridge: Cambridge University Press)

Wittgenstein, Ludwig 1953 *Philosophical Investigations* (Oxford: Basil Blackwell)

Wolf, Susan 1990 *Freedom within Reason* (Oxford: Oxford University Press)

7 *Pro-Tempore* Disjunctive Intentions

Luca Ferrero

1. Introduction

1.1 In planning for the future, we face two basic challenges. First, we must organize our steps to deal with the uneven temporal distribution of the opportunities and resources to make progress in our pursuits. Second, we need to deal with our limited knowledge of the future and with the changes that this knowledge undergoes over time. We hardly ever plan while being certain of all the relevant future circumstances.

Our intentions have a built-in protection against the unexpected: it is in principle always possible to give up an intention in the face of unanticipated future circumstances. Yet this should be a strategy of last resort, to be used sparingly and judiciously. Overreliance on changes of mind, even when justified, would undermine the *stability* of intentions, which is a distinctive feature of the standard operation of intentions and a major source of the appeal of planning abilities.

1.2 Our plans need to balance stability and flexibility. We must avoid two opposite responses to uncertainty and risk, just giving up the intention altogether as we come to learn more, or a reluctance to undertake any plan at all when we are not certain about the future.

There are two obvious strategies to make intentions more flexible, while preserving their stability. First, as Bratman (1987) shows, our plans are *partial*: they need not specify all the details of implementation. We make appropriate adjustments as we progress and acquire relevant knowledge.

Second, most intentions are conditional. Even when expressed categorically, as intentions to φ, we are not usually committed to φ-ing *no matter what*. Most ordinary intentions are implicitly conditional upon circumstances we are uncertain about—they are of the form "I intend to φ if C," where C has not yet been ascertained by us.[1]

1.3 Here I present an additional strategy: the contemporaneous pursuit *for the time being* of two or more eventually *incompatible* projects. Consider a not yet fully resolved traveling plan: I am settled on either going to a conference in Italy or to a festival in Austria. The two projects are incompatible because the events are contemporaneous. Although I have not yet made up my mind between the two, I have already determined that I am going to one of them. Hence I can legitimately say: "I *intend* to go either to Italy *or* to Austria."

Let's call these intentions *pro-tempore disjunctive* (PTD, for short). They are usually undertaken when we don't *yet* know which alternative is preferable while expecting that we will gain relevant information later, at which point we will continue to pursue only one of the alternatives.

1.4 PTD intentions are a familiar strategy to trade off stability and flexibility. But one that has been neglected by philosophers, with the exception of Holton (2009). They are worth of a closer look. There is much that we can learn from them about the nature and dynamics of planning agency.

I first discuss simple disjunctive intentions, where one is indifferent between the alternatives. I then show that PTD intentions are different from simple disjunctive intentions. I discuss how PTD intentions meet the rational pressures of intentions and how they settle practical matters. PTD intentions are a pervasive feature of diachronic agency: they are the simplest tools to agglomerate potentially conflicting pursuits without rejecting them outright. They are the basic strategy for balancing both rigidity and flexibility, and stability and responsiveness to changing circumstances. I close by rejecting Holton's *partial* intentions. We should not introduce a novel kind of attitude: all-out intentions with complex contents, such as PTD, can do all the necessary work.

2. Disjunctions of Indifference

2.1 When one is indifferent between two incompatible courses of action, one can acquire a disjunctive intention of indifference. Someone approaches me with a tray with two cookies. I only want one and I am indifferent between them. I just need to form the intention to either get cookie A *or* get cookie B, that is, to make true the disjunction (get $A \lor$ get B). Because of my indifference, I need not be more specific. No need to settle for "getting the closest cookie," say. After all, by the time I reach the tray, the cookies might be equidistant from me. It only matters that I get one.

2.2 Disjunctive plans of indifference are common. But their incompleteness might be troubling. An intention is supposed to settle what to do, but a disjunctive plan leaves open what the agent is to do once she reaches the "conflict point" (c-point, hereafter) where the two alternatives become incompatible. Even so, a disjunctive intention has practical import. Prior to the c-point, it guides one's conduct to reach the c-point (i.e., to get within the reach of the cookies). And it continues to guide even at the c-point by telling what *not* to do then: not to refrain from reaching for one of the cookies. It directs the agent away from the complement of the disjunction, which is the practical import that Buridan's ass missed.

The disjunctive intention does not tell one *which* cookie to pick. But this is unproblematic. The intention fulfills its guiding role as long as it secures the presence, at the c-point, of some mechanism that moves one to pick one cookie. The resolution of this practical underdetermination does not need to come from an *intention* with a non-disjunctive content (i.e., directed at a specific cookie). The intention is only for making a disjunction true, which is something that at the c-point can be accomplished by whatever psychological or physiological mechanism might make one grab a cookie.

2.3 To sum up, given one's indifference, the job of an intention *as a planning attitude*, is fully discharged by a purely disjunctive intention to $(A \vee B)$. This intention is fully adequate to discharge its characteristic job: settling on and stably guiding future conduct, framing further deliberation, and coordinating conduct transtemporally and interpersonally.

This intention needs to be paired with the power to pick one disjunct. But this is just one of the fundamental executive capacities that all agents need in order to carry out any of their intentions.

2.4 Purely disjunctive intentions are familiar but they apply to a limited range of circumstances. An intention to $(A \vee B)$ is adequate when one is indifferent between A and B but the two options do not call for distinctive preparations and the choice between them does not have distinct long-term effects. In the cookie example, no specific preparation is required. Upstream of the c-point, one is only required to keep the possibility of future choice open. And once the choice is made, one does not expect it to have divergent downstream effects.

2.5 This analysis applies also to cases of normative underdetermination (including those due to incommensurability and parity). When facing the prospect of two incompatible options that one cannot rank, as far as planning is concerned, one might simply settle for making their disjunction true. This is not to deny that, given the specific source of the underdetermination, the agent might have other resources that could help choose or pick between the options when she reaches the c-point. But the resources that come from intentions as planning attitudes run out in the face of this underdetermination, if one anticipates that the underdetermination is going to persist up to the c-point. This is so even if one takes into account the needs for advanced coordination with one's future choice between the two options.

Obviously, the choice at the c-point is much more momentous in cases of underdetermination than in those of mere indifference, including the potential for massively divergent downstream effects. But in a genuinely *persistent* underdetermination, even the anticipation of divergent downstream effects cannot make a difference to what the planning capacities might contribute. The best we can do is to settle for making a disjunction true.[2]

2.6 A disjunctive intention puts no pressure on keeping the options open. If the opportunity to carry it out by going for one of the disjuncts arises earlier than the anticipated c-point, one might be strongly advised to take it (but not required to do so, if the opportunity for future success remains available). Likewise, if one expects future success in executing one of the disjuncts, it is permissible to let go of taking means necessary for the pursuit of the other. Nonetheless, if the disjuncts do not make distinct and costly preparatory demands, one might keep both open until the c-point. Incommensurability over momentous choices might favor delaying the choice but this delay is rationally demanded only if one suspects that additional information might become available and remove the incommensurability.

3. *Pro-Tempore* Disjunctive Intentions

3.1 Let's return to the original scenario about my summer vacation and consider how it differs from a plan to make a disjunction true. First, I am not indifferent between Italy and Austria. It is rather that I am *presently* unable to tell which among these two incompatible options is better. I expect to be in a position to rank them later, on account of either new information or the additional time I can devote to deliberation. I might even expect one of the two options to turn especially bad, possibly one to be avoided at all costs (if, for instance, I were to learn that my archenemy is expected to be at one of those locations). Nonetheless, I have already determined that I should restrict my travel options to these two locations and I am already under pressure to start making the appropriate arrangements.

In this scenario, I am not intending to make the disjunction (going to Italy ∨ going to Austria) true, as if I were indifferent between the options. In particular, I am committed to keeping them *open* for the time being, since I do not yet know which one will turn out to be preferable at a later time. This makes a real difference if, as it is often the case, the preparatory arrangements for the two options are substantially different. When so, to keep both of them open carries a price. In addition to the distinct preparatory steps, I have to pay various costs, including those associated with the delaying of the final determination and with the forgoing of opportunities for alternative investments. But the circumstances might be such that it would still be reasonable for me to incur these costs, rather than risking to make either option impossible or being forced to make an early choice under more limited information.

The plan is not to leave the matter unresolved, in an unattended manner. This strategy might be best in those cases where delaying the preparation carries limited costs, if any. But this is not so in the present scenario. Hence, the plan is not to let the events run their course, make no commitment, and revisit the question later at a hopefully more favorable time. Here I am rather undertaking the intention to sustain the viability of the options until I am in a better position to make a choice, either because I will have gained more relevant information or I will face an unavoidable c-point.

3.2 What I have is an *intention*. It is the intention to sustain the viability of two options (Italy and Austria) *for the time being* in spite of their eventual *incompatibility*. This is not a case of simply intending to make a disjunction true. It is rather a plan that comes in two stages: at the earlier stage, while the implementation of both options is still possible and I am still waiting to determine which is best, I am pursuing *both*. At the later stage, once I have determined which is best or I am faced with an inevitable c-point, I only pursue one (although, prior to that moment, I do not yet know which one). This is a familiar kind of plan, whose form underlies many of our disjunctive expressions of intention. Oftentimes, when we say that we intend to do either A or B, we are not expressing indifference but rather a *pro-tempore* disjunctive intention.

Here is a more formal characterization of the content of a *pro-tempore* disjunctive intention (PTD intention). Given two eventually co-impossible goals A and B, let's call *c-point* the time when the two goals inevitably conflict and it is no longer possible to continue to pursue both; let's call *d-point* the moment prior to (and inclusive of) the c-point when the subject determines, if ever, which option is better than the other.

A *pro-tempore* disjunctive intention to pursue either A or B is the intention:

1. To pursue A at least until d-point (or until c-point if there is no d-point), *and*
2. To pursue B at least until d-point (or until c-point if there is no d-point), *and*
3. To continue pursuing A past d-point only if A is deemed better than B at that time, *and*
4. To continue pursuing B past d-point only if B is deemed better than A at that time, *and*
5. To pursue $(A \lor B)$ as a matter of indifference at c-point if there has been no prior or contemporaneous d-point,[3]
6. While believing throughout that A and B are co-impossible past the c-point.

3.3 Because of the two stages, a PTD intention does not run afoul of the demand for agglomerativity. As Bratman (1987, 134) has compellingly argued, "given the role of intentions in coordination, there is a rational pressure for an agent to put his various intentions together into a larger intention." The "larger intention" has usually been interpreted as the *conjunction* of the objects of the agent's intentions. For instance, if I intend to go to Italy and I intend to go to Austria, I am under a rational pressure to intend to go *both* to Italy *and* to Austria.[4]

The principle of agglomerativity formalizes the intuitive idea that a rational agent is not to embark on a self-defeating course of action. The rational agent must avoid the contemporaneous pursuit of co-impossible goals. But this does not entail that the agent is to give up one of the goals altogether, at least not from the very beginning. Given that the pursuits are co-impossible only *past* the c-point, it is not irrational to continue pursuing both, up until that point, *in the form of a PTD plan*. That is, with the knowledge that one is going to face a choice between them no later than at the c-point.

3.4 The mutual consistency among one's plans is something to be achieved *dynamically*, as one progresses in their pursuit and one acquires (and sometimes loses) additional relevant information.

The simplest method to secure consistency is to give up one of the co-impossible goals. But this might be too radical and premature a response, especially if one expects to gain relevant information later.

Another useful strategy is to *subordinate* one pursuit to the other. I might settle on Italy and intend to go to Austria only if going to Italy becomes impossible or unadvisable. When so, I am no longer under a demand to take steps to go to Austria, as long as the Italy option is still feasible. Hence, I take on the risk of incurring additional costs if going to Italy turns out later to be unfeasible or unadvisable (given that, by that time, the trip to Austria might have become much more expensive, if not impossible).

3.5 Alternatively, I could pursue both plans in a conditional form. I might list the conditions CI under which I intend to go to Italy, and the conditions CA under which I intend to go to Austria. If CI and CA are mutually exclusive, I might be able to agglomerate the two intentions without irrationality, provided that the preparatory steps do not encroach on each other (by making one pursuit possible at the expense of the other). I still cannot expect to succeed at both, but the failure would not count as self-defeating conduct. Rather, because of the mutually exclusive conditions, at least one of the plans will turn *moot* once its conditions are no longer believed to be possible.[5]

The conditional strategy is a more sophisticated form of the dynamical consistency secured by PTD: it requires the articulation of conditions that qualify the alternative options. By comparison, a PTD intention only requires that one is committed to keeping the options open while one is trying to figure out which one is preferable. As one learns more about the alternatives, however, the PTD intention might transform into a combination of two conditional plans, with mutually exclusive antecedents. This kind of metamorphosis is part of the adjustments that a rational agent makes in securing the mutual consistency and continuous progress of her plans over time.

3.6 All these strategies for mutual consistency are in principle equally available. Which strategy is preferable depends on the specific circumstances.

But the PTD strategy is always the simplest strategy to dynamically handle competing demands of eventually incompatible projects—short of renouncing one's goals altogether.

As the simplest strategy, it does not impose a specific structure on how the plan is to unfold. It does not articulate in advance the specific conditions under which one option rather than the other is to be pursued. Nor does it institute a structure of subordination. But it is a useful starting point, which occasionally might lead to subordination or articulation of more specific conditions. Because of its simplicity, it might be abused and become a tool for massive procrastination. An agent who is reluctant to make choices might continue to undertake more and more eventually incompatible projects in the PTD form, but later face the dreaded moments of choice at much higher costs. On the positive side, learning about PTD structure and its characteristic demands helps us become more proficient in handling the balance between fixity and stability, on the one hand, and flexibility and responsiveness to changing circumstances, on the other. It offers the basic tool to keep concurrent projects open but also to develop strategies to reduce the risks and costs associated with concurrent pursuits (for instance, by inviting us to settle on projects with a larger overlap in preparatory steps or to increase reliance on all-purpose means).

PTD plans are the most basic tools for managing the temporal agglomeration of distinct projects and for achieving dynamic consistency by limited agents like us, who need to handle the uneven temporal distribution of resources and opportunities for action, and their limited information. It allows for the simplest, even if only temporary, integration. This integration is a better strategy than the rigidity of agents who either refuse to undertake incompatible plans altogether or undertake them while ignoring their eventual incompatibility, thereby setting themselves up for the high costs of the unanticipated later encounters with c-points.

3.7 There is still a concern with PTD plans. The characteristic benefits of planning derive from its power to settle one's conduct in advance, but a PTD intention might still leave wide open what the agent is going to do. Can one really *settle* a practical question in a stable manner by acquiring a PTD intention?

Consider stability. A PTD intention is subject to the pressure for stability with the same force and character as any other intention. That a rational agent is expected to drop one of the disjuncts does not make her intention more unstable. Dropping one of the options is not the same as giving up the PTD intention altogether. Rather, it is a rational transformation that the intention is *supposed* to undergo as time goes by and more information is acquired. This is similar to the metamorphosis that occurs to a conditional intention, when the conditions are taken to hold and the antecedent is thereby discharged. When so, one is not giving up the original conditional intention, one rather continues to carry it out in its new categorical shape.[6]

Likewise, once the agent drops one disjunct, she continues to carry out the same intention in its new non-disjunctive form.

This conclusion holds for all rational pressures constitutive of intending (such as means-end coherence and agglomerativity), not just stability. They apply with the same force regardless of the content of the intention. In this sense, a conditional intention is not weaker than a categorical one. Likewise for PTD intentions, which are neither weaker nor more unstable than categorical ones. This is why an agent can be *settled* on a PTD intention as much as she is on a categorical one.

3.8 At least, this is so for what might be called "structural" settledness. A rational agent is structurally settled on what she is going to do in the future by way of taking a distinctive attitude—an intention—toward her future conduct. There is still a substantive question: *what* is she settled on? This depends on the *content* of the intention rather than the nature of her attitude. In the substantive sense, a PTD intention seems to determine much less about one's future conduct than a categorical intention, for much is still open about what one is going to do both before and after the c-point. This might seriously reduce the contribution of PTD intentions to transtemporal intra- and interpersonal coordination—which is a distinctive benefit of our reliance on intentions.

But PTD intentions still help with coordination. Like any intention, a PTD one *restricts* the range of conduct expected of a rational agent. At the very minimum, a rational agent is not to make impossible for her to carry out her intention. For a PTD intention, prior to c-point, the range of expected conduct might actually be more limited than for a categorical one to pursue just one disjunct. For instance, before reaching the c-point, not all the ways of my going to Italy are compatible with my going to Austria. Hence, if you need to coordinate with my position prior to the c-point, there are situations in which you might be better off if I have a PTD intention rather than an unconditional one. Prior to the c-point, there are fewer courses of action that I might rationally take in the PTD case than in the unconditional one. For prior to the c-point, I am not expected to make progress toward my going to Italy in a way incompatible with my going to Austria. When you consider coordination after the c-point, however, my PTD intention is less useful than an unconditional one, because of the divergent unfolding of the ways in which I might rationally pursue the plan that I have now chosen at the expense of the other. In any event, even after the c-point, having a PTD intention still provides a better guide for transtemporal coordination than having no intention at all.

3.9 Coordination only requires that one be able to determine, at least roughly, the size and shape of the area where the rational agent is supposed to be on account of her intention. A rational agent is not supposed to get outside of this area, since this exit would amount to a failure of her project.

This is not the same as coordinating with an agent who is settled on a very specific goal which can only be implemented in a univocal way. When so, one could anticipate each and every move of this agent and reduce the area for coordination to a single trajectory. But this is a limiting case. Much coordination, especially for the longer term, is done by securing that one stays within a certain *area* rather than on a specific trajectory.

The smaller the area, the easier it is to anticipate where the agent will be and to coordinate with her based on her expected future location. The extent of this area, however, is not simply a matter of the form taken by the content of the intention. A PTD intention might actually delimit a smaller area than a categorical one. After all, my PTD intention "to go either to Italy or to Austria" is more helpful, for purposes of coordination, than a non-disjunctive but more generic intention "to go to Europe."

4. The Pervasiveness of *Pro-Tempore* Disjunctive Intentions

4.1 The structure of PTD intentions is not just a curiosity. It is a pervasive feature of our intentional diachronic agency. It provides the basic strategy to secure the dynamic consistency of our plans in their temporal unfolding, given that we need to respond to our previous successes and failures, and to changes in relevant information. PTD intentions are not limited to brand new plans about options we are yet unable to rank. They might also be adopted to agglomerate a new goal with some old ones. PTD intentions work as placeholders for future choices, while we continue to move on with all of our concurrent (although not all eventually co-possible) projects.

4.2 In addition, the PTD structure underlies the "partiality" of plans (see Bratman 1987). Many details about implementation are initially left unspecified and they are filled over time. As one gets closer to the need to take more specific steps, additional and more accurate information is usually more easily available to help determine the details of implementation.

Partiality is just a special instance of the PTD structure. Let's imagine that one intends to φ at a later time t_3. Consider the step that she has to take at t_2 to progress toward her later φ-ing. This step might take very different specific shapes, $s_1, s_2, \ldots, s_m$. These shapes are mutually incompatible pieces of local conduct. When the agent at t_1 thinks about what she is going to do at t_2, she might be in no better position than to intend to take one of these steps, without knowing yet which one. But she is also committed to keeping them open for the time being. For she is committed to taking whatever steps will help her progress toward her future φ-ing. In other words, at t_1 she intends *pro-tempore* to either s_1 or s_2 or . . . or s_m at t_2 (which are ultimately co-impossible implementations). As she gathers additional relevant information, she might decide before reaching the c-point at t_2 which of these disjuncts to pursue. But her decision might wait as late as t_2. Either way, this shows that any partial plan has the same structure as a PTD intention.

4.3 The only notable difference is that in the case of the partiality of implementation, the PTD sub-plan is subsumed under a larger and longer plan, which need not be disjunctive. In this scenario, the PTD sub-plan about what to do at t_2 is subsumed under the plan to φ at t_3. As a result, the effects of the choice among the disjuncts is not expected to amplify downstream, but rather to be "reabsorbed" soon, since it is a matter of *local* implementation of one's continuous advance toward the future φ-ing.

4.4 The disjunctive character of PTD intentions becomes more dramatic when the choice is supposed to amplify after the c-point. But this amplification is not a necessary feature of PTD intentions. Likewise, in many cases of partiality, the costs of keeping open the mutually incompatible disjunctive sub-plans might be much less severe than in the standard PTD intentions, since the sub-plans are ultimately alternative ways of implementing the same goal, which is set by the master plan. Even so, the differences between paradigmatic cases of partial plans and paradigmatic instances of PTD intentions are not due to a different underlying structure. Both cases share the basic structure of *pro-tempore* disjunction, even if the structure can be differently instantiated depending on the specific forces that one's goals and circumstances exert on both the upstream and downstream stages of one's pursuit.

5. Partial Intentions

5.1 Richard Holton (2009, ch. 2) has recently suggested that a distinctive kind of attitude—a *partial* intention—is required to handle scenarios similar to the one discussed here. In Holton's example, I intend to remove a tree brought down by a storm. There are several possible ways of achieving this end E: I can lever the tree with a crowbar, saw it with a chainsaw, drag it with a rope, or call the local tree company to move it. These options are eventually incompatible. But I do not yet know which is better. For the time being, I want to keep all of them open, which might force me to bear the costs of multiple preparatory steps.

This scenario falls between the paradigmatic cases of partial plans and those of PTD ones. The options are ways of implementing a single master end E (like in a partial plan), but the costs of keeping them open are higher than in simple partiality, whence the resemblance with standard PTD intentions. As I have argued earlier, the difference between these two scenarios is a matter of degree. Hence, I will consider Holton's proposal as it applies more generally.

5.2 Partial intentions are intention-like attitudes. As Holton (2009, 35) writes, partial intentions

> play the same roles in curtailing deliberation, resolving indeterminacy, and enabling coordination that intentions play: you fix on a small number of plans from the many that occurred to you and that you might

have pursued, and as a result of this you can coordinate around your other plans [. . .] and with other people.

What is distinctive is that partial intentions, unlike all-out ones, are not subjected to the pressure for strong consistency. An all-out intention is supposed to be consistent both with the agent's beliefs and her other intentions. But a partial intention is only subjected to a weaker requirement: if one partially intends to φ, one is required to have a *partial* belief that one will succeed in φ-ing (or at least one is required not to all-out believe that one will fail at φ-ing)—see Holton (2009, 41–46).

5.3 In support of his account, Holton often appeals to the parallel between partial beliefs and partial intentions. But given the controversial nature of partial beliefs and the limitations of arguments from analogy, I will set these considerations aside.

Additional support comes from an argument by elimination. Holton claims that all-out intentions with complex contents cannot account for these scenarios. First, we can't appeal to all-out conditional intentions since no specific conditions are attached to alternative options. Second, all-out disjunctive intentions lack appropriate explanatory force. It is only when the all-out disjunctive intention is broken down into its components (the alleged partial intentions) that the agent's specific actions can be explained. As Holton (2009, 38) writes: "It is my partial intention to get the tree company to move the tree that causes me to phone them; if we are limited just to all-out disjunctive intentions, we can give no explanation of this."

5.4 I agree with Holton that explanatory force is crucial. The issue arises even when the incompatible options are not subsumed under a single end *E*. For a standard PTD intention, Holton's concern is that an action such as my making flight reservations to Austria could be explained by the *partial* intention to go to Austria, but not by the all-out PTD intention to go either to Austria or to Italy.

5.5 To see where the problem might lie, consider the explanatory force of a categorical all-out intention. Imagine that I categorically intend to get the tree company to move the tree. It is uncontroversial that, if I am rational, this intention might explain my phoning the tree company now. This is so when the categorical intention is non-deviantly combined with my current belief *B* that (1) phoning the company now is a necessary means to get the company to move the tree, and that (2) I have now the ability and opportunity to phone the company.

Consider now a *partial* intention to get the company to move the tree. This intention, when non-deviantly combined with the belief *B*, appears as apt as the categorical intention at explaining my phoning now. Its partiality makes no difference to its explanatory power with respect to my phoning. The partiality rather affects which belief is rationally justified on account of my intention: I am not justified to all-out believe that I will get the company to move the tree, since I might end up handling the tree in another manner. But, Holton would claim, I am justified in holding the *partial* belief that the company is going to move the tree. So far, so good.

5.6 Consider now the PTD intention to either call the tree company or saw the tree. Does this intention lack explanatory power? If I phone the company, this action can be explained by combining the PTD intention with the same belief B that I now have the ability and opportunity to take the necessary means. The disjunctive and *pro-tempore* character of the intention does not seem to make any difference to its explanatory force. Given the instrumental fit between phoning the company and getting it to move the tree, it seems that the PTD intention together with B can explain one's action in the same way in which a categorical intention (either all-out or partial) does.[7]

5.7 Even so, one might suggest on Holton's behalf that in a PTD plan the explanation often goes through an intermediate step: in order to implement the plan, the agent might first focus on one of the disjuncts and *only then* considers what she is to do to make the disjunct advance. For instance, I might be moved to phone the company not directly by the PTD intention but by my focusing on the sub-plan of getting the company to come. It is only because I am now thinking about getting the company to come that the belief B together with the intention explains my action of phoning the company. Or so one might argue.

If this is the correct explanation, one might then ask what kind of attitude as a rational subject I have toward the disjunct that is at the focus of my attention. It seems that I cannot say that I all-out intend to get the company since this intention would not meet the demand of global consistency. Holton's suggestion would come to the rescue: what I have is just a *partial* intention, which is not under the pressure of global consistency.

5.8 This is a radical solution. It asks us to acknowledge a novel *kind* of attitude. Before going this route, however, we must make sure that we cannot accomplish something comparable by relying on familiar all-out intentions, albeit ones with a complex content. If I have to explain why I am phoning the company, it seems that I can appeal to my intention to get the company *as* one of the alternative options that I am still pursuing right now. This is the option on which I am focusing my attention right now without having thereby acquired a novel kind of attitude. I have a standard all-out intention with a content that is qualified as one of the open sub-plans of a larger project.

This is a very familiar kind of sub-plan, one in which we often engage when implementing a PTD plan. But this sub-plan does not commit me to believing that I will succeed in it, since I appreciate how the continuous pursuit of the sub-plan ultimately depends on the fate of the other options.

The content of my present all-out intention directed at getting the company has a complex structure. But this complexity just reflects the contribution of this sub-plan to the PTD intention. I intend all-out to get the company *as* the alternative on which I am presently focusing among the various eventually co-impossible options (which I am still trying to keep all open for the time being) to remove the tree. This a perfectly fine content for an all-out intention. One that we are very familiar with even if not necessarily one that we would usually express in this form.

5.9 Hence, it is no objection to my analysis in terms of a complex content of an all-out intention (rather than of simple content of a partial intention) that the intentions toward a sub-plan of a PTD plan are often expressed categorically. How much complexity one is going to express is ultimately a matter of the pragmatics of communication. If all that I am trying to communicate to you is how my phoning makes any instrumental sense, it might be sufficient to say: "I intend to get the tree company." But if you are the one who is supposed to provide me with the saw that I need for the alternative sub-plan, I will be more inclined to spell out the complexity of my plan. I want to avoid any troubling misunderstanding, including your giving the saw to someone else. So I might say to you something like:

> I am phoning the tree company because I want to keep open the option of their coming to get the tree but I have not yet made up my mind about which is the best way to remove the tree . . . so please still lend me the saw.

5.10 I have two additional serious reservations about Holton's proposal. First, there should be a methodological presumption against the proliferation of novel kinds of attitudes, especially on the face of the availability of alternative accounts in terms of complex contents. If my previous considerations are correct, there is no reason to favor introducing a distinctive kind of attitude—a partial one—directed at an unqualified content rather than sticking with standard, all-out attitudes directed at a complex content.

5.11 Second, turning Holton's objection on its head, I want to argue that partial intentions are explanatorily weak. Although a partial intention might explain some of the steps of implementation, it cannot explain how these steps figure into any larger plan and how this might affect their nature and timing. Phoning the tree company might be explained by the partial intention to get the company, but the fact that I am phoning now rather than later, say, might have to do with how this action fits with my other concurrent sub-plans. As far as getting the company to come, I might have called at a different time. But I have to call *now* given that at the other times I will be busy preparing for the alternative options I am still keeping alive.

The partial intention cannot explain the timing of my call, since these other sub-plans do not figure into the content of the partial intention.

True, the partiality of the intention hints at the existence of other eventually incompatible sub-plans. But it does so only in a generic form that does not help explain the specific unfolding of the agent's conduct in response to the actual character of these sub-plans. Likewise, a partial intention offers no explanation of why one might find advisable to give up that sub-plan in favor of another one. A partial intention, at most, warrants only a partial belief in its future success but it is silent on what might stand in the way of its being carried out, that is, it is silent on the specific character of the alternative sub-plans. Finally, a partial intention is not under any rational pressure to handle the competing sub-plans. Hence, except for those times when the agent does not face any immediate interference from the sub-plans, a psychology that operates through partial intentions rather than all-out intentions is in danger of being ineffective.

5.12 The partiality only shows in the irrationality of an all-out expectation about future success but it does not affect the rationality of the agent's conduct in the pursuit of the partially intended goal. This deprives the partial attitude of its explanatory force. If the psychological work were really done by the partial attitude, we should expect the rational agent to be oblivious to the existence of the competing projects while carrying out any specific partially intended goal. But this is exactly what we do not expect of such an agent. The only way to give a full explanation is to revert to an all-out intention directed at a complex disjunctive content. It is only at that level that the pressure for global practical consistency is exerted, a pressure that plays an indispensable role in any perspicuous psychological explanation of the agent's conduct.

5.13 To sum up, I see no benefits in preferring an account in terms of a partial attitude toward a simple content, rather than an all-out attitude toward a complex content. The resources to explain the conduct of the rational agent are to be found in the complexity of the content. Much work still needs to be done to spell out the implications of the complex contents of intentions both for practical rationality and for psychological explanation. But I have not yet found compelling reasons to despair that this job can be successfully carried out without introducing any *partial* practical attitude.[8]

Notes

1. For my view on conditional intentions see Ferrero (2009, 2015). See also Yaffe (2004), Klass (2009), and Ludwig (2015).
2. For a discussion of intentions and underdetermination, see Ferrero (2010), Bratman (2012), and Ferrero (2014).

3. Normally, in undertaking co-impossible projects in the *pro-tempore* mode, an agent expects that if she is unable to determine which is better by the c-point, she will be indifferent between them at that point. When that happens, she can resolve the indifference by relying on any of the mechanisms that operate for standard disjunctive intentions of indifference. In some special cases, the agent might qualify her PTD intention by specifying what she would do at the c-point if she has not been able to determine which disjunct is best by then. For instance, she might even intend to pursue neither at that point.

4. For further discussion of agglomerativity see Bratman (2008) and Yaffe (2009).

5. For a discussion of how conditional intentions become moot, see Ferrero (2009, 705–707) and Ludwig (2015).

6. For a discussion of the transformation of a conditional intention into what I call a "circumstantially unconditional" one, see Ferrero (2009, 710).

7. If the intention is disjunctive because of my indifference between getting the company to move the tree and sawing the tree myself, phoning might not be *necessary* to the intention's success. It is only when it is no longer possible for me to saw the tree that phoning the company becomes necessary for the success of the disjunctive intention. However, this does not entail that one is unable to explain the taking of non-necessary means to the success of a disjunctive intention. One could explain this in the same way as one explains the taking of non-necessary means for either categorical all-out intentions or for partial intentions. Whatever additional elements, if any, might be required to complete the explanation for these intentions, these elements appear to be available for disjunctive intentions as well, both all-out and PTD ones. In any event, unlike the case of indifference, if I have a PTD intention, phoning the company might be a necessary means to the success of this intention even if sawing the tree myself is still an available option. For phoning is a necessary means to getting the company to move the tree, which is something that I am presently committed to, given that I am committed to keeping *both* options open for the time being.

8. An earlier (and much longer) version of this chapter was presented at talks at the University of Southern California, Queens University at Kingston, and the University of Toronto. Thanks to the audiences, and especially to Philip Clark, David Hunter, Rahul Kumar, Lewis Powell, Sergio Tenenbaum, Gary Watson, George Wilson, and Gideon Yaffe. Thanks to Ray Buchanan and Luis Cheng-Guajardo for stimulating conversations. Thanks to Roman Altshuler and Michael Sigrist for their comments and editorial work. A special thank you to Gideon Yaffe who first prompted me to think about disjunctive intentions.

References

Bratman, M. E. (1987). *Intentions, plans, and practical reason.* Cambridge, MA: Harvard University Press.

——— (2008). Intention, belief, practical, theoretical. In Timmerman, J., Skorupski, J., & Robertson, S. (Eds.), *Spheres of reason* (pp. 29–61). Oxford: Oxford University Press.

——— (2012). Time, rationality, and self-governance. *Philosophical Issues, 22*, 73–88.

Ferrero, L. (2009). Conditional intentions. *Noûs, 43*(4), 700–741.

——— (2010). Decisions, diachronic autonomy, and the division of deliberative labor. *Philosophers' Imprint, 10*(2), 1–23.

——— (2014). Diachronic structural rationality. *Inquiry, 57*(3), 311–336.

——— (2015). Ludwig on conditional intentions. *Methode, 4*(6), 61–74.

Holton, R. (2009). *Willing, wanting, waiting.* Oxford: Oxford University Press.
Klass, G. (2009). A conditional intent to perform. *Legal Theory, 15*(2), 107–147.
Ludwig, K. (2015). What are conditional intentions? *Methode, 4*(6), 30–60.
Yaffe, G. (2004). Conditional intent and mens rea. *Legal Theory, 10,* 273–310.
—— (2009). Trying, acting and attempted crimes. *Law and Philosophy, 28*(2), 109–162.

8 Evaluative Commitments

How They Guide Us over Time and Why

Monika Betzler

1. Introduction

Consider the following two examples:

1. A couple has been operating a mom-and-pop store for more than two decades. They have come to value their own business. Even though it is sometimes frustrating to work long hours and not make much money, they refuse to consider other, possibly more promising career options. Their own business means a lot to them.
2. A mother has decided to work part-time to make more time for her daughter. She loves her daughter, giving her a lot of attention and doing many things for her. Even though she occasionally feels exhausted, she would never want things to be different. After all, she only has one daughter and she is very important to her.

These admittedly stylized examples bear important similarities. They are both about particular kinds of engagements that seem to be of the utmost significance to the people involved. These engagements can be individuated as personal projects and close relationships. Personal projects, such as causes, vocations, ambitions, hobbies, and other comprehensive goals, and close relationships to one's friends, spouse, partner, children, and parents, and sometimes even to one's colleagues and neighbors, share important normative features: the people involved not only take themselves to have reasons to do more to realize those projects or relationships (and to disregard other reasons that are in conflict with those for the realization of these engagements). They also seem to be doing something wrong if they do not act on these (both positive and negative) reasons. That is, once one has come to pursue a particular project, or once one has a close relationship, one faces the possibility of rational criticism if one fails to do as much as one can to realize them, and if one does not exclude reasons for doing things that could undermine these engagements. Henceforth, I shall use the term "evaluative commitments" to cover this class of particularly important and normatively significant engagements.

The term "evaluative commitment" covers two intuitions underlying the phenomenon I have just highlighted. First, personal projects and close relationships have important *evaluative* dimensions: they connect an agent with many values or evaluative properties that are likely to make her life more meaningful. Being able to respond to the various values that personal projects and relationships manifest gives content to one's life. Second, and relatedly, by providing the agent with evaluative content to which she comes to respond, evaluative *commitments* form (at least in part) a practical standpoint from which she is able to guide her life. This is because they feature in many areas of our lives and have a wide effect.[1] Let me call this the *content and shape hypothesis*. Without evaluative commitments, our lives would be markedly different, that is, devoid of meaning and structure, less good, and less autonomous. This explains, in part at least, why a good human life without evaluative commitments is but an empirically rare conceptual possibility.[2]

Evaluative commitments not only provide content and shape to a life when considered third-personally. They also exert normative pressure on the agent that confronts her first-personally. Reflecting on the aforementioned examples, we see that an agent who is committed to another person or to a project is under a specific normative relation. Let us suppose, for example, that a new supermarket threatens the profitability of the mom-and-pop store mentioned earlier, so that the couple's friends and children strongly advise them to give up the business. But against this well-meant advice the couple is adamant, pointing out that the store is their life, and arguing that even though they make very little money, that is not enough to make them give up on their project. It is "their" store, after all.

Thus the normative relation under which evaluatively committed agents find themselves does not seem to be entirely dependent on the value-based reasons they have for their commitment. In addition, the relation also seems to be partially independent from the attitudes that typically support the agent's commitments. For instance, if the store owners feel frustrated or have other negative attitudes, they could still regard themselves as being committed to their enterprise. Hence, evaluative commitments exert a normative pressure that is neither entirely dependent on the value-based reasons that ground them nor on the attitudes that underwrite them. But this does not entail that they are entirely decoupled from these value-based reasons and attitudes. To be sure, if the store owners were to go out of business, they would face overwhelming reasons to give up on their valued project. The normative relation that evaluative commitments put us under is thus not absolute.

What I want to highlight is that there are cases in which we are bound to some project, despite the existence of reasons for the contrary. This is what being evaluatively committed seems to imply. Call this the *normativity of evaluative commitments hypothesis*. In this chapter I conduct an inquiry into this hypothesis by addressing the following question: how can we explain the normative relation that evaluative commitments entail?

The structure of this chapter is as follows. After offering some clarifications in Section 2, in Section 3 I will examine how the normative relation to which we are subject in virtue of our evaluative commitments can be captured. In Sections 4 and 5 I examine four possible explanations that, in the end, I find wanting. In Section 6, I outline and defend an alternative explanation, according to which there is a distinct self-referentiality at work in evaluative commitments that explains our reasons for continuing with them and thus accounts for their particular normative force.

2. Four Clarifications

Before I proceed, four clarifications are in order. First, when I refer to close relationships and personal projects as prime examples of evaluative commitments, I do not mean to neglect the important differences between the two. Neither do I mean to suggest that there aren't significant variations within different kinds of close relationships and projects. These differences and variations notwithstanding, I hope to have made it clear that these kinds of engagements fall under one class of evaluative commitments in that they put us under a similar normative relation that, so far, I have spelled out only intuitively. It is not ruled out that there are other kinds of evaluative commitments in addition to close relationships and personal projects, but I take these to be the most typical and pervasive kinds that put us under the normative relation I examine in the following subchapter.[3]

Second, the term "evaluative commitment" lacks a clear definition; while the differences between different kinds of commitments are not discussed here, I take evaluative commitments to be pervasive and important normative phenomena that, at first glance, bind us over time. They do this via the particular values or evaluative properties to which we have come to respond by way of engaging in close relationships and personal projects. This makes them different to other kinds of commitments, such as moral, ontological, or logical ones. They do not shape our lives by binding us to moral norms, ontological convictions, or logical connections, but to valuable properties or things. Whatever other terms have been used to account for such engagements—such as "attachments" (Raz 2001; Wallace 2013), "involvements" (Owens 2012), or "robustly demanding goods" (Pettit 2015)—the aim of this chapter is to provide a better understanding of what it is that binds us in our commitments to people and projects and the evaluative properties those commitments manifest. The term "commitment" captures the way in which a committed agent is under a normative relation. The qualification "evaluative" accounts for the fact that an agent proves to be committed in virtue of having come to respond to particular evaluative properties. Contrary to other kinds of practical commitments, such as intentions, this neither presupposes an act of will before the fact nor any other conscious or deliberative effort.[4]

Third, there is an important yet poorly understood connection between normativity and time that requires further elucidation. Evaluative commitments are noteworthy in that they call for a particular valuing response, thereby guiding a person over time in that they bind her in some specified way and let her identify with them. In short, they are diachronic ways of engagement in virtue of putting the agent under a particular normative relation.

Fourth, in claiming that evaluative commitments allow us to respond to values or evaluative properties by coming to love or value them, I do not wish to defend any metaphysical view of values. The way I characterize evaluative commitments might convey realist leanings. But this is just a mundane way of describing how we typically try to make sense of our projects and relationships. In fact, I try to stay neutral—as far as I can—with regard to any further claims about what values actually are.

In light of these clarifications let me now explain the normative relation manifested by evaluative commitments.

3. Evaluative Commitments as Normative Relations

So far, I have tried to make it clear *that* evaluative commitments put us under a particular normative relation. My aim now is to explain that relation in more detail. It seems to be an analytical truth that evaluative commitments—the fact that we are evaluatively committed in virtue of valuing a particular project, or in virtue of entertaining a particular personal relationship—entail that we take ourselves to have particular commitment-dependent reasons to which we should respond. In short, once we take on a particular evaluative commitment, we are confronted with reasons that we would not have if we did not have that particular commitment. This does not entail, however, that we should respond to every commitment-dependent reason that we encounter. All it takes for the relation between evaluative commitments and commitment-dependent reasons to hold is that we respond to a sufficient range of commitment-dependent reasons to qualify as being evaluatively committed.[5]

But what are commitment-dependent reasons? They can be described in terms of both substantive and formal features.[6] With respect to their substantive features, evaluative commitments generate reasons for specific sets of action-types, as well as forward-looking and backward-looking (mostly emotional) attitudes.[7]

More precisely, these commitment-dependent reasons are reasons to engage in various meaningfully related action-types that jointly manifest the existence of a particular personal relationship or project. The mother mentioned earlier will respond to such relationship-dependent reasons, for example, by calling her daughter on a regular basis, showing interest in her, and being empathetic. Similarly, she will take herself to have reasons to meet up with her daughter, buy her birthday presents, and cook her favorite meals (to name but a few possible action-types). That is, the mother will respond, repeatedly and over time, to reasons to engage in a set of different actions that jointly manifest her relationship to her daughter. These action-types are indirectly related via substantive norms, which more or less determinately prescribe what constitutes a mother-daughter relationship.

In addition, the mother will be liable to a set of forward-looking and backward-looking emotions with regard to her relationship with her daughter. On the one hand, these emotions respond to distinct features of her relationship: how it unfolds over time, and how it fares in changing circumstances and contexts. For example, if her daughter stops reciprocating her efforts to maintain the relationship, the mother will worry about the relationship. And if her daughter is not well, she will feel upset and concerned. She might feel resentment if someone fails to behave respectfully towards her daughter; she could be liable to joy when her daughter, and her relationship with her daughter, flourish. On the other hand, the mother will be disposed to feeling reflexive emotions. These let her perceive salient features about herself given that she is in a close relationship with her daughter. For instance, she may feel proud of her daughter if the daughter is successful. By contrast, she may feel ashamed if her daughter behaves badly.

As a result, being evaluatively committed to a person or a project entails that a person thinks of herself as having reasons for performing meaningfully related action-types and various kinds of forward-looking and backward-looking emotions with regard to a person or project, their circumstances, and herself in relation to that person or project.

Commitment-dependent reasons can also be described in terms of their formal features. That is, they are reasons of particular weight, with a particular role in deliberation, and with a particular scope. As a result, we can say that they are (1) modally stringent, (2) non-instrumental, (3) agent-relative, and (4) diachronic.

In terms of the first feature, commitment-dependent reasons are modally stringent in that they continue to be reasons in various and not too distant possible worlds. For example, being committed to a person or project generates reasons to value or love her or it, even if one experiences frustration or if it undermines one's self-confidence. In sum, these reasons remain in force by default. It remains open as to how distant these possible worlds, in which the normative force of such reasons remains intact, can be. But even if it is a matter of debate (and context), it seems obvious that these reasons have *some* modal stringency. And this entails, in part at least, that an agent who is committed to a particular project or to a friend gives priority to it (or her) over other possible projects and persons and over other momentary countervailing reasons.[8]

Furthermore, with regard to the second feature, being committed to a particular project or person entails reasons that are, to some extent, non-instrumental. That is, they do not play a role in means-end deliberations and thus do not bring about a commitment-independent outcome. Rather, they are, in part at least, reasons to manifest one's already established relationship to a particular person or project. In other words, evaluative commitments provide reasons to love or value things or people for their own sake and not with regard to their comparative value.[9]

These reasons are agent-relative—the third feature—in that commitment-dependent reasons are not reasons for everyone; they are only reasons for the person who is evaluatively committed to a particular project or person.[10] They explain why *she* behaves and acts in a particular way towards *her* friend or project, provided that she is committed to her project or to her friend. The scope of these reasons thus depends on the pronominal reference. As for the fourth feature, once an agent is committed to a particular person or a particular project, he has reasons to pursue his relationship with that person or to engage with his project over time by undertaking recurring activities and by valuing or loving that person or that project by default. This implies that he will ignore reasons that speak for prioritizing other people or that favor other projects. It also involves him taking himself to have a practical standpoint from which to understand himself and from which to think about past and future goals.[11]

Our definition of commitment-dependent reasons clearly needs to be refined and explained in more detail. To start with, however, it seems plausible enough to suppose that evaluative commitments put the person in question under a distinct normative relation in virtue of the commitment-dependent reasons that apply to her once she finds herself evaluatively committed.

What remains unclear is how to account for the force of that normative relation. What exactly does the normative work of commitment-dependent reasons consist in? How far does it reach and what are its limits? In the following two sections I will examine various possibilities that could account for its force.

Even if we agree that having an evaluative commitment entails taking oneself to have commitment-dependent reasons, the stringency of these reasons is not as conceptual a matter. Similarly, it is not clear what explains their stringency. After all, how can we account for their stringency in case the value-based reasons undergirding them are bad or underdetermined reasons? And how can we make sense of their stringency if the agent in question is liable to attitudes that *do not* underwrite them? Furthermore, if we consider evaluative commitments and the commitment-dependent reasons to which they give rise as entirely independent from value-based reasons and attitudes, we face a problem of self-justification. The fact that a person is evaluatively committed does not sufficiently explain why she should be so committed, countervailing reasons and attitudes notwithstanding.

The challenge, then, is not only to explain where the normative force of evaluative commitments comes from. It is also to explain how the value-based reasons for or against a particular commitment and the fact that one takes oneself to have commitment-dependent reasons relate. After all, commitment-dependent reasons are *modally* stringent, and this means that they can lose their force in a world in which strong reasons speak against them. But they are still modally *stringent*, which precludes any reasons, however slight, from tipping the balance against them.

In short, there are two issues that need to be tackled. First, is the normative work performed by commitment-dependent reasons entirely independent of the value-based reasons that ground them? Call this the *Independence Claim*. If the Independence Claim is correct, we need a further explanation of why commitment-dependent reasons put us under a normative relation and change our normative landscape just because we find ourselves committed. Or, second, do they, in some to-be-specified way, interact with the value-based reasons that ground them? Call this the *Dependence Claim*. If this is the correct view, we need a further explanation of the additional weight they provide. The diagnosed stringency of commitment-dependent reasons might be thought to bolster the Independence Claim. The fact that their stringency is only modal speaks for the Dependence Claim. In what follows I will discuss these two possibilities in more detail[12] and then propose a third explanation.

4. The Independence Claim

The clearest way to summarize the Independence Claim is to say that there is a different normative dimension at work once we have become evaluatively committed and thus respond to commitment-dependent reasons. Provided we have come to value or love a project or person and thus respond to commitment-dependent reasons, we might maintain that a rational requirement is at work. That is, once we have come to respond to some commitment-dependent reasons—which we might characterize as having developed certain attitudes with regard to the particular evaluative properties that define a particular commitment—we are rationally required to have further attitudes that are coherent with the former. Rational requirements are very specific kinds of relations that govern our attitudes and differ from reasons. Call this the *Rational Requirements View*.

a. The Rational Requirements View

One might think that commitment-dependent reasons are reasons, in part at least, for taking particular attitudes towards a particular commitment. They encompass reasons for being liable to a range of forward-looking and backward-looking emotions, but also reasons for being susceptible to beliefs and judgments about one's emotions and the value of one's commitments, their fate, and their relation to oneself. These attitudes—which could be called loving and valuing—could be thought to be governed by independent principles of rationality, that is, so-called rational requirements. Rational requirements reflect valid inferences between propositions and are captured as conditionals.[13] Rationality is thus thought to forbid patterns of attitudes that are incoherent. If anything, rational requirements are stringent. That is, on pain of irrationality, if one harbors a certain attitude, one *must* harbor further attitudes that cohere with the first attitude. One might therefore think that rational requirements could account for the stringency of commitment-dependent reasons: once we have a particular attitude with regard to a project or person, we are rationally required to have further attitudes that cohere with it. One might think that this sits well with the idea that commitment-dependent reasons are reasons by default. And this proposal seems to make sense of the normative relation that evaluative commitments put us under even in light of countervailing reasons disfavoring them.

To fully understand this proposal, we first need to clarify which principles of rationality are at work in the case of evaluative commitments. Two principles seem suggestive in the cases of loving and valuing:

(EMOTION–JUDGMENT COHERENCE) Rationality requires [if you are liable to a range of emotions with regard to a project PRO or person PER, and take these emotions to be appropriate, you judge that PRO or PER to be valuable].

(EMOTION–INTENTION COHERENCE) Rationality requires [if you are liable to a range of emotions with regard to project PRO or person PER, take these emotions to be appropriate, and believe that A is expressive of them, you intend to A].

What these principles tell us is *not* that a person has a reason for any of these attitudes. Quite the contrary, they allow for any of the attitudes to be based on bad reasons. They only require—in virtue of the validity of inference patterns—that if an agent has a particular attitude, then further attitudes, on pain of irrationality, have to cohere with it. The agent is thus under a requirement to make the conditional true.

There are, however, two possible objections to the idea that the stringency of evaluative commitments can be captured as a rational requirement that governs the attitudes that commitment-dependent reasons call for.

The first objection pertains to the question of whether rational requirements can really account for the stringency of evaluative commitments. All these principles do is require an agent to make a conditional true. But there are two ways of attaining the truth of such conditionals: an agent can choose to have attitudes that are coherent or give them up altogether. Rational requirements do not tell the agent which to choose. The phenomenology of having an evaluative commitment, however, suggests that an agent takes himself to already be under a normative relation. That is, he takes himself to have and continue to have reasons for his evaluative commitment even if he happens to have incoherent attitudes. This suggests that it is the fact of having evaluatively committed himself to a particular person or project that puts him under a normative relation—incoherent responses to commitment-dependent reasons (to some extent at least) notwithstanding. Since rational requirements do not tell an agent how to make the conditional true, they do not account for the normative relation that the agent takes himself to be under even in light of incoherent attitudes.

This is related to a second concern, namely that rational requirements do not provide any independent authorization for what they require.[14] Thus they cannot explain why an agent who happens to respond to commitment-dependent reasons by having coherent attitudes should be so committed. As a result, rational requirements seem ill-equipped to explain why we should—to some extent at least—conform to our evaluative commitments independently of coherent or incoherent attitudes. We need such an explanation, however, to make sense of the normative work that evaluative commitments seem to do (since this is partly independent of the attitudes that underwrite them). After all, there can be overwhelming reasons to discontinue a particular project or end a relationship with a particular person. If we try to conceive of commitment-dependent reasons for attitudes in virtue of the rational requirements that govern the relations between these attitudes, the stringency is absolute. What we need, however, is an account that makes sense of the *modal* stringency of commitment-dependent reasons. That is, we need to explain why they are *pro tanto* reasons despite their particular, yet non-absolute stringency. Rational requirements cannot provide such an explanation. We need a better understanding of why we are committed to *not* considering reasons disfavoring a particular project or relationship with a person. The interaction between our value-based reasons favoring the choice of a particular commitment and the commitment-dependent reasons that make us react to those initial value-based reasons is a natural place to start.

Let us therefore examine how the normative force of evaluative commitments and the reasons they generate are linked to the value-based reasons that undergird them. What speaks for this Dependence Claim is that it could make sense of the *modal* stringency of evaluative commitments. It might thus be able to explain both why we sometimes have stronger reasons once we are evaluatively committed and why we sometimes have strong reasons to question the normative force of our commitment-dependent reasons—bearing in mind that they could have been overruled initially by reasons disfavoring our taking on the commitment in the first place. If we can show that commitment-dependent reasons interact—in some to-be-specified way—with the value-based reasons supporting a particular commitment, we might finally be able to explain why even if newly arising reasons disfavor our commitment there is still reason to stay committed. We might thus be able to make sense of the fact that commitment-dependent reasons are both stringent and yet only modally so.

5. The Dependence Claim

There are various ways of developing the Dependence Claim. First, evaluative commitments could be thought to generate additional, however lightweight reasons. These might figure as addenda to the balance of reasons that initially influenced our choice of a particular evaluative commitment.

Second, and relatedly, one might think that commitment-dependent reasons are second-order pragmatic reasons. They are thus negative reasons, that is, reasons to disregard other reasons undermining one's evaluative commitment. And third, commitment-dependent reasons could be interpreted as intensifiers of the value-based reasons underlying a particular commitment. They would then figure as positive reasons, adding further weight to the reasons favoring them. Let me discuss these possibilities in a little more detail.

a. Additional Reasons

As for the claim that commitment-dependent reasons are merely additional lightweight reasons, this might do justice to the fact that evaluative commitments cannot normatively get off the ground in case they are based on bad reasons. The reasons underlying abusive relationships, for example, carry such negative weight that the fact that a person is committed to such a relationship and therefore takes herself to have a range of commitment-dependent reasons cannot counterbalance that negative weight.

This way of normatively capturing evaluative commitments, however, does not do justice to the phenomenology. Once we are evaluatively committed, we consider commitment-dependent reasons to be pretty heavyweight. If we took them to be lightweight reasons, we would not think it wrong to refrain from acting on them. They would instead be take-it-or-leave-it reasons; but this is not how we respond to commitment-dependent reasons. However, if we describe them as heavyweight reasons we cannot make sense of the fact that evaluative commitments that are based on bad reasons should be given up. We would always have a heavyweight reason to continue. As a result, it seems that we cannot account for the normativity of commitment-dependent reasons by characterizing them as additional reasons with a particular weight. It is hard to explain their weight independently of the reasons favoring a particular commitment. And if we need the latter to explain their weight, commitment-dependent reasons, as either lightweight or heavyweight reasons, become redundant.

b. Second-Order Pragmatic Reasons

Let's assume that commitment-dependent reasons are not merely an addition—however light or heavy—to the balance of reasons, but are more directly linked to the value-based reasons supporting a particular commitment. This might allow us to make better sense of the intuition that their stringency (or weight) is grounded in the reasons supporting them. In addition, we might be able to come up with an explanation of their function that does not render them redundant.

One way of highlighting the connection between commitment-dependent reasons and the reasons supporting them is to regard commitment-dependent reasons as second-order reasons. That is, once an agent finds herself evaluatively committed, the reasons that apply to her change. The fact that she is evaluatively committed is not an additional reason. This would just be bootstrapping a reason into the space of reasons in virtue of her having that commitment. Rather, the fact that she is evaluatively committed provides second-order reasons that pertain to the status of other considerations as reasons for or against a particular evaluative commitment. What makes these reasons second-order is that they involve ignoring first-order reasons that favor other evaluative commitments. Once John has become a partner to Mary, he has reasons to ignore other reasons favoring other partners. That is, this particular evaluative commitment alters what a person like John should treat as a reason.

What ultimately explains evaluative commitments as second-order reasons, however, is the further pragmatic fact that it would be costly and ineffective to continually reconsider one's evaluative commitments.[15] It is these pragmatic considerations that explain why we continue to respond to commitment-dependent reasons and regard them as normative. After all, simply staying committed in light of countervailing reasons will make it more likely that one will ultimately be successful in one's project or relationship. This certainly supports the proposed view, at least.

However, on closer inspection, this proposal is not compelling. Both the rationale of second-order pragmatic reasons as well as the content of those reasons proves limited. It can, in the end, be far more costly *not* to revise one's evaluative commitments in light of changing circumstances (Brunero 2007; Kolodny 2011). For example, there might be very few or poor reasons favoring a particular choice. If an agent nonetheless proves committed and hence takes herself to have reasons to disregard other reasons disfavoring her commitment, she might miss out on strong reasons for changing her commitment. Hence, evaluative commitments cannot merely generate pragmatic reasons. Furthermore, commitment-dependent reasons are not only about disregarding other reasons, but also about reasons for valuing and loving non-instrumentally. As a result, they do not just generate negative, but also positive reasons. If John is committed to Mary, for example, he will disregard other partners; but he will not stay committed to Mary unless he also regards her as special and cherishes her value.

c. *Intensifying Reasons*

In light of this diagnosis, a third suggestion presents itself: one might concede that commitment-dependent reasons are second-order, but that they are not (exclusively) negative in that they are not just about which other reasons *not* to consider. In addition, they are not merely pragmatic, but actually mainly non-pragmatic. That is, they are positive, and hence they are "about" the reasons that actually bolster the particular evaluative commitments. Positive second-order reasons can be referred to as "intensifiers" of first-order reasons.[16] Accordingly, once a person has become evaluatively committed, thereby responding to commitment-dependent reasons, the fact that she is committed intensifies the reasons favoring the commitment itself. This is non-pragmatic in that, since they intensify the first-order reasons favoring a particular evaluative commitment, these second-order reasons could be thought to be about continuing to respond to those first-order reasons and thus about valuing them for their own sakes. This helps to explain why someone who is evaluatively committed—in virtue of having second-order reasons intensifying the reasons supporting his commitment—withholds further comparison, since he is now confronted with an over-whelmingly strong reason that makes other reasons fade away. This way of capturing the interaction between reasons for evaluative commitments and commitment-dependent reasons seems roughly on the right track. What is still missing, however, is the rationale for this intensification.

Philosophers who adhere to an intensifying view, however, are "primitivists." That is, they take this effect to be unquestionably given and therefore think that no further explanation can be provided as to why evaluative commitments intensify the value-based reasons that underscore them—they just do.[17] But if no explanation is provided, the normative relation is simply redescribed, and ultimately remains elusive. The intensifying view therefore leads to a similar problem to the additional reasons view: without a more informative explanation as to why commitment-dependent reasons intensify, we seem to be drawn to the most ready rationale at hand, namely that their intensifying force co-varies with the weight of the reasons underscoring them. But if we need those first-order reasons to explain their intensifying force, we risk making commitment-dependent reasons redundant. If their force, however, is derived from some other source, the intensifying view is simply incomplete: we cannot make sense of that source.

Let me therefore offer a new proposal as to why evaluative commitments have an intensifying effect on the first-order reasons that undergird them.

6. The Self-Referentiality Claim

So far we have seen that both the Independence Claim and the Dependence Claim suffer from complementary problems. If the normative relation that evaluative commitments put us under is thought to be independent from the reasons favoring or disfavoring those commitments, its particular stringency might be more easily accommodated. Commitment-dependent reasons can be conceptualized as exerting pressure—countervailing reasons notwithstanding. The price, however, is that this leaves them largely unjustified. There is no further rationale that could sufficiently explain why commitment-dependent reasons should not be overruled by strong reasons against them. Also, and relatedly, we cannot do justice to the fact that, despite their stringency, evaluative commitments are *pro tanto*. We are able to step back from our evaluative commitments. Revising them sometimes seems like a very reasonable thing to do. So we still need an account of commitment-dependent reasons as modally stringent, non-instrumental, agent-relative, and diachronic.

Allow for a moment that evaluative commitments are dependent on the value-based reasons favoring or disfavoring them. In this case, we need to render intelligible the fact that there are possible worlds in which they are *pro tanto* and thus conducive to comparative assessment. However, as the Dependence Claim is spelled out, their particular stringency remains obscure. What precisely is it that evaluative commitments bring to a comparative assessment and why?

The challenge is to explain why the value-based reasons favoring or disfavoring a particular evaluative commitment and the normative force ensuing from a particular commitment, once entertained, can come apart. How can the normative force of evaluative commitments be made intelligible? To capture this normative stringency we have to make sense of the fact that there is a normative relation that holds (to some extent, at least) independently of momentary attitudes and the value-based reasons grounding a particular commitment.

My claim is that once we have come to respond to the evaluative features supporting particular kinds of relationships and projects (and thus to value-based reasons), we come to ascribe value to that response. Call this the *self-referentiality of evaluative commitments*. This view holds that an evaluative commitment to a particular project or a particular person consists, by default, in valuing the valuing of that project or valuing the loving of that person.

Evaluative commitments could thus be thought to reinforce reason mechanisms, since the success they manifest—and which consists in having come to evaluatively respond to various kinds of values—entails a demand to carry on. The underlying rationale for this is the intrinsic significance of having been able to respond to value-dependent reasons.

This does not imply that the person in question explicitly reflects on her valuing states and consciously ascribes value to them. Rather, she finds herself responsive to the evaluative features of a particular project or person in virtue of being liable to a range of emotions and by performing various kinds of action-types expressing that valuing response. Since this process is a success—after all, she could also fail to track evaluative features and not become attached to anything—it carries further value. The satisfaction condition of loving and valuing is that one maintains and continues with that to which one has managed to evaluatively respond, given that there is value in having developed such successful responses.

One might worry that the self-referentiality claim has the implausible consequence of multiplying value in an infinite hierarchy of value. If there is value in valuing or loving, isn't there also value in ascribing value to valuing, and so forth? The self-referentiality claim, however, is not liable to this charge because it entails that the value we find in valuing a particular project or loving a person is not distinct from valuing that project or loving that person.[18]

It is the self-referentiality of evaluative commitments that explains the normative pressure they exert. Since there is value in having come to value particular projects and in having come to love particular persons in virtue of the evaluative features they possess, this generates reasons, by default, to continue with these responses.

To further explain this value let me refer to some analogical achievements. There is value in knowledge, for example, to the extent that we successfully managed to acquire justified and true beliefs. Whatever it takes to render our beliefs justified and true, there seems to be additional value in having thereby come to know something (Williamson 2000). Similarly, in the practical realm, there is additional value in having managed to climb to the top of a mountain that goes beyond *being* on top of a mountain. Otherwise it would be equally valuable to have gotten there with the help of a helicopter. The value here consists in having overcome some difficulty (Bradford 2013).

In the case of evaluative commitments, it is independently valuable to have succeeded in connecting with the evaluative features of projects and persons, and thereby having come to value or love them. We thus acquire meaning and a basis from which to steer our life. It is this value, and the reasons it grounds to continue valuing or loving, whose force we experience even when our momentary attitudes are countervailing, or if we realize that there are value-based reasons disfavoring our evaluative commitment.

One might wonder how this could be, given that our attitudes do not underwrite this force. The answer is that it is part of valuing and loving that we are also prone to negative emotions. To the extent that those negative emotions continue to manifest our valuing a project or loving a person depending on context and circumstance, there continues to be value in valuing or loving even if this is not momentarily underwritten by positive emotions. In addition, if a change in reasons now questions our commitment to a particular project or relationship we can make sense of the fact that there continues to be value (in still) successfully valuing it. The success of our response to the valuable features of a particular project or person has not incidentally changed because there are new value-based reasons in the vicinity that question that response. To the extent that our response has proved successful, it continues to carry value that exerts normative force. It is the success of that valuing response, once achieved, that explains the recalcitrance, and thus the stringency of the value connected with it.

Consider again the owners of the mom-and-pop store. There are various correct descriptions of the content of their self-referential evaluative commitment: the fact that they own a business, their interactions with clients, being part of a small community, contributing to its nourishment, and so on are things they have come to value, become vulnerable to, and identify with. Since there is value in valuing their project they have reasons to continue realizing that value even in light of the fact that they make less money than they could. The value-based reasons for starting to disfavor their project are therefore overruled (or even disregarded) to the extent that there remains value in valuing their own business.[19]

So far I have tried to locate the normative force of evaluative commitments, but how can we explain their limits? The self-referentiality of evaluative commitments does not entail that the value of valuing and loving, and the reasons they generate, is absolute. They come with a strong weight, but they are *pro tanto*. This is the case when a person is no longer liable to manifesting the attitude of valuing, or if the reasons disfavoring her valued project or the person she loves become so strong that her responses ultimately prove ungrounded and hence lose their value.

It has recently been suggested that commitments are a distinct normative category. They come with a particular strictness and therefore differ from reasons. But they also differ from rational requirements in that their strictness is not such that they can never weigh considerations with reasons. Their normative force is *pro tanto* (Sphall 2014, 2013). I think this is a helpful and illuminating way of categorizing evaluative commitments. Their being normatively distinct, due to the self-referentiality of valuing and loving, allows us to make sense of the idea that we are under a particular normative pressure even in the face of strong countervailing reasons or attitudes. But we can also explain why there might be stronger reasons to give up on our evaluative commitments.

Finally, the Self-Referentiality Claim provides an explanation of why the value-based reasons favoring a particular evaluative commitment are intensified. They are intensified to the extent that the person in question proves successful in relating to those values. To the extent that she is successful there is value in her success. This also undergirds both the content and shape hypothesis and the normativity of evaluative commitments hypothesis. If a person proves successful in relating to values and thus comes to value a project or love a person, she acquires meaning (Wolf 2010) and a normative identity. If she remains ambivalent, and does not manage to value projects or love people, an important dimension of value that would make her life meaningful and factor in her normative identity is lost.[20]

Notes

1. On the one hand, these kinds of endeavors let us strive for something; on the other hand they enable us to disregard other goals on which we do not settle. See Goodin (2012, 64–65).
2. Calhoun (2009) argues for such a conceptual possibility.
3. For example, someone might claim to be evaluatively committed to a particular religion or other kinds of complex values or value systems. To the extent, however, that these values are not part of her more particularized project it remains unclear how they give rise to reasons that guide her in a unified way over time.
4. I therefore think that the term "volitional commitment," as Sphall (2014) mentions in passing, is less appropriate.
5. Many thanks to Ed Hinchman for pressing me on the alleged analytic truth of evaluative commitments and commitment-dependent reasons.
6. See Betzler (2013), especially pp. 105–111, where I describe these features in detail.

7. Cf. Scheffler (2004, 257).
8. In a similar vein, Pettit (2015) has analyzed "robustly demanding goods," such as attachments, by way of their modal stringency.
9. Cf. Scheffler (2004, 257), in which he emphasizes that project-dependent reasons are reasons "not to maximize aggregate value."
10. See Keller (2013, 33).
11. In another paper I demonstrate that these formal properties substantiate the claim that personal projects are a unique category of practical reason that cannot be reduced to other more familiar sources of reasons, such as desires, plans, or personal ideals. See Betzler (2013, 116–124).
12. Here I draw from Betzler (2015, 73ff.) where I spell out the Independence Claim in more detail.
13. Cf. Broome (1999, 406), who initiated the debate on rational requirements (which he initially called "normative requirements"): "The relation of normative requirement that holds between the beliefs mirrors this relation of inference that holds between the belief's contents. If one proposition follows from others, then believing that proposition is normatively required by believing others."
14. Some, therefore, suspect that such authorization cannot be had without a serious regress problem. Cf. Chang (2013); Kolodny (2005).
15. Cf. Scanlon (2004, 231), for such a view with regard to intending a goal. Raz calls such reasons negative second-order reasons, or "exclusionary reasons." Cf. (Raz 1990 [1975]).
16. This terminology was introduced by (Dancy 2004, 41–42).
17. Keller (2013, 132ff.), makes use of Dancy's idea of reasons being intensifiers in the case of close relationships. Keller adheres to a primitivist view of commitment-dependent reasons (or "reasons of partiality," as he calls them). The only explanation Pettit provides in his analysis of robustly demanding goods is our need for security. See Pettit (2015). But this, again, is a pragmatic rationale.
18. I benefitted here from Roth's work on the self-referentiality of intentions. Cf. Roth (2000).
19. Matthew Anzis suggested to me that the fact that we take meaning from our evaluative commitments explains the normative force of commitment-dependent reasons. I think that what underlies meaning in this context is that there is further value in having come to value a project or love a person.
20. I am grateful for very helpful comments that I received on earlier versions of this chapter. Many thanks to audiences at the Humboldt University Berlin, the University of Bern, the University of Bremen, the University of Vienna, the University of Munich, Constance University, Erlangen University, Tulane University, the University of Wisconsin–Milwaukee, and George Washington University for penetrating questions. Written comments by Matthew Elias Anzis, Claus Beisbart, Alison Denham, Carl Hammer, Jörg Löschke, Nicole Osborne, and Jacob Rosenthal, as well as discussions with Sarah Buss, Luca Ferrero, Hannah Ginsborg, and Stan Husi, helped me clarify and improve my arguments. Any remaining errors are entirely my responsibility. I am thankful to Roman Altshuler and Michael Sigrist for inviting me to include my chapter in this volume.

References

Betzler, M. (2013). The Normative Significance of Personal Projects. In M. Kühler & N. Jelinek (Eds.), *Autonomy and the Self* (pp. 101–126). Dordrecht: Springer.

———— (2015). Evaluative Bindungen und bindungsabhängige Gründe: Eine Herausforderung für den metaethischen Realisten? In D. von der Pfordten (Ed.), *Moralischer Realismus?* (pp. 71–92). Münster: Mentis.

Bradford, G. (2013). The Value of Achievements. *Pacific Philosophical Quarterly,* *94*, 202–224.

Broome, J. (1999). Normative Requirements. *Ratio, 12,* 398–419.

Brunero, J. (2007). Are Intentions Reasons? *Pacific Philosophical Quarterly, 88,* 424–444.

Calhoun, C. (2009). What Good is Commitment? *Ethics, 119,* 613–41.

Chang, R. (2013). Commitments, Reasons, and the Will. In R. Shafer-Landau (Ed.), *Oxford Studies in Metaethics* 8 (pp. 74–113). Oxford: Oxford University Press.

Dancy, J. (2004). *Ethics Without Principles*. Oxford: Oxford University Press.

Goodin, R. E. (2012). *On Settling*. Princeton, NJ: Princeton University Press.

Keller, S. (2013). *Partiality*. Princeton, NJ: Princeton University Press.

Kolodny, N. (2005). Why be Rational? *Mind, 114,* 509–63.

——— (2011). Aims as Reasons. In S. Freeman, R. Kumar, & R. J. Wallace (Eds.), *Reasons and Recognition: Essays on the Philosophy of T. M. Scanlon* (pp. 43–78). New York: Oxford University Press.

Owens, D. (2012). *Shaping the Normative Landscape*. Oxford: Oxford University Press.

Pettit, P. (2015). *The Robust Demands of the Good: Ethics with Attachment, Virtue, and Respect*. Oxford: Oxford University Press.

Raz, J. (1990) [1975]. *Practical Reason and Norms* (2nd ed.). Princeton, NJ: Princeton University Press.

——— (2001). *Value, Respect, and Attachment*. Cambridge: Cambridge University Press.

Roth, A. S. (2000). The Self-Referentiality of Intentions. *Philosophical Studies, 97,* 11–52.

Scanlon, T. (2004). Reasons: A Puzzling Duality? In R. J. Wallace, P. Pettit, S. Scheffler, & M. Smith (Eds.), *Reasons and Value: Themes from the Moral Philosophy of Joseph Raz* (pp. 247–68). Oxford: Oxford University Press.

Scheffler, S. (2004). Projects, Relationships, and Reasons. In R. J. Wallace, P. Pettit, S. Scheffler, & M. Smith (Eds.), *Reasons and Value: Themes from the Moral Philosophy of Joseph Raz* (pp. 247–68). Oxford: Oxford University Press.

Sphall, S. (2013). Wide and Narrow Scope. *Philosophical Studies, 163,* 717–736.

——— (2014). Moral and Rational Commitment. *Philosophy and Phenomenological Research, 88*(1): 146–172

Wallace, R. J. (2013). *The View from Here. On Affirmation, Attachment, and the Limits of Regret*. New York: Oxford University Press.

Williamson, T. (2000). *Knowledge and Its Limits*. Oxford: Oxford University Press.

Wolf, S. (2010). *Meaning in Life and Why It Matters*. Princeton, NJ: Princeton University Press.

9 Updating the Story of Mental Time Travel

Narrating and Engaging with Our Possible Pasts and Futures

Daniel D. Hutto and Patrick McGivern

It's a poor sort of memory that only works backwards.

—Lewis Carroll, *Through the Looking Glass*

We need to account for the ways in which narratives of selfhood look forward in anticipation and imagination, as well as backward in memory.

—MacKenzie (2008, 122)

1. Everyday Adventures in Mental Time Travel

"Mental time travel" refers to the everyday ability to think about events in one's past or possible future as events in one's *own* past or possible future. Through mental time travel one "wanders" away from the present to travel to other points within one's own timeline (Corballis 2012). Thus:

> We exercise the capacity for mental time travel whenever we revise for this year a class we gave last year—remembering what worked and what didn't—whenever we reflect on what kind of career or job would best suit us, whenever we plan a holiday or a shopping trip, arrange a meeting, organize a party, or commit ourselves to a course of study, an exercise program, or a marriage.
>
> (Gerrans and Kennett 2010, 17)

Despite being utterly ordinary, the capacity for mental time travel is fundamental to the kind of reflective, evaluative agency that typifies human practical reasoning. Mental time travel apparently underwrites our ability to reflect upon, evaluate, and appraise our past actions and to make decisions—some of the greatest import to us—in light of those actions when planning and motivating future actions. It has been argued that this linking of our past, present, and future is the main adaptive advantage of mental time travel: thinking about the past confers benefits for dealing with challenges in the present and future rather than simply providing a special form of experientially charged access to our pasts (Tulving 1985; Suddendorf and Corballis 2007). Mental time travel provides an emotionally hot yet reflective way of connecting our pasts, presents, and possible futures—a way that increases our chances of making better choices about and, thus, of acting better in the present and future. The activity is dramatic and emotionally engaging.

The kind of affectively charged episodic remembering of personal life events (e.g., revisiting what happened on one's visit to Istanbul) afforded by mental time travel is quite distinct from the sort of recall afforded by semantic memory of facts (e.g., recalling that Istanbul was once Constantinople). Semantic memory certainly plays some part in mental time travel: general knowledge can help frame its special forms of remembering and imagining (Schacter et al. 2012). Even so, the sort of episodic memory that does the heavy lifting in mental time travel is a different animal than factual recall (see Goldman 2006, 47). Indeed, it is because they are distinct capacities that the experiential and affective dimensions of mental time travel can aid in remembering facts by linking them to emotionally charged stories about those specific happenings in our lives that matter to us.[1]

From the perspective of sciences of the mind, the curious feature of mental time travel is that many of the same processes appear to be at work both when we think about our actual pasts and when we imagine our possible futures. It will seem counterintuitive that this should be so for anyone who assumes that thinking about our pasts is really a matter of accurately recalling actual prior happenings, while thinking about our futures is really a matter of entertaining counterfactual possibilities. The former is a matter of remembering—which requires getting things right to succeed—whereas the latter is a matter of suppositional imagining—which serves an entirely different and seemingly more speculative and preparatory cognitive function. However, in the context of mental time travel, the gap between memory and imagination looks smaller once we recognize that thinking about actual past episodes in such cases requires active selection and ordering.[2] Thus an "essential" feature of mental time travel is that it exercises the "ability to create and recreate these experiences under voluntary control rather than via the presentation of an eliciting situation or object" (Gerrans and Kennett 2010, 15; see also Schacter et al. 2007, 2008).

Seen in this light, mentally time traveling to our own pasts begins to look more akin to imagining our possible futures than it would otherwise. Both forms of thinking involve actively constructing scenarios in ways that are quite unlike the mere involuntary reliving of past events or the mere retrieval of previously recorded memories. Thinking about our pasts is not simply a matter of replaying archived records of previous happenings; it involves the active construction of scenarios. This is precisely why, in mental time travel, episodic memory and imagination must "combine under executive control" (Gerrans and Kennett 2010, 5).

2. Theory of Mind and Mental Time Travel: The Common Network Hypothesis

Focusing on the constructive activity that lies at its heart, mental time travel has inspired many scientists to conjecture that there may be a common font to our related capacities to wander retrospectively or prospectively within our own timeline (Buckner and Carroll 2007; Schacter et al. 2007; Addis et al. 2009; Buckner 2010; Kwan et al. 2010). A number of researchers have been attracted to the idea of a common mechanism or network, formulating this working hypothesis both in more narrow and broader ways. For example, they posit:

> A common set of processes and an underlying neural network that includes the hippocampus.
>
> (Kwan et al. 2010, 3183)

> A common core network that includes the hippocampus in addition to other medial temporal, parietal, and prefrontal regions.
>
> (Gaesser et al. 2013, 1150)[3]

There is a good deal of empirical support for the hypothesis that there is a common cognitive basis for acts of memory and imagination involved in mental time travel. Since 2007, novel scientific work has focused on investigating mental time travel and has repeatedly confirmed the existence of some strong similarities in the patterns of neural activity associated with the sorts of cognitive procedures employed in thinking about our pasts and imagining our possible futures.[4] Behavioral studies suggest that the same cognitive processes are involved in both tasks and neuroimaging experiments have consistently found a "striking overlap in the brain activity associated with remembering actual past experiences and imagining or simulating possible future experiences" (Addis et al. 2007; Szpunar et al. 2007; Schacter et al. 2012, 677).

It has also been found that deficits in remembering the past correlate with deficits in imagining the future. For example, recent work involving elderly adults reveals that these two capacities are strongly correlated, as well as establishing that older adults suffer from age-related declines in both capacities, tending to remember the past and imagine the future with less episodic detail than younger adults (Schacter et al. 2013; see also Addis et al. 2008). Other research has found deeper connections, revealing not only that there are parallel deficits affecting both backward-facing and future-facing mental time travel capacities but that the deficits in these capacities are developmentally linked. As Kwan et al. (2010) report, the "one ability does not develop independently of the other" (3185). Early hippocampal damage impairs both capacities even when the injuries occurred early enough to allow for the possibility of neural reorganization. Like many others, Kwan et al. (2010) take these findings to lend strong support to the idea that early hippocampal injury may interfere with the proper development of the common mechanism, process, or network that is required for dealing with the constructive aspects of mental time travel.

Taken together this body of empirical findings has made it plausible that a common mechanism, network, or set of processes is called into play whenever one thinks about one's past and future by means of mental time travel. In this regard, Gerrans and Kennett (2010) speak of "an emerging body of evidence that mental time travel does not exploit different systems for memory and imagination" (15).[5]

While the empirical work proceeds apace, it proves useful to critically assess the theoretical speculation, popular in the literature, that theory of mind capacities and mental time travel may have a common cognitive basis. Quite a number of authors propose such a connection. For example, Suddendorf and Corballis (2007) propose that the increase in brain size of members of the genus *Homo* was tied to "selection for such interrelated attributes as theory of mind, language, and . . . mental time travel" (312). Mental time travel, on this view, comes as part of a larger cognitive package—it is one of a suite of interrelated capacities that "included the ability to entertain alternative future scenarios (mental time travel), to read others' minds (theory of mind), and to communicate (language)" (Suddendorf and Corballis 2007, 312).

What connects all of these capacities? Suddendorf and Corballis (2007) maintain that "*To represent self and others realistically* requires declarative knowledge of individuals and some folk psychology (i.e., theory of mind) to predict how they act (e.g., knowing that they typically act on the basis of their beliefs to fulfill their desires)" (308; emphasis added). This, however, would make theory of mind capacities importantly not just linked to mental time travel but a fundamental basis for it.[6]

According to these proposals, a capacity to construct and adopt divergent but realistic perspectives on events is what allows us to escape the present and "wander mentally not only in time, but also into the minds of others" (Corballis 2012, 877). Pressing this point, Corballis (2012) also suggests that the default network enables such mental movements and thus is "the common basis for both mental time travel and theory of mind" (Corballis 2012, 880).

There is a compelling theoretical reason for thinking that folk psychological abilities for understanding divergent perspectives are minimally required for engaging in mental time travel. The logic is simple: when engaging in such exercises "One sees one's past or future self as another" (Goldie 2012, 42). All sort of adjustments of perspective and attitude are required if I am to recall the episode of the scraped knee that was absorbing to my six-year-old self, or to imagine what it might be like for my 80-year-old self to cope with the daily routines of retired life. The point is, as MacKenzie puts it, that even though

> the focus or apparent focus of the imagining is one's self, the "I" of the imagining need not be the empirical me. I can imagine living all sorts of possible lives, and I can imagine these lives from the inside, as it were. But, in doing so, I may not be really imagining *myself*.
>
> (MacKenzie 2008, 124; emphasis in original)

Successful mental time travel cannot be a matter of just thinking about "myself" just as I am now, without adjustment. A major part of that adjustment is to get in touch with the relevant perspectives and attitudes of our past or future selves.

That a basic folk psychological competence is needed for mental time travel is consistent with empirical evidence that shows that folk psychological abilities and capacities for mental time travel unfold in tandem. In their full-fledged forms folk psychology and mental time travel appear to be uniquely human, late developing and rooted in capacities for ever more sophisticated perspective taking (see, e.g., Corballis 2012 for a discussion). It is only after children can securely pass false-belief tasks that they show the rudiments for the sort of perspective taking that is needed to understand another's reasons for acting or to practically deliberate about their own reasons for acting in the ways mental time travel demands. Suddendorf and Corballis (2007) propose that "it may require *this level of folk psychology* to be able to identify with one's future self, understand that this future self may have mental states that differ from one's current states, and care about them" (308; emphasis added).

Despite the plausible suggestion that mental time travel and folk psychological abilities are strongly linked, to date there has been no serious attempt to provide a detailed account of just how we should understand the sorts of folk psychological capacities that are involved in mental time travel. Highlighting the limitations of the two standard proposals in the theory of mind literature—theory theory and simulation theory—the remainder of this chapter makes an initial case for thinking that the root folk psychological competence needed for mental time travel is better understood as a special kind of narrative engagement.

3. From Theorizing to Narrating

Classic theory theory assumes that "the mind contains a single mental faculty charged with attributing mental states (whether to oneself or to others)" (Carruthers 2011, 1). Moreover, it assumes that when we understand minds we use the same basic sorts of tools we use when understanding other phenomena—the same sort of tools we use in natural sciences: theories. Tradition has it that such theorizing takes the form of deductive-nomological explanations—such that in the case of folk psychology mental states serve as theoretical postulates and the general principles of human psychology play the role of laws of nature. We succeed in understanding others if we manage to infer the correct mental states by applying the laws of folk psychology to particular cases.[7]

Trouble starts for this idea as soon as it is noticed that any adequate application of such laws to particular cases would require appeal to auxiliary hypotheses. Additional theorizing about what particular people are likely to do in specific situations seems necessary if we are to apply folk psychological theory. To get any interesting predictive or explanatory results a core theory of mind would need to be seriously augmented, for if we only knew that general laws of folk psychology apply to the targets of our theorizing then, as Botterill (1996) admits, "we wouldn't know what to expect of them" (115).

A serious problem for classic theory theory is that it needs to explain the origin of these auxiliary hypotheses and how we become so accomplished at applying them sensitively in actual cases. Without that supplementary account classic theory theory would have little or nothing to say about how we manage to understand the mindset of particular individuals—ourselves or others—in particular situations.

To be clear, even small sets of relatively simple laws can give rise to extraordinarily complex ranges of phenomena—and, used in the right way, those relatively simple, very general laws can explain diverse phenomena very specifically. So the problem for theory theory is not one of a simple mismatch between the generality of laws and specificity of understanding. The trouble is that even if the mental states of others could, in principle, be specified in the way just described, it still would not be at all credible that this is how we actually come to understand ourselves and others in daily life: our capacity to recognize the consequences of a set of theories is in general much more limited than our ability to recognize the mental states of others.

A much more promising idea is that our folk psychological competence centrally involves creating and consuming narratives. Bruner (1986, 1990) highlights the ways in which narratives provide a special kind of understanding—one that is distinct from general ways of understanding people and things. Whereas theories enlighten by focusing on what is abstract and law-bound, narratives focus on particulars, casting specific happenings in a certain light in order to reveal their significance.

A narrative understanding of people and events gives us a handle on relevant details, presenting these in a meaningful way. Illustrating this, Bruner (1986) says of these two styles of understanding, "One leads to a search for universal truth conditions, the other for likely particular connections between two events—mortal grief, suicide, foul play" (Bruner 1986, 11–12). An illuminating narrative identifies "which events are significant and why" (Roth 1991, 178). It does so by selectively focusing on "the elements that are relevant to our interests, and to the interests of our audience" (Goldie 2012, 16). Narrating involves situating events in a certain light against a larger frame and, usually, also characterizing them within a certain genre as opposed to subsuming them under general laws. For this reason, Currie (2010) contends that "general theories are simply not narratives at all" (35).

It is precisely this focus on providing contextualized meaning that makes narratives so perfectly suited for supplying insight about particular attitudes and perspectives. Exemplary narratives are detailed accounts of the events and doings that feature in the lives of particular people and how such happenings shape how they see things. These are "rich, particularized, and unified histories of cycles of thoughts, actions and contingencies" (Currie 2010, 36).

Typically narratives do not just give an account of how particular happenings unfold, they also give insight into the various perspectives taken toward such happenings (László 2008, 3). Emphasizing this, Goldie (2012) provides the following rough and ready definition: "a narrative is a representation of events which is shaped, organized, and coloured, presenting those events, and the people involved in them, from a certain perspective or perspectives" (Goldie 2012, 8). He observes:

> Narratives often provide explanations of why someone had a particular motive, or why someone has a particular character or personality trait, or why someone was drunk, depressed or angry. And the explanations we get are narrative-historical explanations, they locate the motive, the trait, the undue influence on thinking, within a wider nexus, in a way that enables us to understand more deeply why someone did the thing they did through appeal to aspects of their personal history or circumstances.
>
> (Goldie 2012, 20)

The distinction between general knowledge supplied by theory, and personal and particular knowledge supplied by narratives, fits with findings about individuals who are impaired in their capacity for mental time travel. For example, Kwan et al. (2010) describe the situation of H.C.—an amnesic person, whose autobiographical episodic memory was compromised relative to her semantic memory. Because of this deficit,

> her future narratives lacked self-relevant information and did not seem to integrate her present life or future aspirations, unlike controls' narratives. For example, when given the cue word "coffee," H.C. described a future scenario in which she took an order at a coffee shop even though she did not currently work in a coffee shop, had no future plans to do so, and at that point in the future would likely have graduated from college as a trained chef given her current schooling. Her future events, which lacked such self-specific relevance, may instead reflect imagination of more generic events that are not personally contextualized and may not require the same level of hippocampal contribution (Addis, Cheng, et al. 2008; Addis, Wong, et al. 2008). Observations from the current study suggest that H.C. may have *difficulty in contextualizing imagined episodes with the use of personal information.* One way to test this possibility is to specify to H.C. and controls that they should generate events within the scheme of their probable future lives vs. generate events that are more generic or unlikely to occur.
>
> (Kwan et al. 2010, 3184; emphasis added)

Narratives provide insight into the peculiarities of personal perspectives in ways that make the *narrative practice hypothesis* plausible (Hutto 2007,

2008). It holds that we first develop our folk psychological competence and further hone it by learning how to create and consume narratives. Plausibly, it is extended practice with narratives that makes us capable of appropriately handling and understanding multiple perspectives on, and attitudes toward, particular events. Such perspectives are not only those described within stories, but also those of people presenting the stories. By developing our narrative understanding through such practice we become sensitive to a variety of possible perspectives that may be adopted on events, including—especially—cognitive, emotional, and evaluative perspectives that may diverge from our own. This puts us in a position to understand others as well as our past and future selves.

This line of reasoning encourages the idea that some general kind of basic theory of mind abilities must already be in place in order for us to appreciate narratives in the first instance (see Zunshine [2006], and especially Ryan [2010, 478] and Apperly [2011, 117]). But, as seen earlier, this gets matters exactly backwards. Narratives focus on particulars and theories on generalities. Thus it is wholly unclear how having a general theory of mind or mind-reading machinery could possibly enable us to cope with *ad hoc* details and idiosyncratic attitudes. Indeed, classical theory theory is at a loss precisely when it comes to explaining how we deal with details; it is uncomfortably quiet on how we fluently come to understand particular people and their situations.

Recognizing this problem, some trace the trouble back to what they take to be classic theory theory's commitment to an outmoded deductive-nomological vision of theories. In the place of that once standard view, many of today's philosophers of science maintain that theories should be

> identified with sets of models, rather than sets of sentences. These models are real structures . . . These structures are meant to be isomorphic to the world in some respect. Though the models of the theory are often described using language, the important linkages hold between models and the world, not between any set of descriptions and the world . . . [accordingly] a theory consists of two parts: a set of models, and a postulation of isomorphism between certain respects of models and parts of the world.
>
> (Klein 2013, 690)

Drawing inspiration from this development, it is possible to advance a model theoretic version of theory theory—one that seeks to explain how we learn to apply our theory of mind in specific cases. Maibom does just this. She claims:

> This form of theory theory has a number of advantages over traditional forms . . . [one major advantage] is that practice falls out of it . . . [it recognizes that] folk psychological knowledge is knowledge

of the (empirical) world only if it is combined with knowledge of how to apply it. *By combining the general and the particular in this way,* model theory gives a deep and explanatorily satisfactory account of the centrality of practice.

(Maibom 2009, 361; emphasis added)

This softer way of construing theory theory, she argues, "accounts *not just as well as, but better than,* narrativity theory for the fact that our folk psychological explanations appear to contain, or form part of, narratives" (Maibom 2009, 361; emphasis added). Her reason for saying this is that model theory theory can be more liberal about which kinds of practice are, in fact, necessary for developing our folk psychological competence. She doubts that the consumption and creation of narratives is in fact necessary for acquisition of that competence. The empirical jury is still out on this question—and we needn't try to decide it here. The important thing to note is this: One might adopt a model theoretic take on theory theory and be cautious about backing the claim that narrative practice is fundamental to folk psychological competence. Still it is possible (and sensible), as Maibom does, to acknowledge the need for practice and to recognize that stories are excellent practice tools for developing or extending our folk psychological competence.

Whichever way we go on the question of how necessary narrative practice is to developing folk psychological competence, there are good reasons to think that something more, and other than general theorizing, is critical to understanding the perspective of others and ourselves. Taken together the aforementioned considerations capture what is right about theory theory and lend credence to the idea that "our mode of access to . . . our own minds is no different in principle from our access to the mental states of other people. Moreover, such access is . . . equally interpretative in character" (Carruthers 2011, 1). The question is: can this be the whole story about how we understand divergent perspectives when engaging in understanding others and mentally time traveling?

4. From Simulation to Enactive Engagement

It might be accepted that narrative understanding is necessary to gain insight into other perspectives and attitudes—both our own at different times and those of others—while denying that such narrative understanding suffices or exhausts what goes on when we mentally time travel. A good reason for thinking there is more to the story of mental time travel is that it is an experientially rich and emotionally charged business. In remembering the past and imagining the possible future we get an insider, truly first-personal take on how we have or will respond to various situations. Narrative understanding, it may be argued, for all its importance in acts of mental time travel, misses out on something critically important when it comes to thinking about our pasts and futures. It has nothing to say about the hot character of such mental exercises. It might be supposed that this essential but missing "something extra" can only be supplied by means of mental simulation.

In the domain of social cognition, Stueber (2006, 2012) has argued that in understanding others we must put ourselves in the person's shoes, identify with the person, and—somehow—replicate or mentally simulate the other's mindset. This, he holds, is the only way to practically and emotionally evaluate the aspects of the situation that the other takes to be relevant.

> Understanding other agents seems to require that I am able to recognize particular aspects of their situation relevant to their actions. And in order to do that *I have to put myself in their shoes* and practically evaluate what aspects of that situation have to be regarded as relevant.
>
> (Stueber 2006, 160; emphasis added)

Stueber (2006) dubs this process *re-enactive empathy* (158). In making the case that it is a necessary component for understanding others he draws initially on arguments by Jane Heal. Heal (1998) holds that a major part of understanding others requires us to co-cognize with them—that is, to think about the subject matter of the other's thoughts for ourselves. For Stueber (2006) this simulative activity involves not just working through chains of rational thought but also applies "to a whole range of psychological phenomena" (159). This, he claims, is what allows us to be moved by the emotional and normative considerations that "are relevant in a particular situation" (159).[8] This expansion of the co-cognizing proposal fits with the assumption that it is only by projecting ourselves in the situation of another—only by seeing and feeling that situation "from the inside"—as the other sees and feels it—that we gain full insight into the other's state of mind.

Many theorists hold that this capacity for simulative self-projection is a central feature of mental time travel. For example, Schacter et al. report that the

> activation of default network regions is observed not only when individuals remember the past and imagine the future, but also when they *engage in related forms of mental simulation that involve taking the perspective of others* (without an explicit requirement for mental time travel).
>
> (Schacter et al. 2012, 678; emphasis added)

Stressing the link with self-projection more strongly, Gerrans and Kennett highlight:

> There seems to be a subtle difference between "pure" episodic memory, which recreates the perceptual or sensory content of experience, and autobiographical memory. In the latter episodic memories combine

with *a sense of self, of being personally present in the episode* (known as autonoesis).

(Gerrans and Kennett 2010, 13)[9]

Hence, on this view, the sort of mental time travel that is involved, for example, in planning

> requires a capacity to *imaginatively project oneself into the future*; this in turn requires both a sense of oneself as the very same individual who will inhabit that future (autonoetic awareness), and also the kind of detailed self-knowledge that is supported by autobiographical memory.
> (Gerrans and Kennett 2010, 17; emphasis added).[10]

Of course, as highlighted in the first section, even if we allow that talk of self-projection is unproblematic it cannot be that we simply project ourselves—just as we are—into our pasts or futures. We must adjust for relevant differences between how we are now and how we were or will be. We must, as Mackenzie (2008) puts it, take up another "notional first-personal perspective" (125). But how do we do this, and how do we know which adjustments to make?

Here it seems we must, once again, call on the personal and particular framing knowledge we have of others and our past and future selves supplied through narratives. We must know how to make relevant adjustments between, say, our selves now and our selves then—even if this is an entirely imaginative exercise. Since general theory won't help, it is plausible that the relevant framing knowledge to engage in the simulation comes through narratives that reveal particulars of the person in question—for example, a narrative that supplies details of the psychological and physical profile of my six-year-old or eighty-year-old selves.

Consequently, even if we allow that self-projective simulation plays a crucial part in mental time travel, "it seems unproblematic to grant that *some degree of narrative understanding is required for simulating a character*" (Kieran 2003, 69; emphasis added). But not only does this requirement make it clear that simulation cannot suffice for the process of self-projection, it also raises the question of whether the relevant sense of simulation should be understood as a kind of self-projection at all.

Currie (2003) agrees with Keiran about this, observing that

> since one claim of [simulation theory] is that mind reading is done by imaginative projection, we might think it tells us this: imaginative engagement with a fiction involves putting ourselves in the shoes of the characters. I agree with Matthew Kieran that this would be a mistake.
> (Currie 2003, 294)

But Currie sees simulation as playing a different sort role for us when engaging with narrative constructions, for he thinks simulation is involved in the way we "imagine the propositions made fictional by the work" (Currie 2003, 294). This echoes the pretend theory proposal of Kendall Walton (1990), which holds that in responding to fictional narratives we adopt an attitude of make-believe, not genuine belief. Accordingly, on this view we only ever pretend to feel genuine emotions in our imaginative exercises when responding to fictional narratives. Thus, in describing the situation of Charles, the moviegoer who appears to react hotly to an imagined filmic threat of green slime, Walton tells us "it is not true but fictional that he fears the slime" (Walton 1990, 242).

In engaging in the relevant acts of narrative-driven imagining, so this story goes, we engage in a kind of make-believe pretense under the direction of the fictive work (see also Goldman 2006, 48). In doing so one does not truly feel fear but represents oneself as being afraid—or angry, or sad, and so on.[11] According to pretend theory the whole process is an intellectually driven activity in which particular representations motivate and direct our acts of pretense—a process that somehow generates quasi-emotional responses in us.

Arguably, going this way, therefore, also leaves us without the right account of the heat we feel when engaged in mental time travel. Medina (2013) rejects these overly intellectualist simulative accounts in favor of an enactive vision of dramatic imaginings. In his view, if we are to understand how and why we respond with real emotions to narratives, we must focus not on the content but on "the manner in which that content is presented and processed" (Medina 2013, 329). His point is whether, and to what degree, we are viscerally moved by a narrative has less to do with "what is imagined" and more to do with "how it is imagined" (see Medina 2013, 328; see also Moran 1994, 83).

What drives our hot narrative engagements is neither self-projective simulation nor simulating or pretending to believe a set or propositions; rather "the central difference between emotionally engaged and emotionally disengaged imagination has to do with the mode of imagining, not with its content" (Medina 2013, 329). What seems right about this proposal is that depending on how it is rendered, I might be unmoved when imagining an episode from my own past, and even fail to identify, or empathize with my own former self. Contrariwise, if presented in the right way, I might be irresistibly moved by the circumstances of someone I know is not myself. This is the sort of thing that happens when watching tightrope walkers, experiencing *Schadenfreude*, and even when watching the excruciating antics of certain purely fictional characters (such as Basil Fawlty of *Fawlty Towers* fame).

It is important to recognize that when enactively engaging with narratives in this way we do not project ourselves into the other's shoes. The process is not transportation but closer to transformation (as Gordon 1995, 1996 held) by means of special kinds of narrative engagement.

However, in mental time travel we do not simply become the other—not even our past or future selves. In thinking about our past and possible futures we can engage with the perspectives adopted by our past or future self to situations affectively without endorsing those perspectives. We can engage with another's situation in a hot way, while still finding it unattractive or repugnant, say, to our current emotional and moral sensibilities. In this respect we can—in an important sense—fail to fully occupy or identify with ourselves at different points in time even when we manage to see the situation from the perspectives of such selves. Crucially, when it comes to thinking about others and ourselves, the possibility of this kind of evaluative distance—which Goldie (2012) calls dramatic irony—always exists alongside our capacity for insider emotional engagement with the situation in question.

As Goldie neatly puts it,

> Narrative thinking thus exploits our capacity for self-reflectiveness. One sees one's past or future self as another, and, in just this respect, one is at the same time both actor and spectator, both narrator and audience, and both agent and judge, judging both the events in which one's past or future self is implicated, and judging the narrative itself.
>
> (Goldie 2012, 42)

Mastery of this capacity seems to be a narrative art, for "narrative . . . requires an external perspective" (Goldie 2012, 40). In particular, it is through mastery of the narrative device of free indirect style—in which third-person narration is intertwined with first-person description—that it is possible for us, at once, not simply to engage with how a situation presents itself to someone from the inside but at the same time to adopt an external perspective on that very engagement (see Goldie 2012, 36 for a discussion). Free indirect style is a crucial tool that, when used in mental time travel, provides us with the means of capturing

> the ironic gap not between the perspectives of two different individuals, but between the two perspectives of one individual—between you then, as someone—a "character"—who is internal to the narrative, and you now—the narrator—who is external to the narrative. It is because of this gap between you then and you now that the ironic gap can open up.
>
> (Goldie 2012, 37)[12]

Free indirect style, although it finds its most developed expression in literary works, "is surely as old as narrative discourse" (Goldie 2012, 36). This is important to note when it comes to assessing the hypothesis that we master the competence of mental time travel through narrative practice.

5. Conclusion

We can learn important lessons about the folk psychological capacities needed for mental time travel by considering the limitations *prêt-à-porter* versions of theory theory and simulation theory encounter when it comes to understanding how we make sense of others and engage emotionally and imaginatively with fictions. The arguments and analyses supplied in this chapter not only warn us against some natural and tempting pitfalls for those who wish to take the familiar theory of mind path; they also offer a positive, narrative-based proposal for going a quite different way. If the analysis of this chapter holds good, then there are strong reasons to explore further the possibility that our capacity for mental time travel may be best understood as a narratively driven form of dramatic engagement. This initial work lays the ground for further research that may provide a richer understanding of the cognitive basis of our special capacities for mental time travel.[13]

Notes

1. This would clearly be the case once the episodes of our past become storied, because stories "are strikingly memorable . . . although the story 'Little Red Riding Hood' is much more complex than a 20-digit number, the story is much easier to remember" (Scalise Sugiyama 2001, 8).
2. This idea is hardly new. As Schacter et al. remind us,

   the general idea that memory is a constructive process . . . rather than a literal replay of the past, dates to the pioneering work of Bartlett (1932), and has been developed by a variety of investigators who have demonstrated the occurrence of memory distortions and theorized about their basis. (Schacter et al. 2012, 681)

3. Schacter et al. (2012) take the common "core" network that underlies both remembering and imagining to be the default network that includes the medial temporal and frontal lobes, posterior cingulate and retrosplenial cortex, and lateral parietal and temporal areas (see, e.g., Schacter et al. 2012, 677).
4. This was a period when work on mental time travel was still new and quite groundbreaking. Indeed, in 2007, the neuroscientific discoveries that revealed strong links between memory and imagination made the journal *Science*'s "list of the top ten discoveries of the year" (*Science*, 21 December 2007, 1848–1849, as reported in Schacter et al. 2012, 677).
5. This conclusion, however, must be tempered by other findings that complicate the story to a significant extent even if they do not defeat the common mechanism hypothesis. Despite the repeated empirical demonstrations of "impressive similarities between remembering the past and imagining the future, theoretically important differences have also emerged" (Schacter et al. 2012, 678). Any fully satisfactory account of the cognitive basis of mental time travel will need to take stock of and accommodate these findings about documented differences between the relevant acts of remembering and imagining. This important constraint on any future theorizing highlights the urgent need for the development of a sufficiently nuanced integrative theory of the cognitive basis of mental time travel (see Schacter et al. 2012, 683). Though we recognize its importance, this issue will not be addressed in this short chapter.
6. By comparison, in their commentary on this proposal, Bischof-Köhler and Bischof contend that these interrelated capacities might be grounded in

   a multipurpose "module" whose adaptive function is much more general . . . [such that] mental time travel and theory of mind need not be externally arranged in a means-end relation, but appear as different outcomes of one and the same competence. (Bischof-Köhler and Bischof (2007, 317)

7. Fodor speaks for this tradition when he claims that folk psychological explanations

exhibit the "deductive structure" that is so characteristic of explanation in real science. There are two parts to this: the theory's underlying generalizations are defined over unobservables, and they lead to its predictions by iterating and interacting rather than by being directly instantiated. (Fodor 1987, 7)

8. In this respect Stueber's proposal is closer to what Collingwood (1946) calls the "egocentric method."

9. Gerrans and Kennett (2010) remind us: "Autonoesis is a term of art intended to capture that aspect of self-consciousness which annexes experience to the self not just at a time, but over time. We might describe it as awareness of diachronic selfhood" (14). Gerrans and Kennett (2010) report empirical findings in neuroscience that apparently lend support to the idea of distinct forms of autonoetic self-consciousness, claiming that "The 'mine-ness' of an episode is experienced together with its content. My episodic memory of a sunny day at the beach is experienced as mine . . . However it seems that 'mine-ness' is a cognitive achievement mediated by the ventromedial prefrontal cortex. While 'pure' episodic memory studies (such as recall of visual scenes) do not activate the ventromedial prefrontal cortex, 'activations of the ventromedial prefrontal cortex are almost invariably found in autobiographical memory studies" (14).

10. Naturally, anyone impressed by these proposals is likely to hold that same simulative "processes used in mental time travel are also recruited when we imagine others' situations" (Gerrans and Kennett 2010, 22).

11. Employing a simulation-based version of the idea of enactment or E-imagining the basic idea of pretend theory can be expressed as follows:

To E-imagine a state is to recreate the feeling of the state, or to conjure up what it is like to experience that state—in a sense to enact that very state. To E-imagine feeling embarrassed involves using one's imagination to create inside oneself a pretend state that phenomenally feels somewhat like embarrassment. (Goldman and Jordan 2013, 456)

12. For these reasons Goldie (2012) plausibly contends that "free indirect style—or at least its psychological counterpart—is central to autobiographical memory just as it is to autobiographical narrative" (39).

13. This vision of the future is in tune with the suggestion by empirical scientists that research on "remembering the past and imagining the future should benefit from establishing closer connections with work on narrative processing and the representation of nonpersonal fictional information" (Schacter et al. 2012, 689).

References

Addis, D.R., Pan, L. M., Laiser, N., and Schacter, D.L. 2009. Constructive episodic simulation of the future and the past: Distinct subsystems of a core brain network mediate imagining and remembering. *Neuropsychologia*. 47. 2222–2238.

Addis, D.R., Wong, A.T., and Schacter, D.L. 2007. Remembering the past and imaging the future: Common and distinct neural substrates during event construction and elaboration. *Neuropsychologia*. 45. 1363–1377.

———— 2008. Age-related changes in the episodic simulation of future events. *Psychological Science*. 19. 33–41.

Apperly, I. 2011. *Mindreaders: The Cognitive Basis of 'Theory of Mind'*. Hove: Psychology Press.

Bartlett, F. C. 1932. *Remembering*. Cambridge: Cambridge University Press.

Bischof-Köhler, D., and Bischof, N. 2007. Is mental time travel a frame-of-reference issue? *Behavioral and Brain Sciences*. 30 (3). 316–317.

Botterill, G. 1996. Folk psychology and theoretical status. In *Theories of Theories of Mind*, ed. P. Carruthers and P. Smith (pp. 105–118). Cambridge: Cambridge University Press.

Bruner, J. 1986. *Actual Minds, Possible Worlds*. Cambridge, MA, Harvard University Press.

——— 1990. *Acts of Meaning*. Cambridge, MA: Harvard University Press.

Buckner, R. L. 2010. The role of the hippocampus in prediction and imagination. *Annual Review of Psychology*. 61. 27–48, C1–C8.

Buckner, R. L., and Carroll, D. C. 2007. Self-projection and the brain. *Trends Cognitive Science*. 11. 49–57.

Carruthers, P. 2011. *The Opacity of Mind: An Integrative Theory of Self-knowledge*. Oxford: Oxford University Press.

Collingwood, R. G. 1946. *The Idea of History*. Oxford: Clarendon Press.

Corballis, M. C. 2012. The wandering mind: Mental time travel, theory of mind, and language. *Análise Social*. 47 (205). 870–893.

Currie, G. 2003. The capacities that enable us to produce and consume art. In *Imagination, Philosophy and the Arts*, ed. M. Kieran and D. M. Lopes (pp. 293–304). London: Routledge.

——— 2010. *Narratives and Narrators*. Oxford: Oxford University Press.

Fodor, J. A. 1987. *Psychosemantics*. Cambridge, MA: MIT Press.

Gaesser, B., Spreng, R. N., McLelland, V. C., Addis, D. R. and Schacter, D. L. 2013. Imagining the future: Evidence for a hippocampal contribution to constructive processing. *Hippocampus*. 23. 1150–1161.

Gerrans, P., and Kennett, J. 2010. Neurosentimentalism and moral agency. *Mind*. 119 (475). 585–614.

Goldie, P. 2012. *The Mess Inside: Narrative, Emotion and the Mind*. Oxford: Oxford University Press.

Goldman, A. 2006. *Simulating Minds*. Oxford: Oxford University Press.

Goldman, A., and Jordan, L. C. 2013. Mindreading by simulating: The roles of imagination and mirroring. In *Understanding OtherMinds: Perspectives from Developmental Social Neuroscience*, ed. S. Baron-Cohen, H. T. Flusberg and M. V. Lombardo (pp. 448–465). Oxford: Oxford University Press.

Gordon, R. M. 1995. Simulation without introspection or inference from me to you. In *Mental Simulation*, ed. M. Davies and T. Stone (pp. 53–67). Oxford: Blackwell.

——— 1996. Radical simulationism. In *Theories of Theories of Mind*, ed. P. Carruthers and P. Smith (pp. 11–21). Cambridge: Cambridge University Press.

Heal, J. 1998. Co-cognition and off-line simulation: Two ways of understanding the simulation approach. *Mind and Language*. 13. 477–98.

Hutto, D. D. 2007. The narrative practice hypothesis: Origins and applications of folk psychology. In *Narrative and Understanding Persons*. Royal Institute of Philosophy Supplement, 82. 43–68.

——— 2008. *Folk Psychological Narratives: The Sociocultural Basis of Understanding Reasons*. Cambridge, MA: MIT Press.

Kieran, M. 2003. In search of a narrative. In *Imagination, Philosophy and the Arts*, ed. M. Kieran and D. M. Lopes (pp. 69–87). London: Routledge.

Klein, C. 2013. Multiple realizability and the semantic view of theories. *Philosophical Studies*. 163. 683–695

Kwan, D., Carson, N., Addis, D. R., and Rosenbaum, R. S. 2010. Deficits in past remembering extend to future imagining in a case of developmental amnesia. *Neuropsychologia*. 48. 3179–3186.

László, J. 2008. *The Science of Stories: An Introduction to Narrative Psychology*. London: Routledge.

Mackenzie, C. 2008. Imagination, identity and self-transformation. In *Practical Identity and Narrative Agency*, ed. C. Mackenzie and K. Atkins (pp. 121–145). London: Routledge.

Maibom, H. 2009. In defence of (model) theory theory. *Journal of Consciousness Studies*. 16. 6–8, 360–378.

Medina, J. 2013. An enactivist approach to the imagination: Embodied enactments and fictional emotions. *American Philosophical Quarterly*. 50 (3). 317–335.

Moran, R. 1994. The expression of feeling in imagination. *Philosophical Review*. 103 (1). 75–106.

Roth, P. A. 1991. Truth in interpretation: The case of psychoanalysis. *Philosophy of the Social Sciences*. 21. 175–195.

Ryan, M.-L. 2010. Narratology and cognitive science: A problematic relation. *Style*. 44 (4). 469–495.

Scalise Sugiyama, M. 2001. Food, foragers, and folklore: The role of narrative in human subsistence. *Evolution and Human Behavior*. 22. 221–40.

Schacter, D. L., Addis, D. R., and Buckner, R. L. 2007. Remembering the past to imagine the future: The prospective brain. *Nature Reviews Neuroscience*. 8. 657–61.

——— 2008. Episodic simulation of future events: Concepts, data, and applications. *Annals of the New York Academy of Sciences*. 1124. 39–60.

Schacter, D. L., Addis, D. R., Hassabis, D., Martin, V. C., Spreng, R. N., and Szpunar, K. K. 2012. The Future of memory: Remembering, imagining, and the brain. *Neuron*. 76 (4). doi:10.1016/j.neuron.2012.11.001.

Schacter, D. L., Gaesser, B., and Addis, D. R. 2013. Remembering the past and imagining the future in the elderly. *Gerontology*. 59. 143–151.

Stueber, K. R. 2006. *Rediscovering Empathy: Agency, Folk Psychology and the Human Sciences*. Cambridge, MA: MIT Press.

——— 2012. Varieties of empathy, neuroscience and the narrativist challenge to the contemporary theory of mind debate. *Emotion Review*. 4. 55–63.

Suddendorf, T., and Corballis, M. 2007. The evolution of foresight: What is mental time travel and is it unique to humans? *Behavioural and Brain Sciences*. 30. 299–313.

Szpunar, K. K., Watson, J. M., and McDermott, K. B. 2007. Neural substrates of envisioning the future. *Proceedings of the National Academy of Sciences USA*. 104. 642–647.

Tulving, E. 1985. Memory and consciousness. *Canadian Psychologist*, 26. 1–12.

Walton, K. L. 1990. *Mimesis as Make-believe. On the Foundations of the Representational Arts*. Cambridge, MA: Harvard University Press.

Zunshine, L. 2006. *Theory of Mind and the Novel: Why We Read Fiction*. Columbus: Ohio State University Press.

Part III

Deliberation, Motivation, and Agency

10 Time for Action[1]

J. David Velleman

1. Prologue: "All-Out Judgement"

[T]he event whose occurrence makes "I turned on the light" true cannot be called the object, however intentional, of "I wanted to turn on the light". If I turned on the light, then I must have done it at a precise moment, in a particular way—every detail is fixed. But it makes no sense to demand that my want be directed to an action performed at any one moment or done in some unique manner. Any one of an indefinitely large number of actions would satisfy the want and can be considered equally eligible as its objects.

(Davidson 1980a, 6)

All we can judge at the stage of pure intending is the desirability of actions of a sort, and actions of a sort are generally judged on the basis of the aspect that defines the sort. Such judgements, however, do not always lead to reasonable action, or we would be eating everything sweet we could lay our hands on.

(Davidson 1980b, 97)

[A]n all-out judgement makes sense only when there is an action present (or past) that is known by acquaintance. Otherwise . . . the judgement must be general, that is, it must cover all actions of a certain sort, and among these there are bound to be actions of which some are desirable and some not. Yet an intention cannot single out a particular action in an intelligible sense, since it is directed to the future. The puzzle arises, I think, because we have overlooked an important distinction. It would be mad to hold that any action of mine in the immediate future that is the eating of something sweet would be desirable. But there is nothing absurd in my judging that any action of mine in the immediate future that is the eating of something sweet would be desirable *given the rest of what I believe about the immediate future.* I do not believe I will eat a poisonous candy, and so that is not one of the actions of eating something sweet that my all-out judgment includes. . . . The point is, I do not believe anything will come up to make my eating undesirable or impossible. That belief is not part of what I intend, but an assumption without which I would not have the intention. The intention is not conditional in form; rather, the existence of the intention is conditioned by my beliefs.

(Davidson 1980b, 100)

What exactly is the problem that is worrying Davidson in these passages? It has something to do with the temporal order of motive and action. The mental states by which action is constitutively caused must precede their behavioral effect, and Davidson thinks that this necessary temporal order precludes the possibility that these mental states can have as their intentional object the particular action that they cause. But why is this temporal constraint problematic in Davidson's eyes?

The constraint is problematic, apparently, because it stands in the way of the evaluative judgment that constitutes an intention to act. Davidson conceives of an intention as an "all-out" judgment about the desirability of an action. An all-out judgment is usually preceded by an all-things-considered judgment, which pronounces an action desirable insofar as the available reasons go. The subsequent, all-out judgment pronounces the action to be desirable, full stop. Davidson thinks that the temporal order between intention and action precludes the agent from making this judgment about a concrete particular action, because a singular judgment would require the agent to be acquainted with the action, as he cannot be with an action that has not yet occurred. The agent's intention must therefore pick out the intended action by description, hence as a kind of action rather than a concrete particular. The problem is that the agent cannot specify the intended kind of action with enough specificity to rule out every undesirable member of the kind (e.g., eating a poisoned sweet). Since the agent will never be able to describe a kind of action with enough specificity to rule out undesirable instances, he seems to be stymied unless he leaps to an all-out conclusion for which he lacks sufficient grounds.

One flaw in Davidson's reasoning—albeit an inconsequential flaw—is that intending a concrete particular action doesn't require that it already be available as an object of acquaintance. A particular action can also be picked out in advance by definite description. More importantly, it can be picked out in advance by what I would call proleptic demonstration.

I perform a proleptic demonstration when I point to something by causing it. For instance, I can tell a kindergartener, "The letter *a* looks like this," fixing the reference of "this" with the movement of writing an *a*, which directs attention to its own result on the blackboard. Or, the referent of "that" can be a stinging sensation pointed out by the slap that causes it: "Take *that*!" Because a demonstration can direct attention to one of its effects, it can point to what doesn't yet exist and cannot yet have been known by acquaintance. Hence the intention to take *this* step can pick out a concrete particular step by pointing to it, because it can point to the step precisely by causing it.

Davidson's point still stands, however, since a singular intention that used proleptic demonstration would still have to be based on general grounds, which would be inadequate to rule out undesirable instances. My practical reasoning, when laid out in Aristotelian fashion, cannot go like this:

I want to take a step.
This is a step.
So I'll take this.

There is nothing to identify as a step until it has been at least intended, and so I cannot intend to take a concrete particular step on the grounds of its being one. If I am going to intend a particular step demonstratively, I'll have to reason something like this:

> I want to take a step.
> So I'll take *this* step.

Although I can intend a concrete particular action, then, I still have to intend it as a member of a general kind—in this case, a step—differentiated from other members only by being the one produced by this very intention, a distinction that cannot support a comparative judgment in its favor and that consequently leaves open the possibility of its being a undesirable member of the kind.

Yet this risk would not be eliminated, as Davidson seems to think, by acquaintance with the concrete particular action, and here is a more consequential flaw in Davidson's reasoning. For even after I have performed an action and become acquainted with it, I cannot canvass all of the respects in which it may be undesirable. If it would be undesirable to alert the burglar in the next room who, unbeknownst to me, is rifling through my desk, then even after I have flipped on the light I may not know everything relevant to the desirability of having done so. The knowledge I gain by being acquainted with the action still leaves open the possibility that it is disastrous, contrary to my all-out judgment of its desirability.

Why does Davidson think that I need to insulate my all-out judgment against falsification? There is no problem with a judgment's being defeasible even after it has been detached from specific grounds or from the generic hedge "all things considered." Even an unqualified or categorical judgment can be revised or retracted after the fact: "Flipping the switch would be desirable, full stop"; "Oops, it wasn't." "Categorically, eating a sweet would be desirable"; "Yuck—not that one!"

What now emerges is a crucial difference between a desirability judgment and an intention, a difference that raises doubts as to whether the one can constitute the other, as Davidson believes. An intention cannot be retracted after having been executed: once the agent has decided to do something and then done it, he cannot change his mind. How, then, can his intention consist in a judgment that can so easily be revised or retracted after the fact? The role of intention in rational thought is fundamentally different from that of a desirability judgment, whose role allows for subsequent revision or retraction. How can intention consist in a judgment that plays a fundamentally different role?

I suspect that Davidson's phrase "all-out judgment" is meant to suggest that the judgment is not just unconditional but final, in the sense that it cannot be retracted or revised. An all-out judgment, in Davidson's mind, is a judgment to which the agent commits himself irrevocably, once and for all. You might think: there is no such thing as a once-and-for-all judgment—no such creature in the mental menagerie. The Davidsonian reply to this is: well, there is such a creature, namely, the desirability judgment that constitutes an intention, which is by its very nature once and for all.

This interpretation not only restores the similarity between judgments of desirability and intentions; it also helps to explain why Davidson is so worried about judging desirable a kind of action that may include undesirable instances. Whereas an unqualified or categorical judgment can be ventured on the basis of incomplete information, because it can still be reconsidered, a once-and-for-all judgment is closed to reconsideration, and so it had better be right the first time.

2. The Temporal Profile of Action

If this problem is the one that worries Davidson, then he has put his finger on a crucial but little noticed feature of practical reasoning as conceived by most philosophers. Philosophers conceive of practical reasoning as proceeding under a deadline. In order to cause action, the agent's intention must precede it. If, alternatively, it consists in the action—a possibility that Davidson envisions—it must be cotemporaneous with it. And in either case, the intention will be an irrevocable conclusion, which will close practical reasoning once and for all. I think that this conception of practical reasoning is mistaken and that it has distorted philosophical views of how practical reasoning works.[2]

I will shortly present an alternative conception of practical reasoning according to which practical reasoning generally supervises behavior that is already underway, having been initiated without any reasoning at all. Before I introduce my conception of practical reasoning, I want to continue exploring how the temporal profile of action is typically conceived.

Consider this famous passage from Christine Korsgaard's *Sources of Normativity*:

> I desire and I find myself with a powerful impulse to act. But I back up and bring that impulse into view and then I have a certain distance. Now the impulse doesn't dominate me and now I have a problem. Shall I act? Is this desire really a *reason* to act? The reflective mind cannot settle for . . . a desire, not just as such. It needs a *reason*. Otherwise, at least as long as it reflects, it cannot go forward.
>
> (Korsgaard 1996, 93)

This passage describes a sequence of events: "*then* I have a certain distance," "*now* I have a problem," "*as long as* [the mind] reflects, it cannot go forward." Action awaits the discovery—or, even more time-consuming, the construction—of a reason for acting.

I don't say that Korsgaard intended this passage to be interpreted literally as the description of a temporal sequence. Maybe no philosopher of practical reasoning consciously envisions it as time-consuming. The fact remains that almost every philosopher of practical reasoning describes it in such terms, sometimes disavowing their implications but, I suspect, never effectively canceling them.

For another famous example, consider Bernard Williams's description of a man who has "one thought too many." The man is standing on the deck of a sinking ship, watching his wife thrash about in the water below. He wonders whether it is morally permissible to save his wife in preference to other victims:

> [T]he consideration that it was his wife is certainly, for instance, an explanation which should silence comment. But something more ambitious than this is usually intended, essentially involving the idea that moral principle can legitimate his preference, yielding the conclusion that in situations of this kind it is at least all right (morally permissible) to save one's wife. . . . But this construction provides the agent with one thought too many: it might have been hoped by some (for instance, by his wife) that his motivating thought, fully spelled out, would be the thought that it was his wife, not that it was his wife and that in situations of this kind it is permissible to save one's wife.
>
> (Williams 1981, 18)

Williams's point is that the thought of moral permissibility would be "one too many" in the sense that, by thinking it, the man would betray less than complete devotion to his wife. That's why his wife might hope that he wouldn't think it. Unfortunately, Williams has couched this point in a story about an emergency, in which an agent must make a split-second choice. His description of the moral thought as one too many therefore steals plausibility from a different interpretation, namely, that the thought would be one more than the man has time to entertain before having to dive in. Under this interpretation, thoughts are discrete units, each of which takes time to entertain. Thinking one more or one less thought entails thinking for more or less time.

There is of course such a thing as overthinking a decision, of deliberating too much. But deliberation is not a mode of practical reasoning; it's a procedure ancillary to practical reasoning, just as procedures of calculation are ancillary to arithmetic. Deliberation is like a checklist, a mental procedure that we use to help us think. It's something we *do*. Indeed, it's something we do deliberately—which cannot mean "on the basis of deliberation," lest it lead to a vicious regress.[3]

Now, if the word "thinking" is being used to *mean* deliberating, then the man in Williams's story shouldn't think at all: a single thought would already be too many, since he shouldn't take time to deliberate. In this sense of the word, he should jump without a thought, though not of course without a reason.

If however "thinking" is not being used for a mental procedure, then no assumptions can be made about how much or how little thinking would fit into the interval between a man's look and his leap.[4] I've never stood on a burning deck, but I am sure that if I did, all sorts of things would cross my mind: the bilious green of the waves; the jaunty bobbing of lifeboats; screams from the swimming pool, of all places; the time I saw someone drown in the college pool; Kate Winslet. They say, in fact, that my whole life would pass before my eyes. How long would any of this mentation take? It's not a serious question.

If a thought is not a step in a mental procedure, then it can be one too many only because it would be irrelevant, not because it would be too time-consuming. And in that case, the phrase "one too many" does not express an argument for the conclusion that thoughts of morality would be irrelevant; it just expresses the conclusion.

I don't say that Williams intended the phrase to express an argument; I say only that the phrase is often read as expressing one. The argument depends on the same assumption that we found in Davidson, namely, that practical reasoning proceeds under a deadline, beyond which it is over once and for all: the man in Williams's story has only so much time to think *before* acting, and so he has only so much time to think. To this assumption, the argument adds the further assumption that the requisite kind of thinking is a time-consuming procedure. According to this assumption, having only so much time to think entails having time for only so many thoughts.

3. Counting the Reasons

The conception of practical reasoning as a time-consuming procedure with a deadline is also implicit in the language that philosophers use to describe the normative force of reasons for acting. They say, for example, that reasons for an action are considerations that "count in favor" of it, and that nothing further can be added by way of explicating the concept of a reason or the concept of "counting in favor." "Counting" is thus characterized as the unanalyzable essence of reasons for acting.

Unfortunately, the nature of practical reasoning is obscured rather than clarified by talk of counting. Counting is not in the first instance something that considerations do; it's something that people do. Counting is an activity, and it is an essentially rule-governed activity. It is governed by rules for incrementing a sum by units corresponding to items of some kind, individuated in some way. When we speak of things rather than people as counting, we mean that those things are *to be counted* under such a set of rules. When the students ask, "Does spelling count?" they are asking whether the rules for grading their work provide for points to be added or subtracted for words spelled correctly or incorrectly. When we say, "Close doesn't count in horseshoes," we're talking about the rules for scoring the game.

The rules under which spelling counts on an assignment need not be fully determinate, but they cannot be utterly indeterminate, either. If bad spelling merely has some probability of lowering a student's grade by putting the teacher into a bad mood, then spelling doesn't count; it makes a difference, but not by "counting." In order for spelling to count, there has to be some rule-governed method for counting it, however vague or imprecise the rules may be.

To speak of considerations as counting in favor of an action therefore raises the question, How do they count? Whether interpreted literally or metaphorically, such talk suggests that there are rules, however vague or imprecise, for scoring or grading actions by enumerating pros and cons. So the phrase "counts in favor of" says significantly more than "is a reason for," unless the notion of counting is explicitly canceled—in which case, nothing of the phrase is left.[5]

The problem, for present purposes, is that gratuitously introducing the notion of counting leaves the impression that reasons are essentially *to be counted*—hence items in a time-consuming mental procedure that ultimately arrives at a bottom line. If reasons are considerations that count, then only so many of them can enter an agent's practical reasoning, since only so many can be counted in time for him to act.

A similar confusion is introduced by the image of reasons as having weights. Here is an especially vivid exposition of that image, by John Broome:

> Suppose you ought to Φ. An explanation strictly analogous to mechanical weighing would be this. There are reasons for you to Φ and reasons for you not to Φ. Each reason is associated with a number that represents its weight. The numbers associated with the reasons to Φ add up to more than the numbers associated with the reasons not to Φ. That is why you ought to Φ.
>
> . . . [W]hen the fact that you ought to Φ is explained by pro tanto reasons, the explanation retains central elements of the mechanical analogy. It includes one or more reasons for you to Φ, and it may also include reasons for you not to Φ. These reasons are analogous to the objects in the left-hand and right-hand pans of the scales. Each reason is associated with a metaphorical weight. This weight need not be anything so precise as a number; it may be an entity of some vaguer sort. The reasons for you to Φ and those for you not to Φ are aggregated or weighed together in some way. The aggregate is some function of the weights of the individual reasons. The function may not be simply additive, as it is the mechanical case. It may be a complicated function, and the specific nature of the reasons may influence it. Finally, the aggregate comes out in favour of your Φ-ing, and that is why you ought to Φ.
>
> (Broome 2004, 36–37)[6]

Strictly interpreted, Broome is speaking here of the weight that reasons have, not of any weighing procedure: he isn't saying that practical reasoning is a process of weighing. But if the guiding force of reasons is analogous to weight, how can it guide an agent unless he weighs them? They have to be loaded into a scale and left untouched while the beam comes to rest. How else can they exert their normativity?

When we ask someone why he has done something, or what he proposes to do, the reply often takes the form of one or two considerations, discrete propositions that might be entertained at a stroke and "counted" or "weighed" on one or the other side of a question. The agent may even tick off these considerations, as if counting them, or mime the action of hefting one in each palm to compare their weights. But these are tropes of practical discourse, not windows into the mentation that actually leads to action.

4. The Supervisory Conception

I now want to present an alternative conception of the relation between practical reasoning and time. On my view, practical reasoning does not necessarily precede action, much less take time in advance.

In his *Principles of Psychology*, William James describes how we get ourselves out of a warm bed on a cold morning. "If I may generalize from my own experience," he says, "we more often than not get up without any struggle or decision at all. We suddenly find that we *have* got up" (James 1950, 524).[7] Truer words were never written about practical reasoning. We think about getting up; that thought is swept away by many others, irrelevant to the moment; and then we are already on our feet, without having been pushed by any occurrent thought.

Here I must insist on the empirical method: argumentation won't do. Perform some careful observation of yourself over the course of the next few days. You cannot plan this observation immediately before acting, but neither can you leave it until after you have acted. Rather, just after you have begun to act—even as you are rising from your chair (or bed)—you must remember to observe yourself acting. Discard the cases in which you remembered to look before you started to rise. If you start to look too soon, you will just enact your prior conception of practical reasoning for the benefit of yourself as spectator. In order to see what actually happens when you act, you have to catch yourself, as it were, unawares. So the relevant cases are the ones in which you remember only after you have started to act, so that you weren't watching at the crucial moment. In those cases, notice what passed through your mind immediately before you stirred. If I may generalize from my own experience (to echo James), you will find that if you had any thought at all of getting up, it was more of a speculation about getting up than a decision to do so. Then suddenly you were getting up.

The articulated mental accompaniments of action are not of the form: "I want a drink; there's beer in the fridge; so I'll get up and go to the fridge." They are rather of the form: "A drink would be nice. Hey, who took the last beer?"

The fact that the intervening behavior didn't follow any practical thinking does not entail that it was automatic or rote. There was practical thinking involved, but it was largely perceptual and supervisory, not deliberative. That is, you perceived yourself getting up and going to the kitchen, which (as you are aware) is the location of the beer, which (as you are aware) is what you had been thinking of fondly. These perceptions and implicit beliefs did not precede and initiate your behavior; they registered it and then followed along with it. Their role was to supervise the behavior, by ensuring that it was something that made sense in the circumstances. It was your body that initiated the action; your conscious mind watched and thought, first, "I get it. OK then," and thereafter, "So far, so good." I defy you to find that anything else has gone through your mind—provided that you remember to look at just the right moment.

In my view, you don't even individuate actions in advance, much less reason about them. You produce a continuous stream of behavior, which practical reasoning supervises and, in the course of supervising, places under action concepts.

As Davidson emphasized, most of the things you do, you do by doing other things or in the course of doing other things. You turn on the light by flipping a switch, which you do by flicking your finger, and so on; you turn on the light in the course of illuminating the room, which you do in the course of looking for your keys, and so on. Actions telescope. But philosophers generally assume that the telescope of actions is assembled in advance out of predefined segments, which are represented in an antecedent action plan. This assumption is what I deny. "Looking for keys, by illuminating room, by turning on light, by flipping switch, by flicking finger" is a *post facto* or perhaps *in medio facto* description of a undifferentiated stream of behavior. Each so-called action is abstracted from a fluid dance in which you also shorten your stride, grasp and release a knob, step and pause, sniff the air, peer into corners, turn your torso and your gaze. . . . To be sure, you don't do all of those things intentionally, whereas you intentionally flip the switch. But again, just after flipping a light switch, remember to ask yourself how that supposedly intentional action stood out from the rest of the dance.

Discrete act-descriptions get attached to behavior at various points in the process. Sometimes practical reason frames an act-description antecedently, but needn't hand it down as an order if behavior spontaneously rises to meet it. Sometimes practical reason frames an act-description simultaneously with the behavior it describes, so that reasoning and behavior proceed in step. Sometimes practical reason lags behind, and you find yourself standing barefoot on a cold floor, or thirsty at the door of the fridge, describing your behavior retrospectively. And sometimes, though not very often, practical reason has to take charge. In that case, you intend your actions in the classical sense, framing a plan and then carrying it out. Otherwise, brain and body go about their business, and your intending takes the form of *super*intending.

I would argue that supervised behavior is what action *is*. What differentiates action from mere behavior is not that it is both caused and rationalized by a desire-belief pair, as Davidson claimed; what differentiates action from mere behavior is that it is supervised.

There are actions that have no desire or pro-attitude behind them, for example, expressive actions such as jumping for joy (see Hursthouse 1991). When my son called to say that he and his wife were going to have a baby, I really did jump for joy, and I was jumping out of joy itself, an emotion whose object was the prospect of a grandchild; I was not jumping out of a desire to jump, much less a desire to vent my joy by jumping. There are also behaviors that are caused and rationalized by desires but do not clearly qualify as actions. Once when I was introduced to someone, I said, "Pleased to meet me"—an utterance motivated, I then realized, by a desire to tell my startled hearer how *he* should feel about the introduction and, by implication, to disavow the reciprocal feeling. I was surprised to hear the words come out of my mouth, and so I felt that, although mine was the mouth they came out of, it wasn't me speaking. My speech act somehow fell short of action. It fell short because I spoke unwittingly, whereas the speech that amounts to action is spoken wittingly. Speech action consists in speech acts knowingly performed.

Note that my saying "pleased to meet me" was not due to a mechanical failure in speech production: it wasn't a slip of the tongue, like saying "meased to pleet you." Saying "pleased to meet me" was a mechanical success at saying something that, at some level, I wanted to say. But my wanting to say it and saying it were not in themselves sufficient for knowing what I was saying. In order to know what I'm saying, rather, I have to say what I am already prepared to hear myself say, so that my saying it will not be a surprise. In the case of my embarrassing blooper, I was prepared to hear "pleased to meet *you*" but ended up saying something else, for which I wasn't prepared by which I was therefore surprised.

I rarely say things for which I am not prepared, and that's no accident. The reason is not that I usually prepare speeches before I make them. Preparing a speech is indeed a way of being prepared for it, but mostly I'm prepared for what I say because my next utterance is both prompted and prefigured by the situation, including my state of mind. The situation provokes the utterance, and on the basis of it I could, if asked, anticipate what sort of utterance it would provoke (though of course I'm not asked to anticipate it, and I don't). I both say what comes next and implicitly know what comes next, so I'm not surprised by my saying it.

When I don't know what comes next, I may have nothing *to* say and, in any case, I couldn't predict what I *would* say if I said anything. And even if I do have something to say, in the sense that there is something I would say if anything, I am usually inhibited from speaking if I feel ignorant as to what it will be. And *that*, in the last analysis, is why I am rarely surprised by what I hear myself say: I don't open my mouth until I have the sense that

I know, if only latently, what is going to come out. I must, then, be aiming to avoid surprises, to know what I'm saying—not as one of the ends, or desired outcomes, of my behavior but as the manner in which I behave.

The sense of latently knowing what's coming next in my behavior often comes from having followed it successfully until now. The word "following" here does not connote temporal succession. When we follow someone's talk in a printed copy of his paper, we read it along with him, simultaneously; we would only confuse ourselves if we followed temporally, by reading each word *after* he said it. I can usually follow my behavior, in the sense of simultaneously thinking along with it, because it is both prompted and prefigured by the same circumstances. And if I have been successfully thinking along with my behavior, then I gain confidence in being able to continue, and so I am no longer restrained by the aim of avoiding surprises. But if my behavior takes a surprising turn, the aim of avoiding surprises kicks in, and I stop doing anything until I have prepared myself for it by thinking ahead. And then I gradually regain my confidence in being able to follow along.

I would argue that this supervisory process is the final stage of practical reasoning, with which it is fully continuous. I cannot rehearse the argument here, but let me offer just one of the intuitions behind it.

When someone acts irrationally, we often say that he doesn't know what he's doing. In challenging him directly, we are just as likely to ask "What are you doing?" as to ask "Why are you doing that?" Behind these remarks is the assumption that a person acting rationally has a high-level, explanatory conception of what he is doing, a conception that sets his immediate behavior in the context of its outer circumstances and inner motives, which needn't be desires and beliefs, since they may be, for example, emotions such as joy. ("What on earth are you doing?" my wife asked, covering the mouthpiece of the phone. If I hadn't thought it was obvious, I would have explained: "Jumping for joy.") Even if we ask someone why he is doing something, he may come back with an answer to the question what he is doing: "Why are you pacing the halls?" "I'm working on my dissertation." These are the descriptions under which the person is following along with his behavior, in the capacity of a supervisor.

To be sure, we usually have in mind some desired end to which our behavior will be conducive. But the conception of our behavior as directed at a desired end is a high-level, explanatory description under which we know what we're doing. Our behavior is indeed desirable to us because we desire the end, and it is indeed guided by our knowledge of its efficacy in producing the end. But the desirability of our behavior is not the same as its rationality. To say that we usually act on a conception of our action as a means to a desired end is not yet to say how that conception rationalizes our action. Does it rationalize our action by showing it to be efficacious, by showing its effects to be desirable, or by giving us a grasp of what we are doing? Well, "jumping for joy" rationalizes an action without doing either of the first two. I say that rationalizing action consists in the third.

Of course, we sometimes come up with a means-end conception before we implement it, at a time when it doesn't yet describe what we're doing. But that conception does prepare us to see ourselves implementing it, so that we know what we're doing when we do so. Our positive motivation for implementing the conception is our desire for the end, but that motive would be restrained if we didn't feel prepared to see what we were about to do, or we were surprised when we began doing it. In addition to the positive motive, then, there is the conception under which the action is allowed to proceed by the supervisory process. In my view, the desire is our motive merely; our reason is the conception under which our behavior is supervised.

According to this supervisory conception, practical reasoning is not in the driver's seat of action; it's in the passenger's seat. Better: practical reasoning is your *spouse* in the passenger's seat, mostly following along silently as you drive; sometimes warning you of upcoming turns; sometimes telling you to take turns that you are already taking anyway, thank you very much; sometimes pointing out that you just took a wrong turn; and just sometimes giving you just the direction you need, just when you need it.

Most of the time, you drive on automatic pilot, as we say. The automatic driver is your skilled, intelligent, goal-seeking mind, which can handle the car by itself most of the time. If you have a live pilot in the passenger seat, he or she plays the role of the human pilots in modern airliners, supervising the automated systems. If you are driving alone, then you play both roles, supervising your inner, automatic driver. You supervise that automated system and intervene only if necessary.

The automatic piloting that is done by your intelligent, goal-seeking mind includes what Davidson identified as practical reasoning. That is, it includes desire-belief motivation. It also includes course-correction in response to perceptual feedback about progress toward the desired outcome. (Davidson left that part out.) In my view, however, practical reasoning should be identified neither with desire-belief motivation nor with the attendant feedback and course-correction. Those processes are what practical reasoning supervises; practical reasoning is the supervisor.

What argument do I have for this supervisory conception of practical reasoning? First answer me this: what argument is there for the more standard, deliberative conception?

I have suggested several debunking explanations for the nearly universal belief in the standard conception—the literal interpretation of metaphors for the normative force of reasons, the conflation of practical reasoning with practical discourse or with its ancillary procedures. To these explanations I would add the sheer force of inertia. The standard conception of practical reasoning carries the weight of unquestioned lore, a weight that has stifled progress in the philosophy of action since Davidson, if not since Hume. The philosophy of action long ago entered the realm of fan fiction, a genre in which volunteer authors spin out the adventures of a character whom everyone knows well enough to distinguish between the fictionally consistent inventions about him and the fictionally inconsistent ones. The Rational Agent has taken his place alongside Harry Potter and Han Solo as a figure whose character is sufficiently determinate to give the ring of truth to some inventions about him rather than others. Unfortunately, no reason has ever been given for believing that he actually exists.

My case for the supervisory conception of practical reasoning is not an argument but an introspective experiment, for which I have provided very specific instructions. Don't watch yourself decide to do something: you'll just see yourself enact the standard conception, in which you too have been enculturated. Rather, catch yourself when you have just started doing something, and cast a retrospective glance at the thinking that preceded your starting to do it. Where you expected to find the mythical Rational Agent, you'll find the Supervisor instead.

If the supervisory conception of practical reasoning is correct, what follows?

One thing that doesn't follow is any conclusion about free will. The moral responsibility of a navigator may be somewhat different from that of a driver, but so long as the navigator has or could have some influence on the route taken, he or she can certainly bear some responsibility for it.

A second implication of the supervisory conception is that automaticity in behavior is not the exception but the rule. Sub-agential mechanisms do not have to wrest control of our behavior away from a rational driver. There is no rational driver, only a rational passenger-navigator, whose task most of the time is to supervise behavior controlled by sub-agential mechanisms.

Finally, the supervisory conception implies that practical reasoning is not a time-consuming procedure that is carried out under a deadline. We don't necessarily think before we act; usually, we think while we behave, and the combination ends up constituting action.

Notes

1. This chapter was originally written for the conference on Time and Agency at the George Washington University (November 2011). It was also presented to a conference on Davidson's "Actions, Reasons, and Causes" in Duisburg/Essen (September 2013); to the Philosophy Mountain Workshop in Jackson, Wyoming (January 2014); and to the Normativity and Reasoning Workshop, NYU Abu Dhabi (January 2014).
2. For another philosopher who questions this conception, see Talbot Brewer (2009, ch. 3).
3. This point is explored at length by Nomy Arpaly and Timothy Schroeder (2012).
4. See Ryle:

   When we think in the abstract about thinking, it is usually reflecting, calculating, deliberating, etc. that we attend to. . . . Indeed it is just because reflecting is what we start off by considering, that we later on feel a strong pressure to suppose that for the tennis-player to be thinking what he is doing, he must be sandwiching some fleeting stretches of reflecting between some stretches of running, racquet-swinging, and ball-watching. (Ryle 1968, 216)

5. There is a temptation to think that the counting rule for reasons is to add up weighted sums for and against an action. But if the weighted sums are sums of reasons, and reasons are considerations that count for or against an action, then the sense in which they count cannot be given by the rule of computing their weighted sum. The rule for counting reasons cannot be the rule of counting them, as they are to be counted according to that very rule. Such a counting rule would be vacuous and could not serve to determine what counts. A substantive counting rule for reasons would have to be the rule of increasing the sum for each consideration that qualifies as a reason in a sense that's independent of counting according to that rule. So the notion of "counting in favor" appears to add nothing to the notion of reasons—except for the misleading suggestion of a time-consuming activity.

6. I've excised the second paragraph of this passage, which reads as follows: "Such a strictly analogous explanation rarely seems appropriate. For one thing, it often seems inappropriate to associate a reason with anything so precise as a number to represent its weight. Secondly, although we can aggregate together the weights of several reasons, to aggregate them simply by adding up also often seems inappropriate. So-called 'organic' interactions between reasons often mean that their aggregate effect differs from the total of their weights" (37). One might have thought that this concession completely undermines the analogy between normative force and weight, but Broome continues, in the third paragraph, as if nothing has been conceded.

7. Unfortunately, James goes on to speak of "an idea" that "flashes across us," which in my experience rarely occurs.

References

Arpaly, N., & Schroeder, T. (2012). Deliberation and Acting for Reasons. *Philosophical Review*, *121*(2), 209–239. doi:10.1215/00318108–1539089.

Brewer, T. (2009). *The Retrieval of Ethics*. Oxford: Oxford University Press.

Broome, J. (2004). Reasons. In R.J. Wallace, P. Pettit, S. Scheffler, & M. Smith (Eds.), *Reason and Value: Themes From the Moral Philosophy of Joseph Raz* (pp. 28–55). Oxford: Clarendon Press.

Davidson, D. (1980a). Actions, Reasons, and Causes. In *Essays on Actions and Events* (pp. 3–20). New York: Oxford University Press.

——— (1980b). Intending. In *Essays on Actions and Events* (pp. 83–102). New York: Oxford University Press.

Hursthouse, R. (1991). Arational Actions. *The Journal of Philosophy*, *88*(2), 57–68. doi:10.2307/2026906.

James, W. (1950). *The Principles of Psychology* (Vol. 2). New York: Dover.

Korsgaard, C. (1996). *The Sources of Normativity*. Cambridge: Cambridge University Press.

Ryle, G. (1968). Thinking and Reflecting. *Royal Institute of Philosophy Supplements*, *1*, 210–226. doi:10.1017/S0080443600011511.

Williams, B. (1981). Persons, Character, and Morality. In *Moral Luck: Philosophical Papers, 1973–1980* (pp. 1–19). New York: Cambridge University Press.

11 Time and the "Antinomies" of Deliberation

John J. Drummond

1. Antinomies of Deliberation

The criminal code of the late German Empire specified what today we might call "sentencing guidelines" for murder. The punishment for homicide without deliberation was a prison term of not less than five years; the punishment for homicide when mitigating circumstances were present was a prison term of not less than six months; and the punishment for homicide after deliberation was in all cases death. Adolf Reinach believed the enormous disparity in the sentences masked what he called "remarkable antinomies" (Reinach 1989, 279) in our understanding of deliberation. He stated them this way:

1. We reproach someone who performs an evil action of great import without "deliberating even for a moment."
2. We judge that same action even more harshly when done "with deliberation."

In the case of homicide, the first proposition is confirmed by the five-year sentence for killing a person without deliberation and the even lesser sentence of six months for killing without deliberation in provocative circumstances, while the second proposition is confirmed by the much harsher sentence of death when deliberation is present.

Similarly, according to Reinach:

3. Meritorious actions count as less meritorious when done "without any deliberation."
4. Meritorious actions also count as less meritorious when done "only after long deliberation."

Reinach argued that these antinomies required an analysis of the notion of deliberation and the related notions of moral and legal responsibility. I raise these antinomies for our consideration not because I think Reinach has properly stated them or because I wish to follow him in exploring the relation between deliberation and moral or legal responsibility. I am struck by the references to time. Deliberation, on the view presupposed by Reinach, takes time at or continuous with the moment of action. The so-called antinomies arise when the agent takes no time or too much time in deliberating. And this suggests, misleadingly, I think, that there is a "Goldilocks-amount" of deliberation and of time—some deliberation and some time—and just enough deliberation and just enough time. Otherwise, the agent acts either impetuously or with an insufficient attunement to what is right to do in certain kinds of circumstances.

When we think about deliberation with respect to time rather than responsibility, the third proposition—namely, that a meritorious action counts as less meritorious when done without any deliberation—strikes anyone with an Aristotelian ear as false or, at least, as importantly misleading. We need only recall Aristotle's somewhat paradoxical descriptions of the brave person. On the one hand, Aristotle tells us,

> the man . . . who faces and who fears the right things and with the right aim, in the right way and at the right time, and who feels confidence under the corresponding conditions, is brave; for the brave man feels and acts according to the merits of the case and in whatever way reason directs.
>
> (*NE* 1115b18–20)

The brave man is guided by reason and must get an awful lot of things "right" to count as brave. Nevertheless, Aristotle also tells us,

> it is thought the mark of a braver man to be fearless and undisturbed in sudden alarms than to be so in those that are foreseen; for it must have proceeded more from a state of character, because less from preparation; for acts that are foreseen may be chosen by calculation and reason, but sudden actions [are chosen] in accordance with one's state of character.
>
> (*NE* 1117a17–22)

So there we have it: the brave person, insofar as he acts from virtue, acts from calculation or deliberation and from right reasons, whereas the braver person acts from a state of character rather than calculation and reason. Acting from the right reasons suggests that at the moment of action one reflects on one's reasons and affirms them as right, that one justifies one's action by an argument that yields as its conclusion a prescription for what to do in the current circumstances. This is just what deliberation seems to do. The braver person, however, just acts without such deliberation, and he does so from a fixed character. So Aristotle's example seems to confirm Reinach's fourth proposition that the meritorious action is less meritorious when done after deliberation—that person is less brave—but it seems at the

same time to disconfirm the third, that the meritorious action is less meritorious when done without deliberation.

Aristotle's claims, of course, are not necessarily inconsistent, and there
is at least one obvious solution to this paradox. When Aristotle speaks of
the braver person, we might think, he is undoubtedly suggesting that the
braver person does not *occurrently* deliberate in a suitably extended sense
of "occurrent." If an agent deliberates about what to do in the current
circumstances and then immediately acts, the deliberation can be thought
occurrent even though the action is temporally posterior to the deliberation. The temporal continuity of the deliberation and the performance of
the action are sufficient for the deliberation to count as occurrent. But Aristotle's braver person does not deliberate occurrently even in this extended
sense; the braver person acts directly from a state of character rather than
"calculation and reason." But insofar as the braver person is brave at all,
it seems he must act from right reasons and, hence, must act deliberately.

This raises the questions of whether deliberation is temporally related
to choice and action, and if so, how. Let us suppose, for the sake of the
argument, that the braver person deliberates non-occurrently. On that supposition there are only two options for when the deliberation occurs: the
deliberation is prior to but temporally discontinuous with the action, or the
deliberation, in some sense, occurs after the action. John Cooper, in introducing the problem of deliberation in Aristotle, has suggested the latter as
a possibility:

> Aristotle's insistence that moral decisions are all of them "choices" (*pro-
> haireseis*), and therefore supported by deliberation, can be defended,
> then, provided one understands by this only that moral decisions are
> always backed by reasons which, when made explicit, constitute a delib-
> erative argument in favor of the decision.
>
> (Cooper 1986, 9–10)

The idea is that if someone demands to know why the braver person acted
as he did, he can produce a list of reasons for his action that would have the
look of a deliberation. As Cooper puts it,

> the attempt to explain what one has done will take the form of setting
> out a course of deliberation by which one *might* have decided to do
> what one has done, and which contains the reasons one actually had in
> acting as one did.
>
> (Cooper 1986, 9)

It is important to stress that the reasons cited in the so-called deliberation
are the reasons one actually had in acting; otherwise, the citation of reasons
would be merely an *ex post facto* rationale and not really the provision of
the reasons governing the action.

In advancing this possibility as a solution Cooper has in mind a narrow, technical sense of what deliberation involves, a sense that Aristotle often seems to advance as *the* account of deliberation and one that Cooper would like to supplant with a broader, more sympathetic reading of Aristotle on practical reason. Against the narrow, technical view, Cooper claims that, given Aristotle's theory of moral development and the inculcation of the habits of virtue in an agent, not all action is the result of deliberation understood as the fashioning of an argument to the effect that some particular means of attaining an end is the preferred course of action or that some particular action is subsumed under a rule. Cooper proposes—rightly, I think—that Aristotle does not mean by practical reason only the production of means-end arguments or a "practical syllogism" endorsing particular actions as falling under a general rule. Instead, Cooper argues, Aristotle is concerned with working out various ways reasons for action might be at work in action—ways that might involve, say, views about the constituents of an end or views about what conduces to an end (but not as an external means to it). Such a determination of reasons gives us reason, according to Cooper, to perform a *specific* type of action in particular circumstances (Cooper 1986, 22–24). Once the agent reaches this state of practical knowledge, no further *reflection* is needed, but only *perception* of the particular situation (Cooper 1986, 58).

On Cooper's view of Aristotle, therefore, practical reason involves an already gained knowledge of reasons determining our sense of specific actions appropriate to different kinds of situations. On seeing her bus pulling away from the bus stop, for example, Jane knows—as a practical matter—that she can start running after the bus or start waving her arms or both in an attempt to get the driver to stop so that she can board the bus. She does not occurrently deliberate about performing these actions; she simply does them. And since virtue demands having a command of right reasons, in the exercise of a virtuous action, for example, returning excess change to a cashier, Mary simply does what is called for, as the braver person does, without deliberation. Nevertheless, she must, since she acts from virtue, have the right reasons and her action must be from those reasons.

This invites reflection on how the practical knowledge of right reasons is gained and the manner in which it determines an agent's present sense of what is to be done. In what follows I sketch a phenomenology of choice in order to illuminate the temporality at work in agency. In brief, I shall claim that the temporality of choice and action is ecstatic in something like the Husserlian and Heideggerian senses. Put more simply, the present moment in which an agent acts incorporates and is informed by both the past and the future.

It is a commonplace that choice aims at an end, but the meaning of the term "end" varies in different accounts of choice and practical reason. An end for Aristotle is in the first instance the completeness, the perfection, the fulfillment, or the maturation of a thing that is achieved through the exercise of its characteristic activity. Hence, the end is internal to the thing and its activity. From ends in this sense we should distinguish purposes. A purpose involves the explicit ordering of an action toward an internal or external consequence that might or might not be identical to the end of the agent or of the activity in which the agent is engaged. Purposes, in other words, are located in our conscious willing and acting, whereas ends belong to the defining activities of the natural thing as well as of the institutions and practices established by beings capable of choice.[1]

Some examples will draw out this distinction. Plants reproduce. Their reproduction involves no purposes, although it does involve an end. Plants when reproducing do not have in mind the purpose of sustaining the species. They do not have minds and do not make choices, but sustaining the species is nevertheless the end of their reproductive activity. Ends can be purposive in and for beings capable of choice, but not every purpose is an end in Aristotle's sense. It is only in medieval theological contexts that the natural changes that realize a being's end can be thought to be at the same time purposive from a divine perspective, and it is only after a teleological understanding of forms is banished altogether from our modern scientific conception of nature that ends and purposes are completely conflated (as in consequentialist theories in ethics).

The notion of an end proper is not limited to natural kinds. Humans, for example, have among their ends the maturation of their biological natures through processes such as nutrition and reproduction as well as the development of their sensory, appetitive, and rational powers so as to acquire and exercise both moral and intellectual virtues. The development of these powers and the concomitant demand to develop the virtues means that humans, by choice, establish institutions and practices that realize their ends, and these institutions and practices have their own proper ends. Medicine, for example, has as its end the restoration and maintenance of the health of the patient. Someone might undertake the practice of medicine to heal her patients, in which case the end of medicine and her purpose in undertaking it coincide. But her purpose might instead be to make a great deal of money or to gain social status. This does not annul the ends of her medical practice; she still treats and heal patients, but another purpose now exists alongside this end as her primary purpose. Similarly, the end of education is learning. But someone might go to college for the purpose of landing a better job or for the purpose of playing football for the further purpose of becoming a pro player. For such a student the purpose for which education has been undertaken can—and often does—subvert its proper end. Again, the end of politics is to secure the common good, but a politician might act so as to ensure reelection or to favor his associates and financial supporters rather than to serve the common good, and in such cases his or her purpose subverts the end of politics. In this way our human ends and the ends of the activities, institutions, and practices we adopt to realize them measure our purposes.

I am not trying to enforce a terminological rigor here, for ends, purposes, and consequences overlap in important ways. I claim only that if one uses the term "end" for all of these, then the term is being used in a systematically ambiguous way. That is not necessarily bad, as long as the differences in usage are marked. Having made my terminological point, I am going to speak of acting for ends, and by "ends" I shall mean ends proper or purposes as measured by ends.

2. Evaluation and Choice

Persons are in the first place intentional beings, and I use the term "intentional" here in its broad sense of "directedness to" rather than its specifically volitional sense. This entails that persons have an end that is the characteristic activity of intentional beings. Intentional beings intend things in both empty and full intentions. An empty intention is one that makes present or re-presents an object or state of affairs that is absent. The most important examples of empty intentions are linguistic expressions in which a speaker makes present to a listener an object or state of affairs in, say, a proposition expressed in a sentence. For example, I can describe my wife to you and thereby direct your attention to her, but both my linguistic presentation and your intending my wife on the basis of understanding this description would be empty intendings without intuitive fullness.

Empty intentions are contrasted with full intentions. Full intentions either present an object intuitively, as in perception, or they present an object that, while not perceptually present, is made present with the aid of a phantasm or an image. A full intention, in other words, has at least some degree of intuitive fullness. Full intentions can fulfill or disappoint empty ones. A full intention is fulfilling or disappointing only in relation to an empty intention. The full intention fulfills the empty one only insofar as it presents the object as it has been emptily intended, thereby "satisfying" or "fulfilling" the empty intention. By contrast, the full intention disappoints the empty one when it presents the object as other than emptily intended. To extend our example, I can show you a picture of my wife and this would fulfill, to a degree, your empty intending based on my description. This is why, in order to remind myself of my wife and children, I put pictures of them in my office rather than nameplates from their desks. Moreover, I can later introduce you to my wife, in which case your perception of her fulfills—and most fully fulfills—both the empty intention based on my description and the partially fulfilled intention in seeing her picture.

A person's rational experience is, in this sense, "teleological" and "evidential." Reason is not so much a matter of arguments; rather, reason involves a striving for evidence, where "evidence" is understood as the experience of the agreement between what is emptily meant and what is intuitively given in a fulfilling intention (Husserl 1976, 334). In the case of a sentence expressing a cognitive judgment, for example, evidence is the act in which

I am aware of what Husserl calls the "congruence" (*Deckung*) between the propositional sense of the sentence—the state of affairs as it is proposed to be—and the sense of the perceived state of affairs itself (Husserl 1966, 102; 1974, 128; 1985, 652). In this awareness I recognize the identity of what was emptily (absently) intended in the judgment as expressed in the sentence and what is intuitively present in perceiving the state of affairs. In truthful encounters with things this congruence is present; in non-truthful encounters it is not. In all three rational domains—the cognitive, the axiological, and the practical—the aim, that is, the end, of experiential life is the same: to live the life of intuitive evidence. While the evidential experiences for which reason strives take different forms in cognition and the theoretical sciences, in valuation and the axiological sciences, and in volition and the practical sciences, the task of reason is always to ensure in fulfilled experiences the "truthfulness" of our judgments about what is the case, about what is valuable, and about what is right to do (Husserl 1976, 290). The end of rational agency, then, is truthfully apprehending things and states of affairs, having appropriate affective and evaluative attitudes towards those things and states of affairs, and acting rightly in response to and on the basis of our truthful cognitions and evaluative attitudes (cf. Drummond 2010).

Against this background we can understand the intentionality involved in deliberation and how we might address Reinach's antinomies. We can summarize—very abstractly!—the intentional structures involved in choice as follows:

> C is a choice that issues in action A (i) as having end E and (perhaps) purposes P, (ii) as grounded in the valuation of E and P as good (or apparently good), and (iii) as conducive to realizing E or satisfying P.

While deliberation, whenever it occurs, is most obviously present in choosing actions that respond to and modify the situations in which we find ourselves, it depends upon axiological reason's evaluation of the choiceworthiness of the ends and purposes we pursue in action and of the action itself. Deliberation is present, then, in both the axiological and practical spheres. Insofar as our purposes are measured by the ends proper to human activity and human institutions and practices, our purposes ought to conform to the ends proper to the characteristic activities of persons and of the institutions and practices they establish in order to realize those ends. This means that we must also deliberate about and preferentially order our ends, although the ordering of these ends can change in response to changes in our personal situation or the circumstances in which we are called upon to act. Our purposes, besides conforming or not conforming to appropriate ends, might be in addition to realizing the ends appropriate to our activity. When Mary receives too much change and returns the money, she realizes the end of being honest in the very act of returning the money, and she might explicitly identify this as one of her purposes in so acting. She might, however, also identify as purposes of her action ensuring that the merchant has received her fair share and that the cashier will not be punished for a short change drawer.

Choice, we have said, is grounded in our evaluating as good or, at the least, as apparently good, the ends of our action and our purposes in acting. Evaluation, according to a widely shared phenomenological view, apprehends the valuable in a moment of feeling or an episodic emotion that is founded on a "presentation."[2] I take this to mean that there must be distinguishable layers of sense within the compound sense of the evaluation such that a "presentational" layer—the layer presenting the merely descriptive, non-axiological features of the object—founds additional, affective layers of sense. Our evaluative experiences, in other words, invariably have a cognitive content. The value-attributes of things are the correlates of feelings and episodic emotions that are the affective response of a subject with a particular experiential history—that is, a subject with particular beliefs, emotional states, dispositions, practical interests, commitments, and so forth—to the non-axiological properties of a thing or situation. The value-attributes are neither separate from nor reducible to the non-axiological properties in which they are rooted, but our valuations—precisely insofar as they are grounded on cognitive presentations—track these non-axiological properties. Conversely, the non-axiological properties provide motivating reasons—and, in some cases, justifying reasons—for the valuation accomplished in the affective response. The intentional feeling or episodic emotion experienced by the subject is appropriate—or, to put the matter another way, the evaluation is correct or truthful—when it is both rationally motivated and justified by the non-axiological properties underlying it and when the underlying apprehension of the non-axiological properties is itself veridical (cf. Drummond 2009, 372–376; 2010, 441–450).

While I cannot argue it here, the appropriateness of a particular feeling or episodic emotion in response to a set of non-axiological properties is determined not simply by the non-axiological facts of the matter but by an intersubjective determination of what feelings and emotions fit what non-axiological properties. By way of example, I evaluate a supervisor's shouting at an employee as rude. My evaluation is immediately grounded in my directly witnessing the behavior or hearing about it from someone whose testimony is reliable. The speaking loudly along with facial or bodily features, say, a reddened face or waving arms, rationally motivate and justify my adverse affective response (say, shock or indignation or anger, depending on my relationships to the parties involved). The behavior and accompanying bodily features are reasons for, say, my felt indignation and the negative evaluation of the behavior. In brief, in experiencing the speaking loudly and bodily features of the employer, I immediately and at once recognize the action as rude and disapprove of it. But this perception and evaluation depend on an understanding of what a conversational situation entails, of what behaviors are appropriate and inappropriate in interpersonal communications, and of the concept of "rudeness." This understanding is contested in and shaped by the cultural communities in which I live. Shouting behavior is inconsistent with what the nature of conversation entails; it subverts what I understand the ends of conversation to be and achieves no good purpose in the context I have described. The evaluative moment that sees the shouting as rude is rooted in the underlying cognitive dimension and so thoroughly united with it that it is, as it were, a "matter of fact" (although not in the sense of "fact" favored by proponents of the fact-value distinction; see Foot 2002, 102–105, although this sense of "fact" is not quite what Foot means either, since it is non-naturalistic). Anyone who fails to recognize it as rude is mistaken and suffers from a misconception of what constitutes polite and rude behavior in conversational contexts.

Experiencing and negatively evaluating the action as rude is, in other words, based on the features intrinsic to the behavior itself and to the ordinary expectation we have about the behaviors appropriate to different kinds of human transactions. The general sense of appropriate behavior is established and modified over time in the light of our untutored, affective responses and the education of the attitudes and emotions that occurs within the communities to which we belong.

3. Time in Deliberation

These brief remarks about intentionality and evaluation connect with our consideration of deliberation in several ways: (1) they provide a fixed, albeit formal, end for all human agency, which end constrains deliberation; (2) they provide a basis for understanding the habituation of judgment that constitutes non-occurrent deliberation; and (3) they provide a basis for our sense that acts done without deliberation or with too much deliberation are less meritorious. The striving toward fulfillment points toward an aretaic notion of the good and reveals the normative dimension of self-responsibility. A rational being strives toward fulfilling her passively acquired judgments, beliefs, and emotional attitudes as well as the empty intentions she forms. When she gains the evidence that allows her to affirm or to revise these judgments, beliefs, or attitudes, she adopts them as convictions and thereby takes responsibility for them, realizing herself as a person having a particular set of beliefs for which the appropriate evidence has been secured and which contribute to defining her character. These convictions inform her subsequent judgments, valuations, and choices. The self-responsible agent who can rightly act without occurrent deliberation toward good ends has already weighed competing goods. She has already considered the rightness or wrongness of actions conducing to these ends in a reflective, deliberative, and self-responsible activity accomplished over time in such a way as to dispose her toward a certain kind of action in certain kinds of circumstances. These "habitualities," as Husserl calls them (Husserl 1963, 100–101), make up our dispositions to see certain features in certain kinds of situations, to pick out what is salient in those situations, to have certain kinds of attitudes toward them, and to act in typical ways. This is not to deny that deliberation can be occurrent, but it suggests that even when occurrent, deliberation is built upon this reservoir of previous, justified judgments. It also confirms Cooper's point that in those circumstances where deliberation is non-occurrent, prior deliberation yields the sense of a *specific* action— that is, a certain *kind* of action—as appropriate for the particular circumstances in which the agent finds herself. It is the perceptual grasp of those circumstances that further determines the particular course of action to be followed in order to achieve one's ends.

The virtuous agent is the one who *appropriately* appraises ends and orders her preferences among them, who deliberates *well* about what specific actions conduce to what ends, who *correctly* grasps and assesses situations, and who acts *rightly* in the circumstances. The virtuous agent lives self-responsibly, judging, valuing, and deciding for herself in the light of evidence rather than passively accepting received attitudes and opinions. The end of self-responsibility, while considered separately is formal, is nevertheless attained only in the course of experiences that have a determinate content. Self-responsibility, in other words, is realized only in the pursuit of first-order goods. The self-responsible agent, acting virtuously in the pursuit of true, first-order goods for herself and others, also realizes the second-order goods of thinking well, feeling well, and acting well—what we might call the goods of rational agency (as opposed to the goods for particular agents) that are rooted in the teleological structure of intentionality itself.[3]

Self-responsible thinking is dependent upon the self-responsible thinking of others against whose beliefs our own are tested in the pursuit of evidence for our beliefs. Hence, the end of self-responsibility constrains our pursuit of first-order goods. Not only is it the case that the different first-order ends available to us must be properly evaluated so as to produce true judgments about choiceworthy ends. Not only is it the case that our deliberation about particular ends can occur only when a first-order end is rightly evaluated. Not only is it the case that we must rightly choose the actions in and through which we attain these ends. It is also the case that we cannot rightly choose ends or actions that would deny the possibility of self-responsible thinking to others upon whom our own exercise of self-responsibility depends. While the first-order goods that are the objects of our first-order valuations and volitions are realized in actions and worldly states of affairs, the goods of agency are superveniently realized in the cognitive, evaluative, and practical achievements of subjects whose experiences both truthfully disclose and rightly fashion the world as morally ordered. In brief, the end of getting it right about what is true, good, and right constrains our first-order deliberations and choices.

This description of how deliberation, including its communal aspects, could be prior to the chosen action reveals the first aspect of the ecstatic temporality of deliberation. We largely inherit our evaluative and moral concepts. In the case of the virtuous agent, her evidential confirmation or disconfirmation of the judgments grounded in these concepts, of her evaluations of ends and purposes, and of her judgments concerning the rightness of the actions she undertakes self-responsibly appropriates or rejects these concepts and judgments. In the present moment of action features of the situation evoke and bring into play in the present these concepts and previous judgments. Our understanding of these concepts is further affected by our experience of how others act in situations in which these concepts are relevant to those actions. To recall our earlier example of an action not occurrently deliberated, in seeing the bus pull away from the bus stop Jane might begin to run or to wave her arms in the hopes that the bus driver will see her, will stop the bus, and will wait for her to board. But it is only because she is already aware of the appropriateness and efficacy of such actions that she undertakes them. Practical reason functions in sizing up the specifics of the situation and in determining what to do in this particular situation. Deliberation in the narrow sense of fashioning an argument yielding a conclusion about what to do is not operative. Were Jane to go through a deliberative process, it might already be too late for the actions decided upon to make the bus driver notice her.

The recall of the past at work in such exercises of practical reason is not memory. Memorial experiences explicitly direct my active attention to the past. When I act in the present on the basis of judgments previously made and self-responsibly appropriated, my past judgments inform my present action without my turning attention to them. The recall of the past into the present occurs passively, without any active reflection on my part. Cooper is correct that if someone were to demand from me reasons for acting as I did, I would identify those reasons by articulating the content of the previously formed judgments about what is valuable and about what sort of thing ought to be done in the present circumstances. Since these previous judgments were not originally intended for the present circumstances, my non-occurrent deliberation yields only a sense of the specific kinds of action that might be undertaken in the present. The particular determination is undertaken on the basis of my perception of the current situation. But all that is needed for that to occur is the current perception. That perception is what we might call a moral perception or a practical perception that is informed by my already achieved sense of what ends are desirable in this kind of situation and of what actions are appropriate to achieving them. Without the prior judgments I have no sense of this apart from an occurrent deliberation, but with them my perception is already informed such that I can act without further active deliberation.

The second aspect of the ecstatic temporality of deliberation is its future-orientation. Deliberate action transforms the current situation in which I value a future possibility more than the current actuality. In effective action an end or purpose is realized or satisfied, and there are four, non-exclusive temporal possibilities for the temporal relation between the deliberated action and the realization and satisfaction of its ends or purposes. The first is that the end is realized in the very performance of the action precisely because that end is internal to the activity itself. For example, in the very act of returning the excess change to the cashier, Mary achieves the end of honesty and fairness. The second is that the end is realized in a future that is continuous with the present. This occurs most obviously in the case of actions whose purpose is to bring about some external effect in the world. For example, in waving her arms to attract the attention of the bus driver, Jane's purpose is to get the bus to stop. Or, to take a different example: in placing a wedge in a log and striking the wedge with a sledgehammer, my purpose is to split the log. The log's being split is an external effect of the action of striking the wedge, and it is temporally contiguous with my striking the wedge. Even Mary's act of returning the excess change, to the extent that it has as a purpose that the cashier will not be charged for the shortage in the cash drawer at the end of the day, has this character, since the drawer is no longer short once the money is returned even though the drawer will not be balanced until the cashier goes off shift.

The third and fourth possibilities involve situations where my deliberate intention involves the realization of an end or purpose in the more distant future. We can think of these cases as involving an element of resolve. So, for example, the third possibility is illustrated by an agent's choice to realize an end that can be attained only by repetition of the same action. Eileen practices the piano, for example, in order to become a good piano player. As a young child just beginning to play the piano, Eileen did not realize that end in the very act of practicing. Nor did she realize it as immediately contiguous with any particular instance of practicing. It is only by practicing over a long time that Eileen becomes a good piano player, and it is only by continuing to practice that Eileen can remain a good piano player. The commitment to realizing that end informs and renews the present commitment to practice.

The fourth possibility occurs when I deliberate and choose now to act later. A student, for example, decides as a sophomore that she wants to apply to graduate schools. She will not actually apply until her senior year. Her choice to apply to graduate schools does not immediately produce an application. In this sense, her volitional intention remains empty, not to be fulfilled until she actually applies. Nevertheless, the future she forecasts with this decision affects her present in a variety of ways. She must plan and begin to do those things that will put her in a position to apply to graduate schools. She must, for example, select the right major and the right courses within the major; she must by the time she is a junior begin to consult her instructors in order to gather information about the schools to which it would be best to apply; she must begin to consider which of the papers she has written would provide a good basis for the writing sample she must include in her applications; and so forth. Although her action of applying to graduate schools is in the future at the time she decides to apply, her resolve to undertake that action in the future ecstatically shapes her present.

The third and fourth possibilities reflect the structure that is at work in many of the commitments around which we organize our lives: my commitments as a spouse, parent, philosopher, teacher, citizen, and so forth. For the committed agent, practical reason recalls the effective force of past deliberation and past actions into the present just insofar as that force is relevant to determining the kind of thing I should do here and now. For the committed agent, the present is conceived in relation to the future-oriented commitments that organize her life, among which the priorities she establishes might sometimes shift as the circumstances of her life change. The temporality of deliberation and choice is ecstatic precisely insofar as the present in which I choose and act encompasses the past and future at once.

Finally, we have noted that Reinach's so-called antinomies apply only in cases of occurrent deliberation. And we have given an account of how a prior, non-occurrent deliberation can illuminate what I took to be the falsity of Reinach's third proposition, namely, that a meritorious act is less meritorious when done without deliberation. We have illuminated, in other words, how the braver soldier can be thought to have deliberated and how his virtue is realized through an exercise of practical reason without occurrent deliberation. If that prior deliberation is absent, the action is merely instinctive or impetuous and, as Reinach noted, less meritorious. But when there has been prior deliberation, the absence of occurrent deliberation makes the meritorious action more meritorious. We can also now understand why Reinach's fourth proposition—that a meritorious act is less meritorious when there is too much occurrent deliberation—strikes a chord.

A clear example of how and when such deliberation might lead us to think the action is less meritorious is when the deliberation is motivated by a disconfirming experience. Mary, for example, receives the excess change and notices she has received more than what is due her, but she nevertheless starts to leave the store. Her affective sense—her dissatisfaction with what she is doing—gives her pause. She recognizes that she does not deserve this extra money, that the merchant does, and that the clerk will have to pay the shortage in the cash drawer at the end of the day. Hence, in response to these articulated reasons, she reverses course, returns to the counter, and returns the money. Here again, Mary realizes the virtue of honesty (even if imperfectly) and the purposes of giving the merchant what is due as well as protecting the clerk from the damage of having to pay the difference in the cash drawer. In the fully virtuous act, Mary noted what was to be done in perceiving the facts of the situation. In the second case, Mary did not do this, or if she did, she chose—at least momentarily—the dishonest act, only to have her feelings reveal the disvalue of it. She then reversed course.

The example reveals that the person who must occurrently deliberate at length is not sufficiently attuned to what the choiceworthy ends in a situation are and what specific kinds of action are most conducive to realizing those ends. Axiological and practical reason have not been developed and nurtured in such a way that allows the agent simply to perceive what is to be done here and now.

Notes

1. Francis Slade and Robert Sokolowski discuss the distinction between natural ends and human purposes, but I have extended the distinction beyond natural ends to include the distinction between the ends of institutions and their practices and human purposes; cf. Slade (1997, 83–85; 2000, 58–69) and Sokolowski (2008, 259–263). For the extension beyond natural ends to the ends inherent to institutions and practices, cf. MacIntyre (1981, 175ff.).
2. The view originates in Brentano (1995, 45, 80, 276) and Husserl (1988, 252) adapts it. To say that *B* is founded upon *A* is to say (i) that *B* presupposes *A* as necessary for it and (ii) that *B* builds itself upon *A* so as to form a unity with it.
3. I have elsewhere called these goods of agency "transcendental goods" and "non-manifest goods" (1995, 165–183; 2002, 15–45; 2005, 363–371; 2006, 1–27).

References

Brentano, F. (1995). *Psychology from an empirical standpoint.* A.C. Rancurello, D.B. Terrell, and L. McAlister (Trans.). London: Routledge and Kegan Paul.

Cooper, J. (1986). *Reason and human good in Aristotle.* Indianapolis, IN: Hackett.

Drummond, J. (2009). "Feelings, emotions, and truly perceiving the valuable." *Modern Schoolman* 86, 363–79.

——— (2010). "Self-responsibility and eudaimonia." In C. Ierna, H. Jacobs, and F. Mattens (Eds.), *Philosophy, phenomenology, sciences: Essays in commemoration of Edmund Husserl* (pp. 411–30). Dordrecht: Springer.

Foot, P. (2002). "Moral arguments." In *Virtues and vices.* Oxford: Clarendon Press.

Husserl, E. (1963). *Cartesianische Meditationen und Pariser Vorträge.* S. Strasser (Ed.). The Hague: Martinus Nijhoff.

——— (1966). *Analysen zur passiven Synthesis. Aus Vorlesungs- und Forschungsmanuskripten 1918–1926.* M. Fleischer (Ed.). The Hague: Martinus Nijhoff.

——— (1974). *Formale und transzendentale Logik. Versuch einer Kritik der logischen Vernunft.* P. Janssen (Ed.). The Hague: Martinus Nijhoff.

——— (1976). *Ideen zu einer reinen Phänomenologie und phänomenologischen Philosophie. Erstes Buch: Allgemeine Einführung in die reine Phänomenologie.* K. Schuhmann (Ed.). The Hague: Martinus Nijhoff.

——— (1985). *Logische Untersuchungen. Zweiter Band, zweiter Teil: Untersuchungen zur Phänomenologie und Theorie der Erkenntnis.* U. Panzer (Ed.). The Hague: Martinus Nijhoff.

——— (1988). *Vorlesungen über Ethik und Wertlehre 1908–1914.* U. Melle (Ed.). Dordrecht: Kluwer Academic.

MacIntyre, A. (1981). *After virtue: A study in moral theory.* Notre Dame, IN: University of Notre Dame Press.

Reinach, A. (1989). "Die Überlegung; ihre ethische und rechtliche Bedeutung." In K. Schuhmann and B. Smith (Eds.), *Sämtliche Werke: Textkritische Ausgabe in 2 Bändenn* (vol. 1, pp. 279–311). Munich: Philosophia Verlag.

Slade, F. (1997). "Ends and purposes." In R. Hassing (Ed.), *Final causality in nature and human affairs* (pp. 83–85). Washington, DC: Catholic University of America Press.

——— (2000). "On the ontological priority of ends and its relevance to the narrative arts." In A. Ramos (Ed.), *Beauty, art, and the polis* (pp. 58–69). Washington, DC: Catholic University of America Press.

Sokolowski, R. (2008). *Phenomenology of the human person.* New York: Cambridge University Press.

12 Habituation and First-Person Authority

Jonathan Webber

How do you know your own mind? Are any ways of doing so available only to you? If so, are claims about your mental states arrived at in this way more authoritative than claims made on other grounds? These traditional questions of self-knowledge and first-person authority have been reinvigorated over the past few decades by the idea of the transparency of belief: you can answer some questions of whether you believe that p by considering whether p is true; but nobody else can discover whether you believe that p simply by considering whether p is true. What does this observation about the attribution of beliefs show us about self-knowledge and about the mind in general?

My central claim in this chapter is that the leading recent account of transparency and first-person authority rests on a form of mind-body dualism that should be rejected. Richard Moran has argued that the first-person authority afforded by transparency can be understood only if we think of ourselves primarily as agents. It is not in our capacity as knowers that we find our most significant access to the contents of our own minds, but in our capacity as doers. Our epistemic authority about our own mental states is derivative of our agential authority over them. This is a significant departure from the philosophical framework that has dominated theories of self-knowledge and the mind generally in mainstream Anglophone philosophy for more than a century. That it nevertheless retains a vestige of dualism seems to me emblematic of a deep problem with that framework, but this larger point cannot be defended here. The broad message offered here is that the Aristotelian idea of habituation is well supported in experimental psychology, which requires philosophers of mind to take the temporal aspect of agency more seriously than it is taken by Moran and the accounts of rationality and self-knowledge with which he engages.

My argument concentrates not on knowledge of one's own beliefs, but on knowledge of one's own desires. Moran's focus is on belief, but he considers his account to cover the cases of desire and intention too. The problem posed to Moran's account by habituation is clearest and best supported by empirical psychology in the case of desires. Having established this problem and traced it back to an implicit mind-body dualism in Moran's account,

I will present some evidence that a parallel problem arises for belief states. But this is not essential to my argument. For if we are to understand first-person authority in the context of ourselves as agents rather than simply as knowers, then the problem posed to Moran's account by the reality of desire is enough to require a fresh approach. Before raising these problems, however, I will outline Moran's account of the nature and limits of first-person authority.

1. Moran's Account of First-Person Authority

Moran's central point is that first-person authority is essential to rational agency. As rational agents, he argues, our minds are responsive to our own deliberations. My beliefs are formed by my consideration of the objects of those beliefs. The authority of the declaration "I believe that p" is not grounded in some special access to the prior fact that I believe that p. It is grounded in the deliberation over whether p is true determining my doxastic state with respect to p. There is a single judgment that can be expressed either as "p" or reflectively as "I believe that p," and this judgment ordinarily causes the formation in me of the belief that p. This is why we have this authority only over our mental states, he argues, and not over the mental states of other people or over other aspects of ourselves such as our health (Moran 2001, 32–35). This is also why first-person authority is not merely a useful way of gaining information, one that might one day be replaced by some more accurate technology, but is rather an ability whose functioning is essential to the agent's psychic health (Moran 2001, 89–94, 148–150).

This is not the only form of self-knowledge, according to Moran. One can also come to know of one's mental states through observation of one's words and deeds. But this form of self-knowledge does not entail first-person authority. Anyone with access to the relevant observational information could draw the same conclusions. One typically has more of this information than other observers, because usually one is the only constant witness to one's words and deeds. But this is merely a contingent advantage that could be cancelled out by memory problems, motivations to see oneself in a positive light, or surveillance technology. In principle, this access to one's mental states is no less available to other people than it is to oneself. Moreover, an agent who was entirely reliant on such evidence to find out about their own mental states could not treat their mind as their own to make up. Such an agent could not deliberate or form a judgment, so could not be fully rational. Thus, the transparency of "Do I believe that p?" is not merely a descriptive fact about human beings, but "a normal rational *expectation* we make of them" (Moran 2001, 68).

Moran does allow that local failures of first-person authority occur within the global context of a well-functioning rational agent. In such cases, the form of self-knowledge grounded in observation takes precedence. Moran gives an example from psychoanalysis. Someone "might become thoroughly convinced, both from the constructions of the analyst, as well as from her own appreciation of the evidence" that the attitude that she has been betrayed by a sibling "must indeed be attributed to her," even though "when she reflects on the world-directed question itself, whether she has indeed been betrayed by this person, she may find that the answer is no or can't be settled one way or the other" (Moran 2001, 85). Deliberation over whether p, in this case, does not lead the agent to the conclusion that p. The ordinary agential route to answering the question whether she believes that p, therefore, leads her to conclude that she does not. Yet the observational evidence shows that she does believe that p.

If this reasoning is right, it provides a tight and sophisticated explanation of some traditional claims of psychoanalysis. It explains what it means for a belief to be beyond the reach of ordinarily authoritative self-knowledge. It explains how such a belief can be uncovered through the therapeutic procedure. And it explains why someone might not only be surprised to find that they have the belief, but might sincerely deny having it. They might do so because when they consider whether they believe that p, they quite naturally fall into the usual procedure of considering whether p. In the example Moran gives, the woman's denial of having the belief that her sibling betrayed her is the sincere result of considering whether her sibling betrayed her. This is why she needs to be presented with the evidence that she does have the belief, rather than simply have the conclusion drawn from that evidence announced by her analyst.

What implications does this explanation of psychoanalytic practice have for normal rational functioning? If this agent reasons to the conclusion that she has not been betrayed by her sibling, she is able to correctly report "I believe that I have not been betrayed by my sibling." But this leaves in place, according to Moran's explanation of the case, another belief whose content contradicts that of the belief she has reported. She correctly declares "I believe that not-p," yet continues also to believe that p. For this to be a failure of normal functioning, therefore, requires more than that the judgment that not-p is normally the formation of a belief that not-p. It requires also that in normal functioning this belief displaces the agent's previous belief on the matter. Indeed, it would seem to be a requirement of rationality that coming to believe something means ceasing to hold any directly contradictory belief. What has failed in this psychoanalytic case is not belief-formation itself, but rather the expected rational effect of such belief-formation on the agent's overall doxastic state.

2. Deliberation and the Displacement of Desire

Should we accept that, in normal psychological functioning, the formation of a new belief through deliberation displaces any previous belief that directly contradicts it? One aim of this chapter is to show that there is empirical evidence against this picture of belief-formation and revision, even though it is presupposed by much recent Anglophone philosophy. As an account of normal psychological functioning, this picture is indebted to a dualism of the rational and the arational that current experimental psychology suggests is false. To reach this conclusion, we will first consider the case of desire. Moran does claim that his account of first-person authority covers knowledge of one's own desires as well as knowledge of one's own beliefs, and many of his examples are cases of desire. Despite this, his theoretical analysis remains closely focused on knowledge of one's own beliefs, with only a few pages devoted explicitly to applying the theory to the case of desire.

This is important because the idea that the deliverance of deliberation is a new mental state that displaces any that directly contradict it faces an obvious difficulty in the case of desire. It is quite common for a desire to persist in opposition to the outcome of deliberation. Where we then act on the desire and against our judgment, this is akrasia or weakness of will. But cases where we go on to act on the judgment and against the recalcitrant desire present the same problem for Moran's account. For the problem is the desire's persistence, not its manifestation in action. Moreover, these kinds of case are not restricted to failures of first-person authority uncovered through psychoanalysis. We are often very well aware that we want to do something other than what we judge to be the best thing to do. And our awareness of this does not seem to depend on observation of evidence that would be directly available to another person. This is why such cases are often depicted in terms of an inner struggle with oneself, the outcome of which is often described using the metaphors of weakness and strength.

Moran's response to this difficulty is to distinguish desires that are sensitive to deliberative judgment from those that are not. The former, which he calls "motivated" or "judgment-sensitive" desires, have not necessarily been produced by reasoning. They are categorized together purely by their being responsive to reasoning about their objects. These are the desires that can be straightforwardly displaced or refined by a deliberative judgment. A motivated desire is justified by a set of beliefs about the desired object in such a way that losing that justification should lead to the loss of the desire. "It is the normal expectation of a person," writes Moran, "as well as a rational demand" that "the question of what he actually does desire should be dependent in this way on his assessment of the desire and the grounds he has for it" (2001, 115). His example is a desire to change jobs, which depends for its justification on beliefs about oneself, one's current job, and prospects for other employment. If these beliefs are revised in light of new evidence, then there rationally ought to be, and we usually expect there to be, a corresponding revision of the desire to change jobs.

Moran contrasts these with a judgment-insensitive kind of desire, which he labels "brute desire" (2001, 115, 116). His examples are desires "associated with hunger or sheer fatigue," "desires of hunger and lust," and perhaps also "mere feelings, including such things as the sensation of thirst" (Moran 2001, 114, 116). These are not the result of deliberation and neither are they sensitive to deliberation. "Like an alien intruder, they must simply be responded to, even if one doesn't understand what they're doing there or what the sense of their demands is" (Moran 2001, 114–115). It is not true to say that we have no control over these kinds of desire. Rather, the kind of control that we do have is not a directly rational control. One can produce these kinds of desires "by training, mental discipline, drugs, the cooperation of friends, or simply by hurling himself into a situation that will force a certain response" (Moran 2001, 117). This kind of control, which can also be exercised over judgment-sensitive desires, is not entirely independent of rationality. One might adopt such strategies for good reasons. Nevertheless, the operation of the strategy itself would not be a rational operation.

This is a neat response to the problem posed to his theory of first-person authority by the familiar phenomenon of recalcitrant desires. For not only would it isolate the problem within a particular category of desires, which Moran considers to play only a minor role in overall behavior (2001, 116). It would also explain why these recalcitrant desires do not exemplify the kind of failure of first-person authority that can be identified only with the help of an impartial observer of the details of one's behavior, such as a psychoanalyst. Given the kind of mental state a belief is, on Moran's view, the failure of a belief to be displaced by deliberation resulting in a directly contradictory judgment is a failure of normal rational functioning. But it is not constitutive of a "brute desire" that it should be sensitive to deliberation, so this is beyond the boundaries of normal first-person authority rather than a malfunction. Moran does not address the question of how we know about these desires, given that our experience is not generally one of inferring them from our behavior. But there seems no obvious reason why this could not be answered consistently with Moran's theory of our knowledge of our own judgment-sensitive mental states (Webber 2016b, §§ 2, 5).

3. Habituation of Deliberative Desire

What would pose a significant problem, however, would be good reason to reject the identification of normally recalcitrant desires with "brute" bodily desires. If normal psychological functioning renders desires formed through deliberation resilient in the face of contradictory judgment, then resilience would not be restricted to cases of "brute" desire and malfunctions to be diagnosed by psychoanalysis. Our own experience does provide us with good reason to think that a desire arrived at by deliberation can become habituated through ordinary functioning to such a degree that the kind of simple rational revision that Moran's theory requires is no longer possible. Moreover, as we will see in the next section, this picture of habituation is supported by the leading model of the consistencies and variations in an individual's behavioral cognition.

Habituation of a mental state has been understood at least since Aristotle, primarily in terms of its repeated employment in reasoning resulting in action (*NE*, 1105b9–18). Moran is right to say that if one learns new information about the development of one's current job or one's prospects for other employment, then a recently formed desire to change jobs will be rationally sensitive to this new information unless one's cognition is not functioning as it should. But compare this to the desire to pursue a particular highly competitive career that requires significant qualifications. One may have decided many years ago that one wants to be an academic philosopher, for example, and employed this desire regularly in making many large and small decisions, including ones concerning a range of financial and personal risks and sacrifices over a sustained period of time. If one then learns, towards the end of this time, that one is unlikely to settle into a stable academic job for many more years after doctoral graduation than one had realized, or that academic life involves far more bureaucracy than one had imagined, then these considerations might lead one to decide that a new direction would be better. But this decision is unlikely to simply displace the old desire. If one does take a new career path, this is likely to be haunted for some time by the old desire. Or one might still pursue the original career despite one's reflection that this no longer seems so wise, given one's other life goals.

Might this case be understood in terms of the distinction between the object desired and the aspects of that object that make it desirable? Could we say that the desire for the academic career as a whole is rationally revised by the judgment, even though the research and teaching continue to be found just as desirable? This move does not capture all the ways that this kind of case might develop. Someone who gives up on an academic career for these reasons might nevertheless later experience envy at the career of a friend from graduate student days who has continued into academia. This would not require forgetting about or reevaluating the original reasons for changing career. The envy might be accompanied by the judgment that leaving academia was the right decision. It is simply that this judgment does not nullify the desire for the academic career. Conversely, in the case where one continues with the original career path, it may be the case that the new discoveries about the profession would have deterred one a few years earlier without there being any change over that time in the degree to which one enjoys or values teaching and research or in the way one would evaluate these new discoveries. All that needs to have changed is the degree to which the goal of an academic career, or indeed of the goals of research and teaching, have become embedded in one's outlook. The more ingrained such a goal is, the greater the force a countervailing consideration would require to remove it.

It is not only in matters of career that a desire can feature pervasively in one's practical reasoning for many years. This strong habituation is perhaps more common in personal relationships. But these kinds of case should not simply be considered as isolated potential counterexamples to Moran's thesis. For the way that they operate indicates an aspect of all desire that presents a deep problem for the idea that in normal functioning a mental state is displaced by a directly contradictory deliberative judgment. These cases exemplify the power of habituation in an extreme and hence easily noticeable way. Since habituation is a matter of degree, however, we should expect this resilience in the face of contradictory judgment to be a matter of degree too. What is more, we seem to be aware of these kinds of desires in a way that does not rely on observation of our own behavior. This kind of self-knowledge, therefore, is not either of the two kinds that Moran describes. The threat this poses to Moran's deliberative account of first-person authority is that all of the work his theory is designed to do might already be done by the correct account of knowledge of one's own habituated desires. Before considering the evidence from empirical psychology that this is indeed the case, it is worth seeing what has led Moran to an account that faces this problem.

A form of mind-body dualism lies at the heart of Moran's strategy for responding to the problem that desire poses for his theory that first-person epistemic authority rests on the power of deliberation to set the mental state that its conclusion announces. Moran divides desires into a pair of mutually exclusive and collectively exhaustive categories. Those resistant to rational revision are "brute" desires identified with bodily needs. The others are categorized together with beliefs as directly and immediately responsive to rational deliberation. This matches the Cartesian idea of the mind as the realm of thoughts held together by rational relations between their informational contents, distinct from the bodily realm that lies beyond its rational reach. Moran is not committed to the metaphysical claim that these realms inhere in distinct kinds of substance, but he is committed to the dualism of the rational and the arational that led Descartes to that metaphysical claim. Moran even implicitly endorses the Cartesian identification of this rational system with the self. He claims that "brute" desires "simply assail us with their force" and are "mere happenings to which the person is passively subject" (Moran 2001, 116). He likens them to intruders coming in from outside and describes strategies for manipulating them as "external means" (Moran 2001, 114–115, 117). However, we should not identify the self, or the agent, with one half of this purported dichotomy between the rational mind and the arational body. For the existence of mental states that have been formed and habituated through reasoning but which, having become habituated, are no longer immediately responsive to reason shows that this is a false dichotomy.

4. Evaluative Attitudes in the Personality System

Habituation is central to the "cognitive-affective system theory" of personality. This was developed as a model of the psychological processing that underlies each individual's pattern of behavior. It is intended to explain why the individual's behavior in response to a particular feature of their situation will vary with changes in some background features of the situation but not others, or might in some cases be invariant across all such contextual changes. As such, it is supported by a very substantial body of empirical research into this aspect of behavior (Mischel and Shoda 1995). The essence of the theory is that this cognitive and affective processing should be modeled as a connectionist system. It is not only the set of beliefs and desires that make up the system that matters, but the associative connections between them and the relative strengths of those connections. We should picture the processing as a flow of activity through the system, where the activation of one mental state causes the activation of those associated with it to a degree determined by the strength of that connection. Each associative connection is strengthened each time this activity flows along it. This is the role of habituation: repeated use of the same associative connection between two mental states increases the strength of that connection, so increases the proportion of mental activity that flows along it.

This theory is not a complete account of the production of behavior, but rather a framework for conceptualizing further research into the psychology of individual behavior (Shoda and Mischel 1996, 415). One area of empirical research that lends itself particularly well to this framework is attitude psychology, of which cognitive dissonance theory is the most famous strand. This tradition of social psychology has converged on a conception of an attitude as a cluster of cognitive and affective mental states that together make up an individual's overall evaluation of some object. For example, someone might have an overall attitude of approval towards democracy, made up of the belief that democracy is the best way to keep the peace, the desire that peace be kept, the belief that current Western models of democracy tend to place too much power in the hands of political parties, and so on. Attitude psychology has also converged on the idea that attitudes have strength as well as content. This strength is not the degree of approval or disapproval, which is included in the content. It is rather the degree of influence the attitude has over the agent's cognitive and affective processing.[1] If the constituents of an attitude are seen as elements in the cognitive-affective personality system, then the attitude's strength is given by the number and average strength of the associative connections between them.

Conceptualizing attitude psychology this way makes clear why attitude strength correlates with consistency of judgment and action across situations. This is well illustrated by an experiment in which people were asked for their attitude towards Greenpeace and asked some questions designed to measure the strength of that attitude, then a week later given the opportunity to donate to Greenpeace and then asked again about their attitude towards Greenpeace (Holland et al. 2002). The attitude content reported at the start of the experiment predicted whether the individual later donated to Greenpeace only where the attitude was strongly held. In these cases, the second report of the attitude generally matched the first. Moreover, where the original attitude was weakly held, the second attitude report was in line with the response to the opportunity to donate even though this did not correlate with the attitude content reported earlier. The experimenters conclude that weak attitudes are subject to situational variation because they are constructed at the time out of whatever relevant beliefs and desires come to mind most easily. If one has just had the opportunity to donate to Greenpeace, then one's knowledge of whether one donated or did not donate is highly accessible and therefore strongly influences one's attitude. But stronger attitudes, comprising a set of strongly interconnected elements in the connectionist personality system, are robust mental states that exert the same significant influence on judgment and behavior across varying situations.

Where an attitude has been significantly habituated, therefore, it exerts a general pressure towards some outcome in judgment and behavior. It is, in short, something that philosophers would classify broadly as a desire. But because it is constituted by a complex set of strong associative connections between its constituent mental states, it can be revised or replaced only through progressively weakening those associative connections or strengthening other ones. It will not simply be displaced by the contradictory deliverance of an episode of deliberation. Moreover, the same theory explains why the same strong attitude has less influence over deliberation than over immediate judgment and automatic behavioral cognition. This is essentially because deliberation is a slower process. The stronger an attitude is, the more accessible it is, where this is measured in terms of the speed with which it is brought to bear on any given episode of processing. When the processing itself is completed in a short time frame, only the most accessible attitudes and other mental states will have any influence. But when the processing takes much longer, many more considerations can be drawn upon. This is how the deliberation that fails to displace a strong attitude can have reached a conclusion contrary to that attitude in the first place.

Thus, attitude psychology and the personality system provide a clear explanation of how desires become progressively less susceptible to immediate revision or displacement by deliberative judgment. The weakest attitudes are not susceptible to such displacement either, for they are not persisting states at all. Instead, we should say that the more an attitude becomes a persisting state, the more it becomes a stable part of the individual's cognitive system, the more resilient it becomes. Habituated attitudes can be revised or displaced, that is, but doing so requires habituation. Therefore, one cannot simply determine one's desire concerning an object by deliberating about that object in the way Moran describes. The outcome of that deliberation will not immediately become a persisting attitude. It may be at odds with an existing attitude, and if so will not immediately displace it. This is not due to any malfunction. It is rather a feature of our cognitive and affective system that it precludes immediate formation of stable attitudes and produces resistance of stable attitudes to immediate change.

5. One Kind of Implicit Bias as Habituated Belief

Is this habituation paralleled in the case of belief? This issue faces a procedural difficulty. For it is part of the usual philosophical understanding of belief that it should be immediately responsive to rational deliberation. A mental state that is not judgment-sensitive in this way is not likely to be classified by philosophers as a belief. Yet it does seem legitimate to raise this question. My strategy is to provide evidence from research into implicit bias that a mental state can behave like a paradigm of belief in all respects except sensitivity to deliberative judgment. The evidence suggests that this resilience is arrived at through erosion of the mental state's judgment-sensitivity through repeated activation of the mental state. This supports the idea that habituation can render a belief beyond direct deliberative control.

The term "implicit bias" currently covers disparate phenomena, so we should not assume that any significant general account of implicit bias could be provided (Holroyd and Sweetman forthcoming). One phenomenon often placed in this category is the tendency to associate young black men with handguns. This tendency might have been a factor in the death of Mark Duggan, who was killed by an armed police officer in north London in 2011. During the inquest, the officer who shot him claimed that Duggan had been holding a gun at the time and described this scene in detail. Another witness, however, claimed that Duggan had been holding a phone. No gun was found in his possession after he was shot. A gun was found some distance away, but was wrapped in a sock. The inquest jury found that Duggan had been unarmed when he was shot. So how should the police officer's testimony be explained?

A wealth of evidence supports the idea that high-speed decisions about whether someone is armed are biased by whether the person is black. After briefly seeing a black face rather than a white face, people more quickly identify guns as guns and when working at high speed more frequently misidentify other objects as guns (Payne 2001). When playing a video game in which one has to shoot all and only the men carrying guns, people making decisions at high speed shoot black men more frequently than white men, whether armed or unarmed. When playing more slowly, people tend to shoot all and only the armed men irrespective of race, but make their decision to shoot an armed man more quickly when he is black and make their decision not to shoot an unarmed man more quickly when he is white (Correll et al. 2002). All of this suggests that people strongly associate black men with handguns in a way that influences the outcome of object identification and decision-making processes executed at high speed, but does not influence their outcomes at low speed (Payne 2006).

Might the armed officer who shot Duggan have misidentified a phone as a gun due to this kind of association? He did have to identify the object and make his decision at high speed. Duggan had been stopped because police had information that he was carrying a gun, and this bias can be exacerbated by recent exposure to information linking black men with handguns (Correll et al. 2007b). On the other hand, police officers have been found to exhibit this bias only in the time it takes to make the decision whether or not to shoot, differing from the overall population in generally not exhibiting a bias in the content of the decision itself. That this finding can be replicated in undergraduate students by training them to recognize whether someone is holding a gun suggests that police officer training makes a difference here (Correll et al. 2007a).

Irrespective of whether this bias was in fact involved in the Duggan shooting, there remains the issue of whether we should think of the cognitive association between black men and handguns as a belief that black men often carry handguns. One recent influential philosophical discussion of this kind of implicit bias argues that we should not. Instead, according to Tamar Gendler, we should think of this association as an "alief," an arational state that can be had by other animals (2008, 557, 574). Why should we agree that the mental state is arational? It does rationalize its cognitive and behavioral effects. If it were true generally that a black man is likely to be carrying a handgun, then that would rationally support the expectation of a handgun that influences perception and decision in each case. Gendler denies this rational relation on the grounds that the agent would deny having the requisite belief (2008, 565). But this shows only that if the state is a belief that rationalizes its effects, then it might be one that the agent is unaware of having.

Gendler's central reason for denying that the mental state is a belief is that "belief aims to 'track truth' in the sense that belief is subject to immediate revision in the face of changes in our all-things-considered evidence" (2008, 585). Why should we accept this denial of habituated belief? Gendler's reason is that belief has to be rationally sensitive in this way "if it is to bear the relation to knowledge and rationality that philosophers require of it" (2008, 563). But this argument can be reversed: if the empirical evidence shows that belief can be habituated in a way that prevents immediate displacement by directly contradictory deliberative judgment, then any philosophical theory that rests on the denial of such habituated belief will need to be rejected.

Not only does the mental state underlying this form of implicit bias rationalize its effects, but it also seems to track experience rationally. Police officers working in communities with both a high crime rate and a high proportion of black residents showed a particularly strong bias in the time it takes to decide whether or not to shoot a character in the video game (Correll et al. 2007a, 1021). In the general population, the bias can be magnified temporarily by increasing the proportion of black characters in the video game who are carrying guns (Correll et al. 2007b, Study 2). Moreover, the bias is strongly correlated with knowledge that the cultural stereotype of black men associates them with violence, irrespective of whether the individual endorses that stereotype (Correll et al. 2002). Since this cultural stereotype is propagated through media imagery across news stories, fictional stories, and music lyrics and videos, knowledge of this stereotype is likely to be due to exposure to this imagery.

The mental state underlying this bias, therefore, shares with paradigmatic cases of belief both its being rationally supported by the individual's experience, however unrepresentative of reality that experience may be, and it in turn rationalizing its psychological and behavioral effects. It is different only in not being immediately sensitive to deliberative judgment that contradicts it. This suggests that receptivity to information that contradicts one's overall experience is limited, since beliefs formed on the basis of that experience are likely to be deeply ingrained. Although it would be more rational to discount the experience of media that one knows to significantly distort reality, it seems that normal psychological functioning precludes the influence of this distortion being counteracted easily. This would explain why people who repudiate the cultural stereotype associating black men with handguns still manifest the bias rationalized by that stereotype. Thus, rather than classify the mental state underlying this bias as an alief, we should consider it a belief that is not revised immediately by any judgment that contradicts it. Since it has been formed through repeated exposure to the stereotype, this recalcitrance seems due to habituation.

6. Knowing One's Own Habituated Beliefs

This analysis of one form of implicit bias as rooted in habituated belief leaves open the question of whether this habituation operates through practical reasoning, as the Aristotelian position suggests, or whether repeated exposure to the cultural stereotype is sufficient. If it does require reasoning, this need not be explicit deliberation, for the background processing of information required for following a news story or fictional narrative is also a form of practical reasoning. But even if this kind of habituated belief comes about through mere exposure to the stereotype, this would at most show that the Aristotelian picture is incomplete, and that reasoning is not necessary for habituation. The same question arises for the case of desire. But consideration of the experimental research into the effect of mere exposure on attitudes and other cognitive states must await another occasion.

Our primary concern here is with the implications of habituated belief for self-knowledge and first-person authority. The common cognitive association of black men with handguns is routinely described in philosophical and psychological literature as beyond ordinary self-knowledge, often as "unconscious" or "not available to introspection." The term "implicit" has come to be used in this sense, even though it originally labeled only a style of measuring mental states. (The phrase "implicit association test" denotes an implicit test of associations, not a test of implicit associations.) This classification generally occurs without serious consideration of what it means. It is usually motivated by the fact that the subjects explicitly stated that they do not think that black men are strongly associated with handguns. But if these statements represent deliberative judgments about the relation between black men and handguns, then we should not be so quick to assume that they express ordinary self-knowledge of the speaker's belief states. For the idea that ordinary knowledge of one's own beliefs is grounded in deliberative judgment about the objects of those beliefs assumes that beliefs do not become habituated as they become stable mental states, which we can now see is not a safe assumption to make.

Might there be another way in which deliberation provides knowledge of one's own habituated beliefs? To deliberate requires one to draw on relevant information. There is evidence that a belief that has been habituated through repetition of some claim will be more easily and rapidly accessible to deliberation than one that has not. Even when the subject has explicit reason not to believe that the claim is correct, this habituation leads to the automatic retrieval and application of the claim, which can be counteracted only through deliberation drawing on the reason not to believe the information (Begg et al. 1992). A belief habituated through exposure to distorted media could therefore be brought automatically to deliberative cognition even though it might then be deliberatively defeated by the knowledge that the media is distorted in this respect. If this is right, then one could know one's own habituated beliefs through their regularly and rapidly coming to mind in deliberation. But this would not entail that all our habituated beliefs can be known in this way. For it would leave open that a habituated belief might be counteracted by another habituated belief in the automatic cognition that subserves deliberation, such that its content never attains the status of being a consideration in deliberation.

This is a speculative suggestion, full consideration of which would require further investigation of the vexed issue of the role of habituated belief in the relation between automatic and deliberative cognition (Thompson 2009). But it does seem possible that belief might mirror desire in that the more habituated it is, the more stable it is and the more resistant it is to being changed by a contradictory judgment. If this is right, then it undermines the idea that first-person epistemic authority over belief consists in the agential authority to form beliefs. For not only would it show that we lack such authority over strongly habituated beliefs, but it would also show that deliberative judgment does not in itself produce persisting beliefs at all. In the absence of habituation, a judgment just reflects the considerations that come to mind in that deliberative process. The considerations that come to mind soonest and most easily would generally be the most strongly habituated, but on any given occasion these would be joined by any considerations that have very recently been thought about. Moreover, the set of considerations that come to mind will depend on the length of time devoted to the deliberation. One would be likely, therefore, to reach different judgments on different occasions. To report a deliberative conclusion would not be to report a persisting belief.

7. The Temporal Dimension of Rational Agency

Moran's theory of first-person authority rests on a dualism of the rational and the arational that does not allow for habituation. We have good intuitive and empirical reason to accept that habituation is central to the normal psychological functioning of desire. There is some empirical support for the idea that habituation plays a parallel role in belief. If there is to be genuine first-person epistemic authority over persisting mental states, therefore, an alternative account to Moran's is required for desire and perhaps also for belief. Ought such an account respect the idea of transparency? To reach a conclusion by deliberation is not in itself to form a persisting mental state, and neither does the conclusion displace any contradictory mental state as a matter of normal psychological functioning. The intuitive appeal of the idea of transparency, moreover, can be explained by its relation to transient beliefs and desires. We should not assume, therefore, that first-person authority over persisting mental states has anything to do with transparency.

The recognition of habituation does not entail a wholesale rejection of Gendler's conception of alief. It requires only that a mental state not immediately sensitive to rational judgment is not thereby entirely outside the realm of reason. We should reject, therefore, Gendler's dualistic opposition of rational and arational, which she maps onto the difference between humans and other animals. This dualism cannot accept mental states that are rational in their formation and effects, but which resist immediate rational revision. Indeed, Gendler herself does not consistently respect her dualism, since she describes one kind of alief as rationally sensitive to statistical evidence in sophisticated ways not often matched by deliberation (2011, 33–36, 54–57). We should instead accept that rationality of a mental state is a matter of degree. We should likewise accept that an animal whose mental states are sensitive to the environment in statistically sophisticated ways is more rational than one whose mental states are not, irrespective of whether that first creature is also capable of deliberative judgment.

Recognizing the role of habituation in our psychology clarifies the sense in which we are rational animals, or imperfectly rational agents. It is not that we are rational angels unfortunately yoked to mortal bodies that assail us with their demands. Neither is it that our rational systems are overlain on the associative alief mentation of our animal bodies. It is rather that our form of rationality inherently involves habituation that limits our deliberative control over our desires, and perhaps also our beliefs. This operates through repeated employment of a mental state in reasoning, whether as a conclusion or as a premise. Moreover, it is rational for a cognitive system of finite capacity to rely on habituation in this way. Progressively embedding a claim in the cognitive system as it is repeatedly employed obviates the need to reconsider one's commitment to it each time it is deployed. It does so in a way that is somewhat sensitive to the claim's degree of rational support without recording that support itself in an accessible format. This is an efficient design.

Rejection of the dualism of the wholly rational and the wholly arational should therefore be accompanied by a recognition that our form of rationality is essentially temporal. Our form of rationality draws on habituated mental states, whose influence over cognition and resistance to immediate change are determined by their degree of employment in past rational processes. This temporal aspect of rational agency needs to be borne in mind when considering how one might aim to remove some habituated belief or desire, or at least prevent its manifestation in action. Manipulating one's environment in order to counteract the distortions of media representation, such as by putting pictures of counter-stereotypical individuals in prominent places, should not be thought of as a merely causal and arational way of changing one's mind, since it operates through the chronic rational sensitivity of habituation. Conversely, the possibility of changing one's habituated mind through deliberative means, such as discussing reasons for decisions or making decisions sufficiently slowly to regularly employ less accessible beliefs and desires, should not be dismissed lightly.

Philosophical and empirical consideration of the extent of our deliberative control over our own minds should be explicitly framed by this point about the essentially temporal nature of our rationality, as indeed should consideration of all forms of agential control over the contents of minds. Likewise, further discussion of our epistemic access to our own minds and those of other people, in relation to our deliberative capacities and more generally, should keep this temporal dimension of human rationality sharply in focus. We are not simply abstract reasoners. We are essentially temporal rational agents. We should keep reminding ourselves of that until the time comes when its significance is automatically taken into account.[2]

Notes

1. For more detailed explanations of attitude psychology, in relation to the philosophical idea of ethical virtue, see Webber 2015 and 2016a.
2. This chapter was developed through presentations at Cardiff University work-in-progress seminar, Kings College London philosophy society, the visiting speaker seminar at Manchester Metropolitan University, and the 2013 conference of the Nordic Society for Phenomenology at University of Copenhagen, and through participation in the Implicit Bias Project at University of Sheffield. I am grateful to the organizers and participants of these for helping to shape my thoughts on this issue. I am also very grateful to Roman Altshuler, Jules Holroyd, and Michael Sigrist for their thoughtful responses to the first draft.

References

Aristotle. 2002. *Nicomachean Ethics*. Translated by Christopher Rowe. Introduction by Sarah Broadie. Oxford: Oxford University Press.

Begg, Ian Maynard, Ann Anas, and Suzanne Farinacci. 1992. Dissociation of Processes in Belief: Source Recollection, Statement Familiarity, and the Illusion of Truth. *Journal of Experimental Psychology: General* 121: 446–458.

Correll, Joshua, Bernadette Park, Charles Judd, and Bernd Wittenbrink. 2002. The Police Officer's Dilemma: Using Ethnicity to Disambiguate Potentially Threatening Individuals. *Journal of Personality and Social Psychology* 83: 1314–1329.

Correll, Joshua, Bernadette Park, Charles Judd, Bernd Wittenbrink, and Melody Sadler. 2007a. Across the Thin Blue Line: Police Officers and Racial Bias in the Decision to Shoot. *Journal of Personality and Social Psychology* 92: 1006–1023.

Correll, Joshua, Bernadette Park, Charles Judd, and Bernd Wittenbrink. 2007b. The Influence of Stereotypes on Decisions to Shoot. *European Journal of Social Psychology* 37: 1102–1117.

Gendler, Tamar Szabó. 2008. Alief in Action (and Reaction). *Mind and Language* 23: 552–585.

——— 2011. On the Epistemic Costs of Implicit Bias. *Philosophical Studies* 156: 33–63.

Holland, Rob W., Bas Verplanken, and Ad van Knippenberg. 2002. On the Nature of Attitude-Behavior Relations: The Strong Guide, The Weak Follow. *European Journal of Social Psychology* 32: 869–876.

Holroyd, Jules, and Joseph Sweetman. forthcoming. The Heterogeneity of Implicit Bias. In *Implicit Bias and Philosophy*, edited by Michael Brownstein and Jennifer Saul. Oxford: Oxford University Press.

Mischel, Walter and Yuichi Shoda. 1995. A Cognitive-Affective System Theory of Personality: Reconceptualizing Situations, Dispositions, Dynamics, and Invariance in Personality Structure. *Psychological Review* 102: 246–268.

Moran, Richard. 2001. *Authority and Estrangement: An Essay on Self-Knowledge.* Princeton, NJ: Princeton University Press.

Payne, B. Keith. 2001. Prejudice and Perception: The Role of Automatic and Controlled Processes in Misperceiving a Weapon. *Journal of Personality Social Psychology* 81: 181–192.

——— 2006. Weapon Bias: Split-Second Decisions and Unintended Stereotyping. *Current Directions in Psychological Science* 15: 287–291.

Shoda, Yuichi and Walter Mischel. 1996. Toward a Unified, Intra Individual Dynamic Conception of Personality. *Journal of Research in Personality* 30: 414–428.

Thompson, Valerie A. 2009. Dual-Process Theories: A Metacognitive Perspective. In *Two Minds: Dual Process Theories and Beyond*, edited by Jonathan Evans and Keith Frankish. Oxford: Oxford University Press.

Webber, Jonathan. 2015. Character, Attitude and Disposition. *European Journal of Philosophy* 23(4): 1082–1096. doi:10.1111/ejop.12028.

——— 2016a. Instilling Virtue. In *From Personality to Virtue: Essays in the Ethics and Psychology of Character*, edited by Alberto Masala and Jonathan Webber. Oxford: Oxford University Press.

——— 2016b. Knowing One's Own Desires. In *Philosophy of Mind and Phenomenology: Conceptual and Empirical Approaches*, edited by Daniel Dahlstrom, Andreas Elpidorou, and Walter Hopp. New York: Routledge.

13 Timing Is Not Everything

The Intrinsic Temporality of Action[1]

Shaun Gallagher

Developmental science traditionally held that postural schemas (understood as mechanisms of motor control) are absent at birth and that their development depends on prolonged experience. It takes time to gain control over one's movement. Infants younger than three months of age seemingly lack proper coordination. Their arms and legs seem not to be in sync, and this may have something to do with adjusting to their newfound gravity (Hopkins and Prechtl 1984; Prechtl and Hopkins 1986). This may be a matter of degree, however. There is more organization in the movements of young infants than the casual glance reveals. Video studies show that close to one-third of all arm movements resulting in contact with any part of the head lead to contact with the mouth directly (14%), or following contact with other parts of the face (18%) (Butterworth and Hopkins 1988; Lew and Butterworth 1995). Moreover, a significant percentage of such movements are associated with an open or opening mouth posture. The mouth *anticipates* arrival of the hand.[2] This should not be a surprise. We've known for a long time that just such temporal organization generally characterizes animal movement. And for the human, as John Gibbon put it, "Timing is everything: in making shots, in making love, in making dinner" (Gibbon and Malapani 2001, 305; see Malapani and Fairhurst 2002).

Timing is not *everything*, however. Timing is something that we can see and measure. Yet timing can be accidental or merely coincidental. When we find consistency in timing, as in the fact that the mouth almost always anticipates the hand, this suggests deeper temporal processes involved in bodily systems capable of such timing. In other words, it is not just a matter of the system carrying or processing temporal information; rather, the important thing is that the system is capable of organizing itself, its processes, and its behaviors in a temporal fashion. In this regard we should distinguish between timing and an intrinsic temporality, or capacity for temporalizing, which characterizes the system.

Human action and conscious experience are characterized by an intrinsic temporality that involves both anticipatory processes and retentional processes. To make sense out of its experience, and to have practical effect in its action, for example, the human animal (like many other kinds of animals) needs to keep track of how previous movement has brought it to its current state. This is especially true if the movement is intentional, and if a conscious sense of movement is generated.

This intrinsic temporality is not objective time that can be measured by a clock, although action certainly does take place in time, and it may be important in various contexts that its duration can be measured. Intrinsic temporality can be found in the dynamics of bodily movement and action, and manifests itself at both the sub-personal and the personal levels of analysis. Phenomenologists distinguish objective time from lived time (e.g., Husserl 1991; Merleau-Ponty 1962; Straus 1966). The latter is time as we experience it passing, sometimes seeming to pass slowly and sometimes rapidly. Intrinsic temporality includes more than lived or phenomenological time; it includes a temporal structuring that shapes action and experience, but might not be experienced as such.

1. The Dynamics of Intrinsic Temporality in Movement

In regard to action and motor control, this intrinsic temporality is expressed in Henry Head's definition of the body schema. According to Head, the body schema dynamically organizes sensorimotor feedback such that the final sensation of position is "charged with a relation to something that has happened before" (Head 1920, 606). Merleau-Ponty, borrowing Head's metaphor of a taximeter, suggests that movement is organized according to the "time of the body, taximeter time of the corporeal schema" (1968, 173). Body schematic processes incorporate past moments into the present:

> At each successive instant of a movement, the preceding instant is not lost sight of. It is, as it were, dovetailed into the present. . . . [Movement draws] together, on the basis of one's present position, the succession of previous positions, which envelop each other.
>
> (Merleau-Ponty 1962, 140)

Such retentional aspects of movement are integrated with the anticipatory or prospective aspects already noted in hand-mouth coordination in infants. Indeed, anticipatory or prospective processes are pervasive in low-level sensorimotor actions. Similar to the mouth's anticipation of the hand, when I reach down to the floor to grab something, my body doesn't go off-balance and fall over; it anticipates that possibility and angles backward in order to adjust its center of gravity as I bend forward (Babinski 1899). Our gaze anticipates the rotation of our body when we turn a corner (Berthoz 2000), and more generally, visual tracking involves moment-to-moment anticipations concerning the trajectory of the target. On various models of motor control a copy of the efferent motor command (efference copy) anticipates the consequences of the action prior to sensory feedback, allowing for fast corrections of movement (Georgieff and Jeannerod 1998). Reaching for an object, for example, involves feed-forward components that allow last-minute adjustments if the object is moved, and the grasp of my reaching hand tacitly anticipates the shape of the object to be grasped. This is not blind automaticity since the grasp is shaped according to the specific intentional action involved (see Jeannerod 2001; MacKay 1966; Wolpert et al. 1995). Berthoz (2000, 25) suggests that anticipation is "an essential characteristic" of motor functioning. Similar anticipations characterize the sensory aspects of perception (see Wilson and Knoblich 2005 for review). Since these prospective processes are present pervasively, even in infants, the "conclusion that [anticipatory processes] are immanent in virtually everything we think or do seems inescapable" (Haith 1993, 237).

A full intrinsic temporality of retention and anticipation helps to structure movement and action. Berthoz suggests that the Husserlian analysis of the retentional-protentional structure of experience is a model that also works for the processes involved in motor control (Berthoz 2000, 16; Husserl 1991).

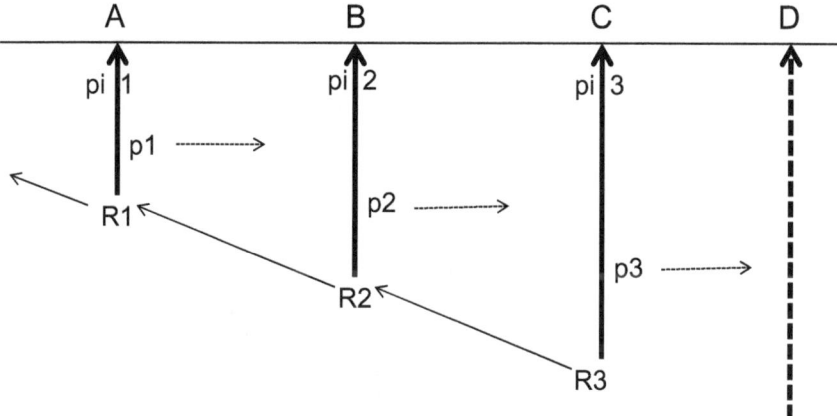

Figure 13.1 Husserl's model of time-consciousness

Figure 13.1 is a diagram of Husserl's model, which he applies to the conscious perception of a melody. As I have argued, it equally applies to action (Gallagher 2011). The horizontal line ABCD represents a temporal object such as a melody of several notes. The vertical lines represent abstract momentary phases of an enduring act of consciousness. Each phase is structured by three functions:

- *Primal impression* (pi), allowing for the consciousness of an object (a musical note, for example) that is simultaneous with the current phase of consciousness;
- *Retention* (R), which retains previous phases of consciousness and their intentional content;
- *Protention* (p), which anticipates experience that is just about to happen.

In the current phase, simultaneous with C, there is a retentioning (R3) of the previous phase of experience, and this just-past phase includes its own retentioning of the prior phase. This means that there is a retentional continuum—R3(R2[R1]), and so forth—stretching back over recent prior experience, on the order of seconds. There is also a double intentionality to this retentional aspect. Retention retains the prior phases of consciousness (longitudinal intentionality), but since those phases include primal impressions of the then current notes (B and A, respectively), retention also retains the prior notes of the melody (transverse intentionality), in the sequential order in which we heard them, which generally reflects the order in which they were sounded. Imagine if that were not the case. If there were no retention of previous notes, then we would not hear a melody. Indeed, if our experience were of only one moment at a time without experiential connection to previous moments, it would be impossible to make sense of the world.

Protention, in turn, provides consciousness with an intentional sense that something more will happen. This protentional aspect allows for the experience of surprise. In listening to a familiar melody, there is some sense of what is to come, a primal expectation of the notes to follow. If someone hits the wrong note I am accordingly surprised or disappointed, in the same way that if a person fails to complete a sentence, I experience a sense of incompleteness as my protention is unfulfilled, or what I experience fails to match my anticipation. The phenomenon of "representational momentum," where movement or implied movement results in the extrapolation of a trajectory that goes beyond what was actually perceived, also provides evidence of protention (Wilson and Knoblich 2005). The content of protention is never completely determinate since the future itself is indeterminate. Indeed, in some cases the content of protention may approach the most general sense of "something (without specification) has to happen next."

I've argued that this model, or its broad structural features, can be extended to non-conscious motor processes as well, and is reflected in Head's description of the retentional aspect of the body schema and the anticipatory aspects of motor control (Gallagher 2011). The same retentional-protentional structure can act as the organizing principle for proprioceptive and efferent processes that give rise to a phenomenal sense of movement and agency. The integration of the different neuronal contributories that would in part account for the organization of action and our consciousness of action arises through the concurrent participation of distributed regions of the brain and their sensorimotor embodiment (Varela et al. 2001). This is consistent with a dynamical systems approach (Thompson 2007; van Gelder 1996; Varela 1999) and is modeled on three different scales of duration (Pöppel 1988, 1994; Varela 1999), the first two of which are said to be directly relevant to the protentional-retentional processes of intrinsic temporality:

1. The elementary scale (varying between 10 and 100 milliseconds);
2. The integration scale (varying between 0.5 and 3 seconds);

3. The narrative scale (involving memory and integration across longer time periods).

The elementary time scale corresponds to neurophysiological processes of intrinsic cellular rhythms of neuronal discharges within the range of 10 ms (the rhythms of bursting interneurons) to 100 ms (the duration of an excitatory postsynaptic potential (EPSP) / inhibitory postsynaptic potential (IPSP) sequence in a cortical pyramidal neuron). Integration of these neuronal processes corresponds to the integration of cell assemblies, distributed subsets of neurons with strong reciprocal connections (see Varela 1995; Varela et al. 2001). Phenomenologically the second scale corresponds to the experienced living present, the level of a fully constituted, normal cognitive operation. In terms of motor processes it corresponds to a basic action, for example reaching or grasping.

Neuronal-level basic events that have a duration on the elementary scale of milliseconds synchronize (by phase-locking) and form aggregates that manifest themselves as incompressible but complete acts on the integration scale.[3] Despite the objective measurements used to characterize these different scales, there is no set or rigid completion time or fixed integration period. Rather, integration is dynamically dependent on activation in a number of dispersed assemblies. Furthermore, the integration does not necessarily preserve an objective linear sequence in the events. For example, when a 50-ms interval is followed by a 100-ms interval the integrated event is not necessarily an additive sum of 150 ms.

> Instead, the earlier stimulus interacts with the processing of the 100 ms interval, resulting in the encoding of a distinct temporal object. Thus, temporal information is encoded in the context of the entire pattern, not as conjunctions of the component intervals.
> (Karmarkar and Buonomano 2007, p. 432)

Importantly, the temporal order that manifests itself at the integration level is the product of a retentional function that orders information according to a pragmatic pattern (a pattern that is useful to the organism) rather than according to some internal or external clock.

One result may be something like the intentional binding that occurs when subjects are asked to judge the timing of their voluntary movements and the effects of those movements (see Engbert et al. 2007; Haggard et al. 2002). Intentional binding refers to the compression in time estimation that occurs when one's action is involved. If A and B are two events separated by one second and you are asked to estimate the temporal interval between them, you are likely to estimate about one second. If, however, A is your action and B is the worldly effect of that action, you are likely to say half a second. This intentional binding effect correlates to and has been proposed as a measure of the pre-reflective sense of agency that is tied to efferent motor-control processes. Note that the temporal window of the integration scale is necessarily flexible (0.5–3 seconds), and consistent with the idea that there is some ambiguity attached to the sense of agency, the integration process depends on a number of factors: context (which may involve physical environmental aspects, but also meaning—the environment as the agent pragmatically perceives it), as well as factors of bodily affect such as emotion state, hunger, fatigue, general physical condition, age of subject, and so on (Bower and Gallagher 2013). Because of such embodied and environmental factors, the integrating neural synchronization is dynamically unstable and will constantly and successively give rise to new assemblies, such transformations defining the trajectories of the system.

The processes that define the integration scale correspond to the experienced present, or to William James's (1890) notion of the specious present, and are describable in terms of the protentional-retentional structure discussed earlier (Varela 1999; Thompson 2007). Whatever falls within this window counts as happening "now" for the system, and this *now* integrates (retains) some indeterminate sequence of the elemental scale neuronal events that have just happened. The system dynamically parses its own activity according to this intrinsic temporality. In each emerging present the preceding emergence is still present in and still has an effect as it traces the dynamical trajectory (corresponding to *retention* on the phenomenological level). This process is constrained by its initial and boundary conditions defined in terms of embodied constraints, the experiential context of the action, behavior, or cognitive act, and the pragmatic interest of the agent (Gallagher and Varela 2003; Varela 1999).

2. Why Action and Experience Are Enactive All the Way Down

That the dynamical analyses offered by Varela and others are not a perfect fit for Husserl's analysis of temporality was critically pointed out by Rick Grush (2006). It is not clear, however, that such analyses confuse vehicle and content, as Grush maintains. He thinks of dynamic processes as neural vehicles, and he equates retention with "aspects of the current contents of awareness." Husserl, however, regarded retention not as the content of awareness or an aspect of that content, but as a structural feature of experience. The important correlation here is a structural one. Indeed, on the kind of dynamical and enactive view that I want to defend here, the vehicle-content distinction that works so well in cognitivist and functionalist approaches is irrelevant. From such perspectives one might say that the dynamical processes described by Varela and others constitute the vehicles that carry information about what has just been happening in the system. That the dynamical processes are retentional is correct, but one can say this without recourse to the vehicle-content distinction, as I'll argue later.

The enactivist view holds that perception and cognition are action oriented, that is, they are characterized in most cases by a structural coupling between the agentive body and the environment, which generates action-oriented meaning (e.g., Di Paolo 2009; Noë 2004; Varela et al. 1991). When, for example, I perceive something, I perceive it as actionable. That is, I perceive it as something *I can* reach, or not; something *I can* pick up, or not; something *I can* hammer with, or not, and so forth. Such affordances (Gibson 1977, 1979) for potential actions (even if I am not planning to take action) shape the way that I actually perceive the world. One can find the roots of this kind of approach in the pragmatists,[4] but also in phenomenologists like Husserl, Heidegger, and Merleau-Ponty. Merleau-Ponty (1962) is most often cited in this regard, but Merleau-Ponty himself points us back to Husserl's analysis of the "I can" in *Ideen II* (Husserl 1952), and to his analysis of the correlation between kinesthetic activation and perception (1973; see Zahavi 1994 and Gallagher and Zahavi 2012 for further discussion).

Consistent with this enactive view, as we've seen in the previous section, the intrinsic structure of consciousness and perception, and more generally, cognition, is such that it integrates with the intrinsic temporal structure of action. Moreover, as I now want to argue, if perception and cognition are enactive, then at a minimum their intrinsic temporal structure should be such that it allows for that enactive character.[5]

One can think of the primal impression, not as the origin and point of departure (as "something absolutely unmodified, the primal source of all further consciousness and being" [Husserl 1966, 67]), but as the "boundary" between retention and protention (Husserl 2001, 4), or the result of a dynamical interplay between retention and protention, the line of intersection between the retentional and protentional tendencies that make up every momentary phase of consciousness. This claim is brought to light in Husserl's idea that the point of departure, rather than being the primal impression, is the empty (i.e., as yet unfulfilled) protentional anticipation (Husserl 2001, 4). That is, the primal impression is conceived as the fulfillment of an empty protention, and the now is constituted by way of a protentional fulfillment (Husserl 2001, 4, 14). The notion of an isolated primal impression is an abstraction and not something that exists in itself.

Primal impression is supposedly the consciousness of the now point of the temporal stimulus—for example, in Husserl's favorite illustration, the note that is currently being sounded, or more precisely, the current moment of the note that is present. It's helpful, however, to return to considerations about the intrinsic temporality of action discussed in the previous section. As we saw, the protention-primal impression-retention model applies to action and non-conscious motor processes as well as to consciousness. Both human experience and human action are characterized by a ubiquitous intrinsic temporality. In this regard, when we look at action we can say that at any one moment the body is in some precise posture—as captured by a snapshot, for example. That momentary posture, however, is a complete abstraction from the movement, since in each case the body is not posturing from moment to moment but is constantly *on the way*, in the flow of the movement such that the abstract postural moment only has meaning as part of that process. One could argue that *objectively speaking*, at any moment the body actually is in a specific posture. But if that postural moment is anything, it is the product of an anticipated trajectory of where the action is heading. Furthermore, we can define that abstract postural moment only when it is already accomplished—but that means, only in retention, and as an end point of what had been a movement characterized primarily by anticipation.

We should think of consciousness in the same way—as Husserl does—as a flow, where it is intentionally directed in such a way that when I am hearing the current note of a melody I'm already moving beyond it, and such protentional/anticipatory moving beyond is already a leaving behind in retention. What we have as the basic datum of experience is a process, through which the primal impression is already collapsing into the retentional stream even as it is directed forward in protention. Hearing a melody (or even a single note in some context) never involves hearing a currently sounded note (or part of a note), *and then* moving beyond it; rather, the "and then" is already effected, already implicit in the experience.

Pre-reflectively, consciousness and action have this structure. There is no impression of the present, or posture in the present, taken as a knife-edge; rather, as Husserl suggests, primal impression is already fulfilling (or not) protentions that have already been retained, and in doing so is already informing the current protentional process. The proposal is not that we should eliminate the concept of primal impression. The point is rather that we should abandon the idea that primal impression is a direct, straight and simple apprehension of some now point of a stimulus that is unaffected by retention and protention. The perception of a currently sounding note, for example, is already modified by my just past and passing awareness of whatever came directly before. In that sense, primal impression is already modified by the retentional performance of consciousness. There is no primal impression that is not already qualified by retention. It is not that in a now phase of consciousness I have a retention of a past phase *plus* a primal impression of a current stimulus (S). It is not an additive function. A description of the full experience of a melody is not well formulated by saying that I first experience (in primal impression) note A, and then (in a new primal impression) note B, as I retain note A. Rather, the primal impression of B is already qualified (impacted, transduced, modulated) by the just previous experience, just as one's present posture is already the result of previous movement. So the primal impression of B is never simply that; it is something that has worked its way through the retention of the previous primal impression of A. Accordingly, the primal impression of B would be a different experience if it were preceded not by the primal impression of A, but by a primal impression of something that was not A. Thus, for example, the primal impression of the note B-minor sounded at a certain point in Bach's *Concerto in B Minor* will be different from the primal impression of the note B-minor sounded at a certain point in Vivaldi's *Concerto in B Minor*. The notes will also sound different in some way.

Protention also has similar dynamical effects. On the one hand, the primal impression of A when occurrent, is producing a determination of what my protentional horizon currently is—for example, a protention of B . . . C . . . D . . . , and so on. Whatever I anticipate must be related to what I am currently experiencing. On the other hand, the primal impression of B, when occurrent, is already qualified by the previous protention (currently retained), whether that was a protention of B (now fulfilled), or something else (now unfulfilled). Generally speaking, then, primal impression constrains the current protention, and is constrained by the previous protention. The occurrent primal impression is partially either the fulfillment or lack of fulfillment of the previous protention and provides partial specification of what I am anticipating.

One objection to this may be that we have confused the content of experience with the vehicles of experience, or its formal temporal properties. Someone like Grush might object that the analysis of intrinsic temporal structure should not be about the difference between how we hear Bach and Vivaldi. But this objection ignores the fact that *what* I experience has an effect on the temporality of my experience. I may find Bach boring and Vivaldi vivacious, in which case Bach's *Concerto in B minor* will seem to drag on—time will seem to slow down—in contrast to my listening to Vivaldi's concerto. Likewise bodily affects such as hunger or pain may modulate the retentional-protentional process so that my painful experience of sitting through the concert will be temporally different from my pain-free listening experience. As Merleau-Ponty suggests, there is an "influence of the 'contents' on time which passes 'more quickly' or 'less quickly', of *Zeitmaterie* on *Zeitform*" (1968, 184).

Temporal masking is another example where the content-vehicle distinction is less than clear, that is, where *what* I experience determines the experienced temporal order. For example, the tonal arrangement of sounds presented in a sequence can affect the perception of that sequence. If in the sequence of sounds ABCDBA the tones A and B are of a particular low frequency, the order of C and D will be masked. That is, you will not be able to distinguish the order of C and D. Variations in the tonal qualities of the sequence A followed by B can lead you to experience D before C (Bregman and Rudnicky 1975). In this regard, it's not simply that the conscious retention of A and B determines the phenomenal order of C and D, since the later sounds of B and A are also required to get these effects. That is, the sounds that follow C and D in the objective sequential order will also determine the way C and D play out on the conscious level.

Of course one can still say that there is some level of formal temporalization that remains invariant—whatever the content, or whatever the phenomenological velocity or experienced serial order, or the implicit temporality of the object itself, I always experience a sequence in which some S precedes another. But what S happens to be experienced, or in what order it appears in the stream of consciousness, or how fast it happens to swim by, makes all the difference in experience.

The primal impression is itself structured by its very dynamic participation in its relations to retention and protention (and vice versa). The content-vehicle distinction doesn't hold firm if, to put it in these terms, what we call content starts to play a role in shaping the vehicle. It is surely true that our consciousness of red is not itself red. It's more difficult to say that our consciousness of a temporal object (like a melody or an action) is not itself temporal or that the two temporalities are not dynamically intertwined. Rather, the intrinsic temporality of experience and action seems to have a fractal character (Gallagher and Zahavi 2014). The structure of intrinsic temporality is protention-primal impression-retention. But each element also reflects this structure again so that any closer examination of primal impression (or retention or protention) finds that same structure repeated—again, not in an additive way, but with a kind of fractal effect—an effect that multiplies itself in such a way that any attempt to define primal impression in itself always finds the effects of retention or protention already included.

If thinking of intrinsic temporality in this fractal way is suggestive, it is more true to the dynamic nature of these processes to think of them as enactive. One can say that there is no primal impression—no current intuition of the present stimulus (S)—without it already being anticipatory (on the basis of what has just occurred), so that my primal impression of the present is already involved in an enactive anticipation of how my encounter with S will work out. Protention, primal impression, and retention are in an *enactive* structure in regard to S in the sense that a certain anticipatory aspect (already shaped by what has just gone before) is already complicating the immediacy of the present. Consciousness is not simply a passive reception of the present; it enacts the present, it constitutes its meaning in the shadow of what has just been experienced, and in the light of what it anticipates.

To be even more explicit on this point, what the primal impression is, and how it relates to retention and protention, are not independent from the intentional nature of consciousness, or from the specifics of what we experience. As in the case of action, the intrinsic temporality of experience should be considered as pragmatically directed towards the meaningful possibilities the agent sees in the world. This lines up well with Husserl's conception of embodied experience as an anticipatory "I can," which draws on my prior experience and my current state. As Husserl put it, "every living is living towards." This anticipatory intentionality is an apprehension of the possibilities or the affordances in the present, of what S *can be* for my experience—possibilities that will be fulfilled or not fulfilled as our enactive perception trails off in retention.

Primal impression, retention, and protention are not elements that simply add themselves to each other. They are rather in a dynamic relation; they have a self-constituting effect on each other. Moreover, together they constitute the possibility of an enactive engagement with the experienced world.

> Just as I perceive the hammer as affording the possibility of grasping it, or in a different circumstance, as affording the possibility of propping

open my window, I likewise perceive the melody as affording the possibility of dancing or sitting in peaceful enjoyment, etc. The point, however, is not about hammers versus melodies. It's about the temporality of affordances and enactive engagements. Nothing is an affordance for my enactive engagement if it is presented to me passively in a knife-edge present; that is, nothing would be afforded if there were only primal impressions, one after the other, without protentional anticipation, since I cannot enactively engage with the world if the world is not experienced as a set of possibilities, which, by definition, involves the not-yet. Moreover, just as nothing would be possible if there were only primal impressions without a retentional-protentional structure, so too if there were no primal impression. If there were only retentions, everything I experience would already have just happened; we would be pure witnesses without the potential to engage. If there were only protentions, there would only be unfulfilled promises of engagement. Meaning itself would dissipate under any of these conditions.

> (Gallagher and Zahavi 2014, 96)

Thus, the enactive character goes all the way down, into the very structure of the intrinsic structure of experience and action.

3. Intentional Action and the Narrative Scale

In terms of temporal structure I've been focusing on the elementary and integrative scales and digging into the micro structures that characterize sub-personal processes and the pre-reflective level of experience as they relate to perception and action. I've been arguing for a dynamic and enactive account of these phenomena. The definition of different time scales (elemental, integrative, and narrative) usefully maps onto three kinds of intention or intention formation (Figure 13.2).

1. The elementary scale (10–100 milliseconds) corresponds to the motor-control processes (including forward and inverse models) that define motor or M-intentions (Pacherie 2007), or what Merleau-Ponty calls motor intentionality (see Gallagher 2012).
2. The integration scale (varying from 0.5 to 3 seconds) corresponds roughly to present or proximate (P-)intentions (what Searle [1983] calls "intentions-in-action"), which involve quick decisions and ongoing monitoring of actions in the detailed parameters of the current environment (Pacherie 2007).
3. The narrative scale involving memory, deliberation, and planning (involving longer-term prospective and retrospective reflection and deliberation) corresponds to future or distal (D-)intentions.

The focus so far has been on time scales that correspond to M- and P-intentions. The claim that intention formation processes are enactive all the way down, however, needs to be complemented by understanding that they are also enactive all the way up.

Intentional action often involves prior intentions, and these may involve conscious and thoughtful deliberations about what to do and what goals to aim at. In these respects intentional action involves the narrative time scale in which we can make sense of actions in terms of reasons. We know, at least from the time Socrates explained it (in Plato's *Phaedo*), that it is not enough, or even relevant, to provide an account of one's actions in terms of motor-control mechanisms if one is asked "Why" one acted thus and so. The question calls for an answer, not in terms of a scientific theory or explanation, but in terms of a retrospective account that in some regard involves narrative structure. "I acted because someone told me this, and someone else told me that, and I thought it was right to do this in the circumstances since my partner was having a rough time." Each part of the account could be fleshed out further to extend the narrative. Likewise, in the case of forming a D-intention, one's considerations are prospectively narrative. In deciding to purchase a car next spring, I may imagine myself driving around in a certain model; I may imagine what my wife will say about my plan; I may imagine taking the bus to the auto dealer; and so forth.

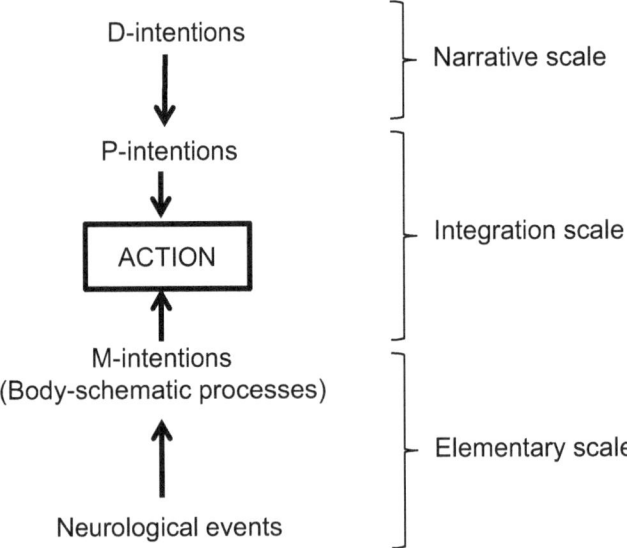

Figure 13.2 Intentions and time scales

Intentions are future directed. They are about the future, since they are about goals, and goals are things to be accomplished. I can, of course, talk retrospectively about goals that I have attained or fail to attain, but to say that I have an intention to do something necessarily means that it is something that is still to be done. It would be odd and nonsensical for me to decide, while writing this chapter, that I intend to meet Napoleon some day, since with regard to Napoleon we are directed only towards the past (unless I believe in some kind of heavenly, or hellish, afterlife). Even after I have formed a properly future-oriented D-intention, as I start to perform my intentional action my P-intention is oriented to moving my present action in the direction of the future, not yet achieved goal. Once the goal is achieved (or once I have given up on the goal or forgotten about it), the intention is no longer operative; it dissipates and fades into the past. I can, of course, remember and talk or write about my past intentions and actions, and about how successful or unsuccessful I have been in carrying them out. I may also have a long-standing intention that I have not acted upon (and may never get a chance to act upon). Directed toward the future, intentions are about possibilities. Even M- and P-intentions are about possibilities for movement or for taking advantage of whatever affordances are current.

This tells us something about the intrinsic temporality of intentions. On the one hand, and unremarkably, an intention, if it exists, exists in the present. At least on one conception of time this is true of everything and anything. Furthermore, once the intended goal is attained (or we give up on the goal), the intention ceases to exist, except in memory or narrative. Again, this is unremarkable. Intentional action takes time; it begins and ends, and takes up some duration in between. The time frame of intentional action may vary from very short to very long, depending on the degree of complexity involved in the action. But again, this is an unremarkable observation in terms of objective measurable time.

What is remarkable is the temporal structure of action that derives from intention. An occurrent action is, *per se*, ongoing towards the future, specifically towards its future end, and this feature is not reducible to the fact that this action requires more time to be complete. As Heidegger (1962, 236) would put it, action is always "ahead-of-itself." Moreover, as a way of being-in-the-world, my action is always and already situated in a particular set of circumstances, and these circumstances are shaped by what has gone before, which includes my own action up to this point. What I *can* do (what my possibilities are) is shaped by those circumstances. Yet my action always transcends these circumstances insofar as I act for the sake of some goal other than the action itself. At the same time, at the same stroke, my action incorporates the situation that has been shaped by *past* actions, and by the projected *future* toward which it is moving, in the *present* circumstances that can both limit and enable it. This is a temporal structure that is not captured by objective time. It's not enough to say that action takes time; there is a time *in* action, an intrinsic temporality or a temporal structure in action, not only at the micro level of motor control, but at the narrative level of acting for reasons.

If timing is everything, in telling a joke, in making shots, in making love, in making dinner, then the implicit temporality of action and conscious experience is even more than that, since it enables the skillful timing that we often find in action and in the very structure of experience as it constitutes meaning.

Notes

1. Research for this chapter was supported in part by the Marie Curie Actions ITN project 264828, Towards an Embodied Science of Intersubjectivity (TESIS), and by the Humboldt Foundation's Anneliese Maier Research Fellowship.
2. This kind of coordination begins even earlier. Ultrasonic scanning on fetuses shows that similar hand-mouth movements, reflecting "a striking similarity between prenatal and postnatal movements" occur between 50 and 100 times per hour from 12 to 15 weeks gestational age (DeVries et al. 1984, 48)
3. See, e.g., Thompson (2007):

   > integration happens through some form of temporal coding, in which the precise time at which individual neurons fire determines whether they participate in a given assembly. The most well-studied candidate for this kind of temporal coding is *phase synchrony*. Populations of neurons exhibit oscillatory discharges over a wide range of frequencies and can enter into precise synchrony or phase-locking over a limited period of time (a fraction of a second). A growing body of evidence suggests that phase synchrony is an indicator (perhaps a mechanism) of large-scale integration. . . . Animal and human studies demonstrate that specific changes in synchrony occur during arousal, sensorimotor integration, attentional selection, perception, and working memory. (Thompson 2007, 332)

4. John Dewey, for example, holds that in perception

   > we find that we begin not with a sensory stimulus, but with a sensori-motor coördination, the optical-ocular, and that in a certain sense it is the movement which is primary, and the sensation which is secondary, the movement of body, head and eye muscles determining the quality of what is experienced. (1896, 358)

   Likewise, George Herbert Mead suggested that what is present in perception is not a copy of the perceived, but "the readiness to grasp" what is seen (1938, 103):

   > the readiness of the organism to act toward [objects]. . . . We see the objects as we will handle them. . . . We are only "conscious of" that in the perceptual world which suggests confirmation, direct or indirect, in fulfilled manipulation. (Mead 1938, 104–105)

5. For a more detailed analysis of this point in terms of Husserl's revision of his phenomenology of time-consciousness, see Gallagher and Zahavi (2014).

References

Babinski, J. (1899). De l'asynergie cérébelleuse. *Revue de Neurologie* 7: 806–816.
Berthoz, A. (2000). *The Brain's Sense of Movement*. Cambridge, MA: Harvard University Press.
Bower, M. and Gallagher, S. (2013). Bodily affectivity: Prenoetic elements in enactive perception. *Phenomenology and Mind* 2: 108–131

Bregman, A. S. and Rudnicky, A. I. (1975). Auditory segregation: Stream or streams? *Journal of Experimental Psychology: Human Perception and Performance* 1: 263–67.

Butterworth, G. and Hopkins, B. (1988). Hand-mouth coordination in the newborn baby. *British Journal of Developmental Psychology* 6: 303–314.

De Vries, J.I.P., Visser, G.H.A., and Prechtl, H.F.R. (1984). Fetal motility in the first half of pregnancy. In H.F.R. Prechtl (Ed.), *Continuity of Neural Functions from Prenatal to Postnatal Life* (46–64). Oxford: Spastics International Medical Publications.

Dewey, J. (1896). The reflex arc concept in psychology. *Psychological Review* 3: 357–370.

Di Paolo, E. A. (2009). The social and enactive mind. *Phenomenology and the Cognitive Sciences* 8/4: doi:10.1007/s11097–009–9143–5.

Engbert, K., Wohlschläger, A., Thomas, R., and Haggard, P. (2007). Agency, subjective time, and other minds. *Journal of Experimental Psychology, Human Perception and Performance* 33/6: 1261–1268.

Gallagher, S. (2011). Time in action. In C. Callender (Ed.), *Oxford Handbook on Time* (419–37). Oxford: Oxford University Press.

——— (2012). Multiple aspects of agency. *New Ideas in Psychology* 30: 15–31.

Gallagher, S. and Varela, F. (2003). Redrawing the map and resetting the time: Phenomenology and the cognitive sciences. *Canadian Journal of Philosophy*, Supplementary Volume 29: 93–132.

Gallagher, S. and Zahavi, D. (2012). *The Phenomenological Mind*, 2nd ed. London: Routledge.

——— (2014). Primal impression and enactive perception. In D. Lloyd and V. Arstila (eds.) *Subjective Time: The Philosophy, Psychology, and Neuroscience of Temporality* (83–99). Cambridge, MA: MIT Press.

Georgieff, N. and Jeannerod, M. (1998). Beyond consciousness of external events: A "who" system for consciousness of action and self-consciousness. *Consciousness and Cognition* 7: 465–77.

Gibbon, J. and Malapani, C. (2001). Neural basis of timing and time perception. In L. Nadel (Ed.), *Encyclopedia of Cognitive Science* (305–311). London: Wiley.

Gibson, J.J. (1977). The theory of affordances In R. Shaw and J. Bransford (eds.), *Perceiving, Acting, and Knowing: Toward an Ecological Psychology* (67–82). Hillsdale, NJ: Lawrence Erlbaum.

——— (1979). *The Ecological Approach to Visual Perception*. Boston: Houghton Mifflin.

Grush, R. (2006). How to, and how *not* to, bridge computational cognitive neuroscience and Husserlian phenomenology of time consciousness. *Synthese*. doi:10.1007/s11229–006–9100–6.

Haggard, P., Aschersleben, G., Gehrke, J., and Prinz, W. (2002). Action, binding and awareness. In W. Prinz and B. Hommel (eds.), *Common Mechanisms in Perception and Action: Attention and Performance* (266–285). Vol. 19. Oxford: Oxford University Press.

Haith, M. M. (1993). Future-oriented processes in infancy: The case of visual expectations. In C. Granrud (Ed.), *Carnegie-Mellon Symposium on Visual Perception and Cognition in Infancy* (235–264). Hillsdale, NJ: Lawrence Erlbaum Associates.

Head, H. (1920). *Studies in Neurology*, Volume 2. Oxford: Clarendon Press.

Heidegger, M. (1962). *Being and Time*, trans. J. Macquarrie and E. Robinson. Oxford: Blackwell.

Hopkins, B. and Prechtl, H.F.R. (1984). A qualitative approach to the development of movements during early infancy. In H.F.R. Prechtl (Ed.), *Continuity of Neural Functions from Prenatal to Postnatal Life* (179–197). Oxford: Spastics International Medical Publications.

Husserl, E. (1952). *Ideen zur einer reinen Phänomenologie und phänomenologischen Philosophie. Zweites Buch: Phänomenologische Untersuchungen zur Konstitution. Husserliana 4.* The Hague: Martinus Nijhoff; English translation: *Ideas Pertaining to a Pure Phenomenology and to a Phenomenological Philosophy, Second Book: Studies in the Phenomenology of Constitution.* Trans. by R. Rojcewicz and A. Schuwer. The Hague: Kluwer Academic, 1989.

——— (1966). *Analysen zur passiven Synthesis,* Husserliana 11. The Hague: Martinus Nijhoff.

——— (1973). *Ding und Raum. Vorlesungen 1907.* Husserliana 16. The Hague: Martinus Nijhoff, 1973. English translation: *Thing and Space: Lectures of 1907,* trans. R. Rojcewicz. Dordrecht: Kluwer Academic, 1998.

——— (1991). *On the Phenomenology of the Consciousness of Internal Time (1893–1917),* Collected Works IV, trans. J. Brough. Dordrecht: Kluwer Academic.

——— (2001). *Die Bernauer Manuskripte über das Zeitbewusstsein (1917–18),* Husserliana 33. Dordrecht: Kluwer Academic.

James, W. (1890). *The Principles of Psychology.* New York: Dover, 1950.

Jeannerod, M. (2001). Neural simulation of action: A unifying mechanism for motor cognition. *Neuroimage* 14: S103–S109.

Karmarkar, U.R. and Buonomano, D.V. (2007). Timing in the absence of clocks: Encoding time in neural network states. *Neuron* 53: 427–438.

Lew, A. and Butterworth, G.E. (1995). Hand-mouth contact in newborn babies before and after feeding. *Developmental Psychology* 31: 456–463.

MacKay, D. (1966). Cerebral organization and the conscious control of action. In J.C. Eccles (Ed.), *Brain and Conscious Experience* (422–445). New York: Springer.

Malapani, C. and Fairhurst, S. (2002). Scalar timing in animals and humans. *Learning and Motivation* 33(1): 156–176.

Mead, G.H. (1938). *The Philosophy of the Act,* ed. C.W. Morris. Chicago: University of Chicago Press.

Merleau-Ponty, M. (1962). *Phenomenology of Perception,* trans. C. Smith. London: Routledge and Kegan Paul.

——— (1968). *The Visible and the Invisible,* trans. A. Lingis. Evanston, IL: Northwestern University Press.

Noë, A. (2004). *Action in Perception.* Cambridge, MA: MIT Press.

Pacherie, E. (2007). The sense of control and the sense of agency. *Psyche* 13/1. http://jean nicod.ccsd.cnrs.fr/docs/00/35/25/65/PDF/Pacherie_sense_of_control_Psyche.pdf

Pöppel, E. (1988). *Mindworks: Time and Conscious Experience.* Boston: Harcourt Brace Jovanovich.

——— (1994). Temporal mechanisms in perception. *International Review of Neurobiology* 37: 185–202.

Prechtl, H.F.R. and Hopkins, B. (1986). Developmental transformations of spontaneous movements in early infancy. *Early Human Development* 14: 233–283.

Searle, J. (1983). *Intentionality: An Essay in the Philosophy of Mind.* Cambridge: Cambridge University Press.

Straus, E. (1966). *Philosophical Psychology.* New York: Basic Books.

Thompson, E. (2007). *Mind in Life: Biology, Phenomenology, and the Sciences of Mind*. Cambridge, MA: Harvard University Press.

Van Gelder, T. (1996). Wooden iron? Husserlian phenomenology meets cognitive science. *Electronic Journal of Analytic Philosophy 4*; reprinted in J. Petitot, F. J. Varela, B. Pachoud, and J-M. Roy (eds.) (1999) *Naturalizing Phenomenology: Issues in Contemporary Phenomenology and Cognitive Science* (245–265). Stanford, CA: Stanford University Press.

Varela, F. J. (1995). Resonant cell assemblies: A new approach to cognitive functioning and neuronal synchrony. *Biological Research* 28: 81–95.

—— (1999). The specious present: A neurophenomenology of time consciousness. In J. Petitot, F. J. Varela, B. Pachoud, and J.-M. Roy (eds.), *Naturalizing Phenomenology: Issues in Contemporary Phenomenology and Cognitive Science* (266–314). Stanford, CA: Stanford University Press.

Varela, F., Lachaux, J. P., Rodriguez, E., and Martinerie, J. (2001). The brainweb: Phase-synchronization and long-range integration. *Nature Reviews Neuroscience* 2: 229–239.

Varela, F. J., Thompson, E., and Rosch, E. (1991). *The Embodied Mind: Cognitive Science and Human Experience*. Cambridge, MA: MIT Press.

Wilson, M. and Knoblich, G. (2005). The case for motor involvement in perceiving conspecifics. *Psychological Bulletin* 131/3: 460–473.

Wolpert, D. M., Ghahramani, Z., and Jordan, M. I. (1995). An internal model for sensorimotor integration. *Science* 269/5232: 1880–1882.

Zahavi, D. (1994). Husserl's phenomenology of the body. *Études Phénoménologiques* 19: 63–84.

Part IV

Phenomenology and the Temporality of Agency

14 Care, Death, and Time in Heidegger and Frankfurt

B. Scot Rousse

1. Introduction

Both Martin Heidegger and Harry Frankfurt have argued that the fundamental feature of human identity is *care*. Both contend that caring is bound up with the fact that we are mortal beings related to our own impending death, and both have claimed that caring has a non-instantaneous, future-oriented, and ultimately circular temporal structure. In this chapter, I argue that Heidegger's conception of the temporal articulation of caring elucidates a misunderstanding at the heart of Frankfurt's view of the relations among care, death, and time. The temporal, existential, and normative significance that Frankfurt finds in a hard-wired instinct for self-preservation, what he calls a person's "love of living," is more compellingly captured by Heidegger's idea that a human identity is lived out in the manner he calls "being-toward-death."

2. Preliminary Remarks

Heidegger gives the name "care" [*Sorge*] to the ontological structure of human existence that he spells out in *Being and Time*. He contends that for me to be a person, or "Dasein," is for me to relate to my life by caring about it. My own being is *an issue* or *at stake* for me. Heidegger writes that "any entity is either a '*who*' [a person or Dasein] . . . or a '*what*' [a mere thing or object]," adding that, for the latter, "their being is [to them] 'a matter of indifference' [*gleichgultig*]; or more precisely, they 'are' such that their being can be neither a matter of indifference to them, nor the opposite" (Heidegger 1962, 71, 68; my gloss in the brackets).[1]

The phenomenon of care, according to Heidegger, cannot be captured by the everyday conception of time as a linear sequence of discrete now points. Care stretches our time; it has an "ecstatically" articulated and ultimately circular temporal structure Heidegger describes as "being-ahead-of–myself-already-in-the-world." Put abstractly, to care is to be related to the present by way of pressing into the future on the basis of what already matters. Caring, Heidegger contends, is rooted in how I *already* find myself, yet it is essentially futural (BT 376).

To care is thus to have final ends "for the sake of which" we carry out our daily activities and in terms of which we organize our time and have a meaningful orientation in our everyday world (see BT 116–119).[2] The final ends that define our identities are not *goals* that we aim to achieve at some discrete now-point in the future and then leave behind in the past. They are that "towards-which" a person always conducts himself. It is helpful here to appeal to an example from William Blattner's discussion of this point: the difference between having the *goal* of getting tenure and having the identity of *being* a teacher (Blattner 1999, 39–43). The goal is something you aim at, achieve at a certain moment, and then "leave behind" in the past. But futural final ends that define a person's identity of *being* a professor remain "always outstanding": they are never "settled" or "actualized" and left in the past, at least as long as the person understands and identifies himself in the relevant way. That is one thing Heidegger means by claiming "Dasein is constantly 'more' than it factually is," and that Dasein "*is* existentially that which . . . it is *not yet* (BT 185–186, cf. 287). The "not yet" here is the ecstatic "stretching out" into the dimension of the future that is characteristic of care according to Heidegger.

Frankfurt has been interested in the same realm of phenomena. With a feigned tentativeness, he writes:

> Perhaps caring about oneself is essential to being a person. Can something to whom its own condition and activities do not matter in the slightest properly be regarded as a person at all? Perhaps nothing that is entirely indifferent to itself is really a person.
>
> (Frankfurt 1999e, 89–90)

Especially since his essay "The Importance of What We Care About," Frankfurt has consistently sought to distinguish caring from the overburdened notion of *desire*. He does so in terms of their differing "temporal characteristics": while desire has "no inherent persistence" and can obtain in a merely instantaneous present "now," caring is essentially "prospective," involving, as Heidegger saw, a *futural* thrust: "The outlook of a person who cares about something is inherently prospective; that is, he necessarily considers himself as having a future" (Frankfurt 1988, 83). Frankfurt contends, again in a Heideggerian vein, that caring provides "continuity and coherence to a life" (Frankfurt 1999c, 162) and prevents it from "being merely a sequence of events" (Frankfurt 1988, 83). Echoing Heidegger's denial that the temporality of care can be understood as a "pure sequence of 'nows' " (BT 377), Frankfurt remarks that the "moments in the life of a person who cares about something . . . are not merely linked inherently by formal relations of sequentiality" (Frankfurt 1988, 83). Caring is a matter of the person's "continuing concern with what he does with himself and with what goes on in his life" (Frankfurt 1988, 84).

Both Heidegger and Frankfurt relate the phenomenon of care and its futural directedness to the fact of human mortality. In his later work, Frankfurt claims that the instinctual avoidance of death, a manifestation of what he calls the "love of living," plays "a comprehensively foundational" role in the constitution of personal identity and in grounding the importance of what we care about (Frankfurt 2006b, 41).[3] Heidegger agrees that human mortality is somehow the source of importance in human life. However, Heidegger would strongly reject Frankfurt's understanding of the significance of death in terms of a naturalistic, instinct-conception of care.

In Frankfurt's efforts to rebut the claims of Kantian rationalism—a still widespread view that sees reflective self-consciousness and rational principles at the core of human identity—he, like Heidegger, holds up care as a phenomenon that does not depend upon and indeed enjoys a certain priority over reason or reflective self-consciousness. Frankfurt, though, does not see how to defend this anti-rationalism without swinging all the way to the opposite dialectical extreme and defending an implausible foundationalist naturalism in his conception of care.

Frankfurt mistakenly sees the priority of care over rational self-consciousness as implying the *priority of the factual over the normative* in questions of human identity. He understands the "love of living" as the "biologically embedded" instinct for self-preservation (RL 30), a fundamental and defining personal commitment that is "determined for us by biological and other natural conditions, concerning which we have nothing much to say" (RL 48). Frankfurt thus conceives the most fundamental commitments definitive of a person's practical identity—what he calls "volitional necessities"—on the model of *natural facts* about which normative questions of justification, appropriateness, or evaluation *cannot* arise at all.

From Heidegger's perspective, this is to misunderstand the significance of the relation between caring, dying, and human identity. Heidegger's emphasis on care is also part of his overall anti-rationalism. Yet Heidegger's view helps us see how Frankfurt's naturalistic interpretation of volitional necessity is motivated by a misunderstanding of the significance of the temporal structure of caring. For Heidegger, human identity cannot bottom out in any fundamental commitments that are hard-wired like natural facts. Against Frankfurt's naturalist foundationalism about ultimate personal commitments, Heidegger would emphasize that the question of who I am can always arise anew and that my identity remains ever at stake. According to Heidegger, there are no substantive commitments that are in principle immune to revision or rejection; not even the ones Frankfurt claims are biologically embedded in human nature. In order to understand all that is at stake in this nest of issues, we need to back up and get more details of these arguments on the table.

3. Frankfurt on Identity and Care

Frankfurt launches his reflections on caring by considering what is involved in posing and answering what I will call "the practical question," the question that asks: "Who am I going to be? What am I going to do?" According to Frankfurt, answering this question takes the form of hierarchically ordering the relative importance of the things a person cares about (RL 23). This is because a person's caring about something "consists . . . in the fact that he *guides* himself by reference to it" (Frankfurt 1999d, 111). The important thing about caring is not captured by thinking of it as some kind of affective or cognitive state. Caring is a matter of having certain practical dispositions; it is about what I am prepared or able to do.

Ranking the relative importance of the things I care about clarifies for me how I can carry out my life in terms of what is of "greatest importance" to me (Frankfurt 1999a, 132) so that I have something worthwhile to do rather than being bogged down in paltry trivialities or being dispersed in activity that is merely "locally purposeful but nonetheless fundamentally aimless" (RL 89, 53).[4] Living in terms of what you care about enables you to live a *meaningful* life characterized by "a fundamental kind of freedom" (RL 53, 97, 99). Caring about something (a person, ideal, cause, etc.) consists in the fact that the person is effectively motivated to pursue its good or interests, structures his activities in accordance with them, and is affectively vulnerable to their diminishment or enhancement (Frankfurt 1988, 83). Hence, caring involves the person both practically and affectively in his own life and activities. To see why this counts as a meaningful life, it helps to consider what a meaning*less* life consists of for Frankfurt. If a person did not care about anything, Frankfurt claims, he would "languish": "the results would be a fragmentation of life, passivity, and boredom" (Frankfurt 1999e, 88). Why is a carefree life boring? Because nothing matters. Again, mattering is not a subjective feeling; it is largely a pragmatic notion. Mattering means there is something *you have got to do*.

The specter of boredom accordingly comes to hold a central position in Frankfurt's middle and late analysis of the self.[5] "Boredom," he writes, "is a serious matter. It is not a condition that we seek to avoid because we do not find it enjoyable. In fact, the avoidance of boredom is a profound and compelling human need" (RL 54).[6] Frankfurt even claims that "avoiding boredom . . . expresses a quite primitive urge for psychic survival . . . [an urge appropriately construed] as a variant of the universal and elemental instinct for self-preservation" (RL 54–55; my gloss).[7]

Reflectively posing the practical question assists one in coming to a wholehearted (i.e., non-ambivalent) *identification with* or acceptance of the motivations that the person finds move him to act. "Being wholehearted means having a will that is undivided. The wholehearted person is fully settled as to what he wants, and what he cares about" (RL 95). But this being "fully settled" presupposes that the person "understand[s] what it is that [he himself] really care[s] about, and [that he is] decisively and robustly confident in caring about it" (RL 28). Asking the practical question is aimed at "articulating what we are to care about" and enabling us "to get clear concerning what is to be important to us" (Frankfurt 1999e, 92). In Frankfurt's picture, as in Heidegger's, to be guided in your action by what you care about is to be identified with your motivations and thus to be autonomous, "self-owned" ("authentic" [*eigentliche*] in Heidegger's term). For both, such autonomy is a response to your receptive sense of the way things *already* matter. However, Frankfurt and Heidegger understand the significance of this "already" and the experience of being identified with it in radically different ways.

4. The Circular Structure of Care

Both Frankfurt and Heidegger see the phenomenon of care as having a distinctive temporal articulation: while being essentially futural or prospective (caring gives us something *to do*; it orients and moves us toward the execution of certain ends), it is rooted in a pre-constituted situation in which a person *already* finds himself. The peculiar interrelationship between the past and futural dimensions of caring shows that it has a temporally *circular structure*, one that comes to light especially when we consider how caring shapes the way a person deliberatively poses and answers the practical question. Both Frankfurt and Heidegger see that this question involves a kind of circularity. Frankfurt writes that it involves "a rather obvious sort of circularity" (RL 24) and when Heidegger writes that "an entity [Dasein] whose own being is itself an issue has, ontologically, a circular structure" (BT 195), he means to capture (in his own terminology and among other things) the circular structure of the practical question.

To say that the practical question has a circular structure is to deny that it can be intelligibly posed from any kind of original position or stance of detached neutrality. Someone can only pose the question of who and how to be—and press reflectively into the future—from out of a substantive prior orientation provided by the way that things *already* matter. This "*already*" captures the temporal dimension of the past. Where you are going is shaped by where you are coming from. Your past does not "follow along after" you, Heidegger claims; it "always already goes ahead" of you (BT 41).

The prior orientation required for any meaningful posing of a practical question reflects and anticipates the answer that will be given to it. Being able to pose the question means you have already got a grip on its possible answer (Frankfurt 2006b, 23). The question cannot be meaningfully posed in abstraction; it seeks a concrete answer, and it does so by an evaluation of a definite range of possible ways of living and final ends that are already intelligible and appealing (otherwise they would not show up in the deliberation in the first place). Frankfurt makes these points helpfully:

> In order for a person to know how to determine what is important to himself . . . he must already know how to identify certain things as making differences that are important to him. Formulating a criterion of importance presupposes possession of the very criterion that is to be formulated.
>
> (RL 25–26)

In deliberating about how to go on, the person does not *decide* what is important to him, but *finds* himself already finding it important (RL 26). Similar to Heidegger's arguments against a conception of freedom as a *liberty of indifference* (BT 183), Frankfurt compellingly criticizes the notion of freedom as unconstrained choice:

> Unless a person makes choices within restrictions from which he cannot escape by merely choosing to do so, the notion of self-direction, of autonomy, cannot find a grip . . . [and] the decisions and choices he makes will be altogether arbitrary. They cannot possess authentically personal significance or authority, for his will has no determinate character.
>
> (Frankfurt 1999d, 110)

According to Frankfurt, the circularity of the practical question shows "that the question is *systematically inchoate*" and that, if it is able to be intelligibly posed and answered, it must be grounded in and oriented by some factually stable aspect of the person's identity about which no normative questions can be raised (RL 25). Richard Moran, in his review of Frankfurt's *The Reasons of Love*, helpfully explains what is wrongheaded about this claim:

> Certainly the inquiry about how to live must begin somewhere . . . And it is also true that any such inquiry must begin with a provisional sense of "the criteria on the basis of which the exploration is to be pursued." But it's not obvious that this makes the question itself "systematically inchoate," any more than ordinary theoretical inquiry is inchoate since it must begin with, and rely on, an initial set of beliefs and standards for making progress. This would amount to begging the question being raised only if the resultant inquiry did not allow for revision or correction of the assumptions with which it began. Similarly, it may be agreed that the normative question of how to live cannot get going without a provisional answer to the factual question of what one indeed cares about, but so far that is just a reason to begin the inquiry there, not to give the factual questions any other priority.
>
> (Moran 2007, 468)

From the Heideggerian perspective, Moran shows in this passage that the circular structure of the practical question is an instance of the more general phenomenon involved in any understanding whatsoever: the *hermeneutic circle*. A brief explanation of Heidegger's understanding of the hermeneutic circle will help us begin to clarify the differences between Heidegger and Frankfurt.

Understanding is a temporally articulated event with both a past and a futural dimension. It is easiest to explain this with reference to linguistic understanding, although Heidegger sees a general structure here that also holds for our understanding of the equipment we deal with in daily life, as well as our understanding of ourselves and our situation of action. Understanding always proceeds out of a pre-given context (the dimension of the past, or what Heidegger calls "having-been-ness" [*Gewesenheit*]) on which a person has a more or less stable and explicit grasp and in terms of which he forms certain anticipations regarding what is coming next (the dimension of the future). The better your prior grasp on (your "pre-understanding" of) the context, the better you will be able to make sense of the content. So, for example, it is much easier to understand a conversation you hear between two people (especially if it is in a language that is not your mother tongue) if you already understand something about the subject matter and have a sense of the positions and commitments held by the interlocutors.

Next, part of the experience of understanding is the satisfaction of a more or less vague anticipation of what is coming next. If you are understanding the words you are reading right now, the words and phrases that immediately follow next will not come as a shock. The more you read of a book or listen to someone speak, the more you know about where he is coming from and the better you can anticipate what he will say next, where he is going; this is why reading a book or watching a movie twice or three times can be such a different experience; the pre-understanding in terms of which you initially approach the text gets enriched and altered with each repetition.

The projective or anticipatory aspect of understanding normally is so much a matter of course that it is not noticed. It becomes evident in cases of breakdown, when your anticipations fail to be satisfied. In such a case of conversational breakdown we might say, "I thought I knew where you were going with that, but now I'm not *following you* anymore." These expressions track the *futural* dimension of understanding, what Heidegger calls "projection" [*Entwurf*]. Gadamer's comment on this is helpful:

> A person who is trying to understand a text is always projecting. He projects a meaning for the text as a whole as soon as some initial meaning emerges in the text. Again, the initial meaning emerges only because he is reading the text with particular expectations in regard to a certain meaning. Working out this fore-projection [*Vorentwurf*], which is constantly revised in terms of what emerges as he penetrates into the meaning, is understanding what is there.
>
> (Gadamer 2004, 269)

When we no longer follow the meanings being presented to us, we need to readjust ourselves and revise the initial grasp on the situation that gave rise to our frustrated anticipation, that is, we need actively to appropriate our "projections" and attempt to figure out why we have misunderstood and formed faulty and unsatisfied anticipations of meaning. As Moran put it, this is the moment in which we tend to a "revision or correction of the assumptions" in terms of which we were initially oriented in the conversation or inquiry. In Heidegger's view, this is the work of interpretation, which he describes as the active development or cultivation [*Ausbildung*], working-out [*Ausarbeitung*], appropriation [*Zueignung*], and making-determinate [*Bestimmen*] (BT 188, 203, 89) of the pre-understanding at work in our initial grasp of our situation. Those interpretive determinations or appropriations accordingly feed back into our sensitivity to the context and then serve to aid the generation of more precise and adequate understanding and anticipations of meaning. This is, abstractly speaking, how the process of coming to a better understanding by way of an interpretation works.

As Frankfurt saw with respect to the particular case of practical deliberation, all processes of coming to an understanding involve a projective (future-oriented) determination of meaning based on one's prior understanding of the context in which he already finds himself (past-oriented). But, according to Heidegger, the fact that one's deliberation concerning what to do (and one's understanding of written and spoken language) is always pre-structured in terms of a prior substantive understanding of oneself and one's situation does not reveal any special problem in need of a solution, as Frankfurt thinks. As a structural feature of understanding and interpretation itself, the hermeneutic circle is not a threat to the intelligibility of the practical question and it is not something to be avoided, straightened-out, or grounded in some deeper fact. In Heidegger's words, "What is decisive is not to get out of the circle, but to come into it in the right way . . . It is not to be reduced to the level of a vicious circle, or even a circle which is merely tolerated" (BT 195). Frankfurt does seek to get out of the circle. He does so by conceiving the most basic personal commitments presupposed in any concrete posing of the practical question on the model of *naturalistic facts* foundational for the normative and existential space of human everyday life.

5. Frankfurt's Misunderstanding of the Circular Structure of Care

From the hermeneutically sensitive premise that practical reflection operates within a prior orientation in which the person already finds himself and which he cannot freely or rationally choose, Frankfurt draws the following unjustified conclusion:

> The most fundamental question for anyone to raise concerning importance cannot be the normative question of what he *should* care about. That question can be answered only on the basis of a prior answer to a question that is **not normative at all,** but **straightforwardly factual—** namely, the question of what he actually does care about.
>
> (Frankfurt 2006b, 23–24; my bolding, italics in the original; see also RL 26)

Conceiving our basic commitments on the model of "straightforward facts" not only allows Frankfurt to plug up what he mistakenly regards as a problematic regress; it supports one of the most important theses involved in his battle against conceptions of human agency and selfhood that are "excessively intellectualized or rationalistic" (Frankfurt 2002, 184). The claim is that a person's deepest commitments ("volitional necessities") are not based in *any*—actual or possible—consideration of the reasons supporting them. Frankfurt asserts that "loving is not the rationally determined outcome of even an implicit deliberative or evaluative process" (Frankfurt 2006b, 41). On the contrary, loving is the factual basis for any practical reasons a person has: "it is the ultimate ground of practical rationality" (RL 56). By construing basic commitments on the model of natural facts, Frankfurt thereby seals them off from being grounded in reasons and secures the foundational status he wants to attribute to them, thus chagrining the rationalist.

In order to support the idea that our prior fundamental commitments are best seen on the model of a factuality and located totally without the space of normativity, Frankfurt appeals to what become in his later writings his central paradigms for the logic (or *lack thereof*) of being identified with one's commitments and motivations: a person's "commitment to the continuation of [his life], and to the well-being of [his] children" (RL 29). Questions of justification or evaluation do not arise here because these most basic commitments are "biologically embedded" and thus "determined for us by biological and other natural conditions, concerning which we have nothing much to say" (RL 30, 48). Driving this point home, Frankfurt adds that the love of living is a "prerational urge" (Frankfurt 2006b, 37) with a special foundational status:

> [O]ur interest in staying alive has enormous scope and resonance. There is no area of human activity in which it does not generate reasons . . . Self-preservation is perhaps the most commanding, the most protean, and the least questioned of our final ends. Its importance is recognized by everyone and it radiates everywhere.
>
> (Frankfurt 2006b, 35)

How is the importance of staying alive supposed to "radiate everywhere"? I can see that love of living gives me reason to look both ways when I cross the street and to avoid drinking bleach, but how does love of living "give rise to the more detailed interests and ambitions that we develop in response to the specific content and course of our experience" (Frankfurt 2006b, 38)?

Frankfurt's point comes across more clearly when we see it in the context of a criticism he makes of Bernard Williams. According to Williams, the primitive drive to go on living that Frankfurt claims is rock bottom itself needs to be sustained by the person's having certain fundamental projects or "categorical desires" that make life worth living at all for him, that is, which settle the question of whether or not they have any interest in just staying alive in the first place (Williams 1973 and 1981). For Williams, life itself is not a meaningful final end. Our interest in staying alive derives from our investment in certain projects and relationships. Frankfurt's view is directly opposed to Williams's:

> Surely Williams has it backward. Our interest in living does not commonly depend upon our having projects that we desire to pursue. It's the other way around. We are interested in having worthwhile projects because we do intend to go on living, and we would prefer not to be bored.
>
> (Frankfurt 2006b, 36–37)

With this, we can now turn to Heidegger's version of the relations between care, death, and time.

6. Heidegger on the Circularity of Care

Like Frankfurt, Heidegger emphasizes the receptive aspect of our identities. Things *already* matter to us. We do not answer the question of who and how to be by standing back and neutrally choosing or endorsing which identities to have. Prior to all of that we already find ourselves with—or, in Heidegger's phrase, we are already *thrown* into—a richly textured practical orientation whose motivational efficacy does not depend on our being aware of it or seeing reasons for it. For Frankfurt, this prior orientation is conceived primarily from the perspective of what matters to *me*, in *the first-person singular*, whereas for Heidegger, a person's prior practical orientation has to do with what tends to matter to *us*, in the *first-person plural* (though, as we'll see, Heidegger uses the third-person singular—what matters to "one"—to make his point).

According to Heidegger, I get the prior orientation in terms of which I am able to pose and answer the question of who and how to be not by a brute instinctual endowment, but by having been socialized into an everyday normal way of doing things, of understanding what is important and appropriate, and by non-reflectively taking over a range of identities that express the normal way *one* does things. "The one" [*das Man*] is Heidegger's substantivized term for the functioning of social normativity; it comes from such phrases as "One eats noodles with a fork; One drives on the right side of the road; One uses chalk to write on the chalk board, not to write on the wall or to throw at students," and so on. Heidegger writes:

> From the world [Dasein] takes its possibilities, and it does so first in accordance with the way things have been publicly interpreted by the one. This interpretation has already restricted the possible options of choice to what lies within the range of the familiar, the attainable, the respectable—that which is fitting and proper.
>
> (BT 239)

The shared everyday way of understanding what is important and appropriate tends to orient us in the world without our being reflectively aware of it as such, behind the back of our self-consciousness; it operates tacitly with "inconspicuousness and unascertainability [*Unauffälligkeit und Nichtfeststellbarkeit*]" (BT 164), and thus with a self-evidence or "instinct-like" immediacy. Heidegger worries that this everyday inherited way of understanding ourselves enables and *encourages* us to avoid a real, *eigentlich*, confrontation with the question of who I am and what finally matters, and tends to pressure one to absorb oneself in more or less trivial and self-instrumentalizing forms of taking care of our "urgent" everyday business and daily grind. As an unavoidable feature of human identity, this tendency is a built-in liability, a tendency towards living out my life, not in terms of my own best sense of what really matters and what this particular situation calls for from me, but in terms of the safe and conventional thing that *one* does.[8]

Heidegger calls this "fallenness": "Dasein has, in the first instance, fallen away from itself as an authentic ability-to-be its self" (BT 220). As such, the person allows the question of who and how to be to be *answered by what "one does"*: "The one . . . supplies the answer to the question of the '*who*' of everyday Dasein" (BT 165–166). This condition is what Heidegger calls "inauthenticity" [*Uneigentlichkeit*]. It is a "socialized" version of heteronomy, a matter of being moved into action by socially imposed and sanctioned motivations with which the person himself does not finally identity.

In the Heideggerian view, a person's own experience of Frankfurtian wholehearted identification with such social motivations cannot be taken at face value. A person may have the experience of being identified with effective motivations of "the one" (they can be so obvious that questions do not arise), yet nevertheless be alienated from these determining grounds of his action (they may not reflect his own best sense of his final commitments and what is to be done in a particular situation of action). According to Frankfurt's later view, the mark of being identified with one's motivations and thus of being autonomous, is "satisfaction," which amounts to a lack of "interest in making changes" (Frankfurt 1999b, 104). Frankfurt compares satisfaction to being relaxed (Frankfurt 1999b, 105n16). Yet, for Heidegger, two features of inauthenticity are (1) tranquilization [*Beruhigung*] (BT 222) and (2) untroubled indifference [*unbehelligten Gleichgültigkeit*] (BT 299). A person with such comfort "has no urge for anything," Heidegger writes, adding that such self-satisfaction "demonstrates *most penetratingly* the power of [self-] forgetting in the everyday mode of concern which is closest to us" (BT 396). Going along with the habits and routines of one's social milieu may feel perfectly natural and relaxing, but that should not be taken as conclusive evidence of being *identified* with them. Such a satisfaction with *what one does* may actually be a symptom of a failure, a "forgetfulness" regarding the challenge to actively appropriate my own individual identity and ultimate commitments.

Hence, although Heidegger and Frankfurt both prioritize a person's receptive sense of how things *already* matter, they conceive of the significance of the preceding dimension of our identities in thoroughly different ways. Frankfurt overlooks the social aspect of human agency that preoccupies Heidegger and that figures into his hermeneutic conception of the circular structure of human identity. Lacking any other resources for combatting the claims of the rationalistic conceptions of the self he opposes, Frankfurt ends up conceiving of the preceding, receptive dimensions of identity on the model of natural instinct, identification with which supposedly provides an ultimate, spade-turning foundation for individual identity and practical normativity. Heidegger conceives of the prior and receptive practical orientation in terms of a tacitly operative and socially shared sense of appropriate and acceptable patterns of conduct: what *one does*. Doing so importantly reframes the issue of whether or not a person identifies with his or her basic, pre-reflective motivations, because the practical orientation provided by our social upbringing can seem so natural and a matter of course that questions do not arise about it. But this is no guarantee of our being identified with such motivations.

Heidegger's view, then, leads him to pose the question concerning the possibility of identification with the preceding social dimensions of identity that mediate our individual self-understanding. Heidegger's response to this issue leads him to a reflection on the basic *futural* dimension of human life: its directedness towards death. This offers a suggestive parallelism and subtle corrective to Frankfurt's appeal to the supposedly foundational role played by an instinctual fear of death.

Whereas Frankfurt argues that the substantive identity of the individual self gets its pre-rational ultimate grounding in an instinctual avoidance of death, Heidegger argues that human identity can never bottom out in such ultimate facts. Yet, in Heidegger's picture too, death is somehow the source of importance in the world. Moreover, appreciating my special relation to my own impending death is what enables me to be released from the satisfying, instinct-like grip of my taken-for-granted, everyday motivations, and thus provides the possibility of being more fully receptive to my own unstable sense of what and how things finally matter so that I may express that in my activities. I qualify this as an "unstable" sense in order to highlight a crucial aspect of Heidegger's conception of death and human finitude: no sense of what matters, no matter how self-evident or satisfying it may be, can be taken as the ultimate, brute ground of identity that Frankfurt is looking for. In order to spell all of this out in more detail, we have to turn to Heidegger's conception of death.[9]

7. Heidegger on Death and the Importance of Being Finite

We should first note the distinction Heidegger makes between "death" conceived (1) as the event of the cessation of biological functions of plants and animals, what he calls "perishing" [*Verenden*] (BT 284); (2) as a significant event that inevitably happens at the end of a *person's* life, what he calls "demise" [*Ableben*]; and (3) death as "a way of being," what he calls, variously, "dying" [*Sterben*], "being-towards-the-end" [*Sein zum Ende*], and "being-towards-death" [*Sein zum Tode*] (BT 291 and §§ 50–51).

Heidegger describes death as the final "end" [*Ende*] of Dasein (BT 276). To be towards this end is to be an essentially *finite* [*endliche*] entity. To *be finite* in Heidegger's sense is to relate to your own being as something that *matters* and calls you to give it significance (a stand on who you are and how you are going to be). In Heidegger's words, to be finite is for your own being to be *an issue* for you, to be ever *at stake* for you. It is our understanding of the fact that our lives end in *demise* that makes it possible for things to be important to us. Importance obtains only in a world of mortals (beings who perish) who *understand* that they are *going to die* (beings who demise); these are the beings who exist in the manner Heidegger calls "being-towards-death."[10]

Heidegger's claim that Dasein's way of being is a finite being-towards-its-death is essentially a reformulation of the basic claim that Dasein's own being is an issue for it, the claim that to be a person is to be always and unavoidably faced with the questions: Who am I? How should I be? This doesn't mean that we are always anxiously going around like a character in a Woody Allen movie in a state of existential breakdown, wondering what the meaning of it all is. Rather, the question of who I am is never settled once and for all, as though it were a question of observer-independent fact. For Heidegger, to be open to the possible revision of my commitments and sense of identity is a sign of hermeneutic health, not of identity-undermining ambivalence.

With his conceptions of demise and death I see Heidegger trying to get at the following train of thought. Coming to appreciate the fact that inevitably my life will come to an end can result in a *relativization of the trivial vis-à-vis* what calls me as important and worthwhile.[11] I get a better grip on what is of final importance to me and actually makes my life worth living, and I experience a call to give myself an identity in terms of that rather than letting my identity be determined by unquestioned ideals and prejudices concerning what is important, what *one* should do. In short, the fact that I will perish and meet my demise gives me something *to do*, for it is because I die that my own being can matter to me in a special way. Living only matters because we die.

From the Heideggerian perspective, the very fact that there are significant differences between what is worthwhile for us and what is trivial is grounded in the fact that we will all, in Raymond Chandler's words, "sleep the big sleep." Life is not in itself a worthwhile final end and source of meaning, at least not without respect to its ultimate termination in death. This is what we can call "the importance of being finite." I will now sketch out the position that I think is behind Heidegger's emphasis on death and the importance of being finite. I'm aware that being convinced by the following "argument from the importance of being finite" would involve some auxiliary assumptions that I don't defend here, but it nevertheless helps to capture the spirit of Heidegger's conception of death and its relation to time for our present purposes.

First, when something matters to you, this is not only an affective state; it involves a temporally articulated pragmatic disposition. Mattering is a matter of *what* you do *when*. Mattering consists in how the commitment in question has an effect on what you do and when, how your projects are prioritized and hierarchized in time. Second, it is because we have an essentially limited amount of time in life—because we will meet our demise—that we have the need to give our time and activities a hierarchical structuring. If you had unlimited time as an immortal you would have no pressure to do something now rather than later. It would not matter *when* you do what, or *what* you do when. You could do it all later.

It is the definite limit imposed by demise that puts the felt pragmatic pressure on our time, the pragmatic pressure to relativize the trivial. Without such limitation significant differences, pragmatically speaking, are leveled; it would become practically impossible for one thing to matter more than another, because it would not essentially matter when you do one thing rather than another. This is why living itself is not the source of value, life is not itself the ultimate source of our interest in having worthwhile projects. Death is. Death is the background against which the distinction between trivial and worthwhile can show up. Death is our deadline.

Heidegger's view of "inauthentic everydayness" as bogged down in instrumental trivialities is directly connected to his view that it promotes a "constant tranquilization about death" (BT 298). The tranquilization is expressed in the everyday attitude to death that presents it as a neutral, third-person event: "One of these days one will die too, in the end; but right now it has nothing to do with us" (BT 297). In such an attitude the impending fact of my own death is avoided, I live life *as if* I were immortal, and can come back to what really matters to me *later*.

The proper "authentic" [*eigentlich*] response to human mortality, the response in which the inevitability and finality of death is grasped as the source of importance such that it pervades and temporally structures one's way of life, is what Heidegger calls "anticipation" [*Vorlaufen*]. Anticipation involves becoming receptive to and getting a grip on your "ownmost" possibilities: those things and activities that finally matter for their own sake according to you (see BT 308 and 435). Living authentically means structuring your everyday activities in terms of such commitments, yet at the same time living in light of the possibility that you may fail at them, or that you may have to reject, revise, or re-hierarchize them. You have no ultimate foundation for your identity.

8. Conclusion

The point of the foregoing account of Heidegger's conception of human finitude was to explain how he maintains a point that is one of Frankfurt's animating concerns, and does so without a misguided appeal to a supposedly foundational instinct for self-preservation. The Frankfurtian concern is to explain how there can be a distinction between final ends that matter for their own sake and define a person's identity, on the one hand, and trivial or merely instrumental ends, on the other. The mode of agency that manifests the former is agency that is both autonomous and meaningful, what Frankfurt calls "self-identified" and what Heidegger calls "authentic."

Heidegger's conception of the importance of being finite provides a compelling alternative to Frankfurt's claim that all importance ultimately "radiates" from biologically embedded foundational instinct for self-preservation. We have seen how Frankfurt was driven to the instinct-model of care because of a misunderstanding of the significance of the hermeneutic circle and the fact that we cannot choose and thus always presuppose a sense of what is finally important. The fact that something is always presupposed in the posing of the practical question does not grant it any special or fundamental priority, only a provisional one. Heidegger builds directly into his view a recognition of the fact that, in the course of living and struggling with the question of who and how to be, a person may be lead radically to revise the most basic aspects of his or her own self-understanding, even perhaps the commitment to staying alive. Previously unquestioned or unquestionable identities might have to be "taken back" (BT 355) or "given up" (BT 443) in accordance with the demands of a particular situation or in favor of what comes to grip the person as a higher end. These are the risks and challenges that accompany the importance of being finite.

Notes

1. I'll refer to *Being and Time* (Heidegger 1962) as "BT" from now on.
2. The "for-the-sake-of-which" [*das Worumwillen*] is Heidegger's name for the substantive final ends or commitments that define a person's identity. See Heidegger (2009, § 11) where he explicitly connects his notion of the "for-the-sake-of-which" [*das Worumwillen*] to Aristotle's conception of final ends.
3. I'll refer to *The Reasons of Love* (Frankfurt 2006b) as "RL" from now on.
4. This is almost exactly how Ernst Tugendhat interprets Heidegger's conception of "inauthentic fallenness," which we will discuss later. See Tugendhat (2001, 81).
5. I divide Frankfurt's work into early, middle, and late periods. The periods can be differentiated according to the way he conceives of the phenomenon of identification and the primary examples he uses in his explanation of it. In the first or early phase, the example is that of a willing addict who is moved irresistibly by his physiological addiction to nicotine but who desires—that is, *prefers*—to be so moved. In the second or middle phase, Frankfurt's favored example is Martin Luther, who is also moved into action irresistibly by means of a necessity, not of physiological addiction, but of personal commitment of the deep kind that Frankfurt calls "volitional necessity." Luther is irresistibly and yet willingly impelled by his most cherished commitments to an act of protest, and in doing so expresses his sense of what is most important. In Frankfurt's third or late phase, Luther drops out and the privileged example becomes the parent who, given his biological makeup, irresistibly loves his children, or the person

who, given his natural instincts, irresistibly loves life and has an uncontrollable urge to stay alive.

6. Heidegger also comes to see boredom as a phenomenon of human being with rich existential implications. This is not a coincidence, but there is no space here to discuss the significance of this overlap. See Heidegger (1995) for his analysis of boredom.

7. For more on the "primitive human need" of psychic self-preservation, see Frankfurt (1999a, 88–89), and Frankfurt (1999d, 139).

8. See Rousse (2013) for a detailed explanation of how Heidegger's focus on the social dimensions of human identity deeply shapes his account of agency, motivation, and answerability.

9. Among recent interpretation of Heidegger's conception of death, the view I present here is close to those sketched by Wrathall (2005) and Richardson (2012).

10. By claiming a pivotal role for the event of demise in Heidegger's understanding of the significance death, I am going against the grain of some of the best recent Heidegger scholarship on this issue. See, for example, Thomson (2013) and Schear (2013). Both Thomson (2013, 217) and Schear (2013, 365) construe Heideggerian death as a world-collapse (collapse of the intelligibility of a way of life), which has little to do with the commonsense understanding of death as the event that happens at the end of life. John Richardson also criticizes other interpreters for construing Heideggerian death in a way that implausibly "takes it out of any direct relation to the biographical event of death" (Richardson 2012, 146–147n15). Richardson, like me, argues explicitly that the impending event of demise plays a crucial role in Heidegger's account (Richardson 2012, 152).

11. Here I am indebted to helpful conversations with Cristina Lafont. I am also grateful for conversations and exchanges with Roman Altshuler, Hubert Dreyfus, Axel Honneth, Michael Sigrist, and audiences at Berkeley, the American Philosophical Association in San Francisco, George Washington University, Northwestern University, and the University of Frankfurt.

References

Blattner, W. (1999) *Heidegger's Temporal Idealism*, New York: Cambridge University Press.

Frankfurt, H. (1988) "The Importance of What We Care About," in *The Importance of What We Care About: Philosophical Essays* (pp. 80–94), New York: Cambridge University Press.

—— (1999a) "Autonomy, Necessity, and Love," in *Necessity, Volition, and Love* (pp. 129–141), New York: Cambridge University Press.

—— (1999b) "The Faintest Passion," in *Necessity, Volition, and Love* (pp. 95–107), New York: Cambridge University Press.

—— (1999c) "On Caring," in *Necessity, Volition, and Love* (pp. 155–180), New York: Cambridge University Press.

—— (1999d) "On the Necessity of Ideals," in *Necessity, Volition, and Love* (pp. 108–116), New York: Cambridge University Press.

—— (1999e) "On the Usefulness of Final Ends," in *Necessity, Volition, and Love* (pp. 82–94), New York: Cambridge University Press.

—— (2002) "Reply to T.M. Scanlon," in *Contours of Agency: Essays on Themes from Harry Frankfurt*, ed. Sarah Buss and Lee Overton. Cambridge, MA: MIT Press.

—— (2006a) *The Reasons of Love*. Princeton, NJ: Princeton University Press.

—— (2006b) *Taking Ourselves Seriously & Getting It Right*, ed. Debra Satz. Stanford, CA: Stanford University Press.

Gadamer, H.-G. (2004) *Truth and Method*, revised translation by J. Weinsheimer and D. Marshall, New York: Continuum.

Heidegger, M. (1962) *Being and Time*, trans. J. Macquarrie and E. Robinson, San Francisco: Harper & Row.

—— (1995) *The Fundamental Concepts of Metaphysics: World, Finitude, Solitude*, trans. W. McNeil and N. Walker, Bloomington: Indiana University Press.

—— (2009) *Basic Concepts of Aristotelian Philosophy*, trans. Robert Metcalf and Mark Tanzer, Bloomington: Indiana University Press.

Moran, R. (2007) "Review Essay on The Reasons of Love," *Philosophy and Phenomenological Research*, 74:2: 463–475.

Richardson, J. (2012) *Heidegger*, New York: Routledge.

Rousse, B. S. (2013) "Heidegger, Sociality, and Human Agency," *European Journal of Philosophy*, doi:10.1111/ejop.12067.

Schear, J. (2013) "Historical Finitude," in *The Cambridge Companion to Heidegger's Being and Time*, ed. Mark Wrathall, New York: Cambridge University Press.

Thomson, I. (2013) "Death and Demise in *Being and Time*," in *The Cambridge Companion to Heidegger's Being and Time*, ed. Mark Wrathall, New York: Cambridge University Press.

Tugendhat, E. (2001) "Über den Tod," in *Aufsätze: 1992–2000*, Frankfurt: Suhrkamp.

Williams, B. (1973) "The Makropulos Case: Reflections on the Tedium of Immortality," in *Problems of the Self* (pp. 82–100), New York: Cambridge University Press.

—— (1981) "Persons, Character, Morality," in *Moral Luck* (pp. 1–19), New York: Cambridge University Press.

Wrathall, M. (2005) *How to Read Heidegger*, New York: W.W. Norton.

15 Merleau-Ponty on the Temporality of Practical Dispositions

David Ciavatta

From the point of view twentieth-century phenomenology, our understanding of time must begin with an account of our lived experience of it. It is in the practical domain that our confrontation with time is most pronounced, and so it is not surprising that for many of the thinkers working within the phenomenological tradition there is an intimate link between our experience of time and our experience of ourselves as agents. In the work of Bergson, for instance, only free agents—beings whose futures are fundamentally indeterminate—demand to be thought of in distinctively temporal terms. A purely deterministic world would make time ontologically superfluous, on Bergson's account, for all future events would be already implied in what is present, and all past events would be simply irrelevant insofar as they no longer had any determinative, causal force in their own right. In contrast, our existence as free agents must be conceived in terms of a continuous temporal duration whereby the past is not simply over and done with, but is prolonged and made effective in the present course of action precisely through the agent's retention of it in memory; moreover, even the future shapes the character of the agent's present, for on Bergson's account, we only ever perceive that portion of the present world that aligns with the range of possible actions that we anticipate performing in it.[1] In the thought of Sartre the link between freedom and the experience of time is likewise brought to the forefront, for on his account it is precisely our basic experience of freedom that introduces a genuine negation into the fabric of being, and as such we as agents become the basic ground of such negative phenomena as "not yet" and "no longer," phenomena that are essential to the experience of time.[2] Even the basic Husserlian notion that our perceptual experience of the present is essentially encompassed by a horizon of potential further experiences seems to recognize a special connection between time and agency. For the horizon of potential experiences consists essentially in an anticipation of a not yet determined future, and our very openness to this future seems to be grounded in an experience of an "I can."[3] The experience of freedom and our experience of time, it seems, are intertwined in a fundamental way in such accounts.

My goal in this chapter is to articulate what I take to be one of Merleau-Ponty's distinctive contributions to this phenomenological account of the link between time and agency. In the *Phenomenology of Perception* (1962), Merleau-Ponty's focus on the fundamentally practical orientation of the lived body, and on its capacity to develop new habits, draws our attention to a basic form of bodily agency that is quite distinct from our more familiar conception of a self-conscious, thinking agent that acts on the basis of reasons and consciously formed intentions. We find this bodily agency at work in such basic natural processes as breathing and balancing and eating, but for Merleau-Ponty it is also the basis of our more developed practical dispositions and the psychological "complexes" that underlie our general characters. Part of what is distinctive of Merleau-Ponty's discussion of the link between time and agency is that he works towards clarifying the distinctive form that our lived experience of this level of agency takes, and then attempts to articulate the experience of time that is peculiar to it. Merleau-Ponty argues that the way we experience time at this bodily or dispositional level is quite distinct from the more familiar experience of time we have as free, self-conscious agents occupied by the various personal decisions we find ourselves having to make on a situation-by-situation basis.

Among other things, in the latter experience agency is essentially episodic in nature, in the sense of having to exert itself anew in each fresh instance of action. As such, this experience is correlated to an experience of a time that can be broken up into distinct episodes, distinct "nows," whose individuation is correlated with that of the individual moment of decision. In contrast, Merleau-Ponty would have us see that the operations of our bodily agency are fundamentally cyclical, and so non-episodic in fundamental ways, with the result that our bodily agency is correlated to an experience of a time that is not so readily divisible into distinct moments. Thus, rather than being alive first and foremost to the unique "now" of some particular practical situation I may be in, at the level of my bodily agency we might speak, instead, of the general now of my life as a whole—a now that is not present to itself all at once, in this or that particular instance, but that is relatively indifferent to particular situations. My goal is to bring this latter experience of time to the fore, and to suggest how our practical dispositions are structured in terms of it.

1. Natural Time and Historical Time

At several points in the *Phenomenology of Perception*, Merleau-Ponty sees fit to make a distinction between two basic kinds of temporality: the temporality specific to historical events, including the biographical events that make up the core of our personal histories as individuals; and the temporality specific to generic, cyclical processes of nature, including those processes of our living bodies that we share with other humans and animals (1962, 82–85; 346–347; 422–424; 449–453). We, as embodied agents, are historical and natural at once, and every phenomenon of human life will to some degree be informed by both these dimensions, and so by both sorts of temporality. However, certain aspects of our lives are clearly more defined by historical time, and others more by natural time.

For instance, our breathing and blinking patterns are almost purely natural, and so help to reveal the sense in which we are thoroughly implicated in the cyclicality of natural time.[4] For Merleau-Ponty, natural time is most essentially grounded in a structure of *indefinite repetition*, and thus in a *lack of meaningful differentiation and individuation among temporal moments*.[5] In the case of breathing, for instance, I breathe now only to breathe again, and then again; my current breath does not mark itself off as significantly distinct from my past breaths, nor does it take itself to build on what my previous breaths have accomplished: with each breath, we are, so to speak, right back where we started the last time; each breath is a wholly fresh start on the exact same process, facing what is from breathing's point of view the exact same situation as the last breath. Each "episode" thus immediately dissolves, leaving no trace and no memory of itself as distinct.

Moreover, the ongoing, cyclical character of breathing is such that there are no decisive, internal beginnings or endings within the overall flow of the respiratory process. While we might be tempted to treat the exhale as the completion of a process begun by the movement of inhaling—as though there were some sort of narrative structure involved in each breath cycle—the exhale is just as much that which calls for the next inhale as it is the end of the previous one. It seems that what really exists here is not primarily the distinct episodes of breathing, each temporally individuated in its own right, but rather the general, *continuous* process of breathing as such. And though, of course, this continuous process only maintains itself in and through distinct episodes of breathing, these episodes *don't matter as such in their distinctness*, but are each in the service of enabling the *continuous flow of breathing* to continue on indefinitely. If we consider that each breath exists in the present, as happening always "now," it seems we must say that each of these "nows" is at bottom *identical* from the point of view of the process of breathing itself; breathing itself carries no historical register within which each breath would take its unique, dated place, but is wholly concerned with making sure that the same thing is happening continuously. In a way, it is more appropriate to speak of the *general "now"* of breathing in its overall continuity—a now that, in our case, lasts the duration of our respiratory lives—rather than an identical now that is *repeated*; for even the notion of repetition seems to be premised upon retaining the temporal individuation of the moments repeated, and this individuation is precisely what seems to be dissolved in such a natural cycle.[6] This privileging of the general now, this stretched-out present, over the more circumscribed, episodic now of the passing moment is, on Merleau-Ponty's account, a constitutive feature of natural time and, indeed, of nature in general.[7]

Historical time, in contrast, is the time specific to eventful unfoldings and narrative developments that begin and lead somewhere, rather than being stuck in an indefinite, repetitive cycle.[8] Within historical time, the now is not wholly indifferent to its past. The historical present is never a wholly fresh start, nor a pure, identical repetition; it carries its past within it *as something past*, and thus *distinguishes itself from it while taking it up into itself* (whether to develop, surpass, depart from, or imitate it). Moreover, the historical present is never wholly cut off from its future, either, for it is in part structured in terms of an anticipation of what is to come, much as, while listening to a melody, our sense of the significance of the note we are currently hearing is informed, in part, by our anticipation of where the melody is leading us.

For Merleau-Ponty, that aspect of our existence that is most clearly structured in terms of historical time is *action*. Indeed, I would suggest that historical temporality is itself a function of the distinctive way in which agents take up time *qua* agents, for action inevitably involves an express differentiation and interrelation of past, present, and future.[9] For instance, if, on a given morning, I commit myself to performing certain tasks that day, I thereby reach out from the present moment toward the future, committing my afternoon in such a way that its temporal unfolding will constantly refer back to, or carry forward, the morning's resolution; it is on this condition, for instance, that the day can come to appear to me as "slipping away," as the fading daylight makes palpable to me that I am "running out of time" in the drama I have instituted. My capacity to make such ordinary commitments is thus founded upon the more basic capacity of my agency to imbricate the temporal dimensions with one another—in this case, by making the future answerable to my resolution in the present, which *will be* this future's specific past.[10]

However, it is clear that for Merleau-Ponty some of our actions—in particular, those that most individuate us as agents—more clearly exemplify the basic structure of historical temporality than others. For instance, many of our most repetitive, routine actions—like brushing our teeth or showering or eating—are basically extensions of our natural cycles, and they are comparably generic and impersonal in character. There was nothing remarkable about my act of brushing my teeth this morning, and I would be hardpressed to single out anything in this act that made it distinct from the thousands of acts of brushing that I have engaged in over the years; and even if a glimpse of my unique personality can be had by observing the particular way I brush my teeth, no one would deem such an action worth mentioning in a biography of my individual life. On the other hand, one's most individual, and personality-revealing, actions—a momentous confession of love, a decision to give up a successful career upon realizing one's heart has never really been in it, a decision to stand up against an injustice despite the fact that doing so exposes one to a threat to one's own livelihood—can claim to warrant chapter-length treatments in one's biography, for such actions seem to make manifest, in an incomparably heightened way, one's most constitutive commitments and characteristics. Such acts are typically "once in a lifetime" decisions, and are typically undertaken with a heightened sense of their singularity and momentousness. They are also typically undertaken with a sense of making a decisive break with the momentum of one's past life, and involve opening oneself to a new and indeterminate future—the embarking into a new chapter, one that may in fact become the decisive, dramatic core of the whole narrative that is one's individual life. Indeed, one can feel as if the *whole* of one's personal narrative is on the line in making such a choice—as if even one's past is not ultimately impervious to being reoriented and redefined by it, given that this past will *from then on* be experienced as the prelude to the particular new future this act alone opened up.[11] It is this momentousness, one's anxious sense as an agent in the midst of deciding, that everything is on the line *right now*—this sense that one's present will not simply "dissolve" into oblivion and be replaced by another present that is indifferent to it, but will from here on in be decisive for understanding all that came before and after it—that ultimately gives the historical present its singular, "once and for all" character for Merleau-Ponty.[12] It is in this respect that the historical present most distinguishes itself from the dissolving, indistinct present of the natural cycle (the present that immediately fades into the general present of an overall continuity).

2. Freedom and the Historical

Within the existentialist tradition that Merleau-Ponty is clearly drawing from, these sorts of heightened decisions are the clearest cases of *free* action. For Sartre, this capacity of an agent to break from the past and inaugurate a new future in the decisive singularity of the present is tantamount to the definition of freedom itself (1993, 21–45); and so in this tradition, and especially for Sartre, the philosophy of human agency comes hand in hand with an account of the distinctively *historical* character of our being, where this is understood specifically as an ability to negate and transcend our merely natural character.

Even leaving aside such heightened decisions, however, it seems that there is an important link between what we generally think of as free actions and the original, individualized character of the temporal present. It seems that, in principle, every practical situation, every distinct episode in which an action is called for, is an original, temporally individuated whole unto itself, and that an agent is free to approach each such situation on its own individual terms. The agent is free to settle upon a way of responding that is unique to the present demands of this situation as they combine with the particular desires, needs, and beliefs she has at the moment.[13] Each episode of free action, each moment of decision is, in effect, a singular and autonomous "now," differentiating itself from all other nows in part by the fact that the agent experiences it as a "live" now—that is, one that actually calls for a response on her part, rather than being a now that is remembered and that is, as such, over and already done with. This now of action, it seems, gets its individuality ultimately by way of the individuality of the agent and her sense that she personally is being called upon to respond, and in particular by the agent's singular *presence to herself* in being thus solicited. As Sartre argued, even if the agent is bringing to bear past commitments or motivations in the response she decides upon, or is anticipating future outcomes, this past and future are *internal* to the decision and have a force in informing it only insofar as the agent is, within the singular moment of decision itself, *actively granting* them their force, thus enabling them to be "live" and "present" concerns.[14] Decisions to act, it seems, necessarily take place in the singularity of the present, and insofar as they claim to be autonomous and free—that is, insofar as they are not *determined by* the past, but claim to be presently *conscious of* the past and *thereby give it* its motivational force in the now—they in effect constitute the present itself as a distinctive and original individuality, *discontinuous* with the past (and with the future).[15]

Against this background it is striking that, in his main discussion of freedom and agency in the *Phenomenology of Perception*,[16] one of Merleau-Ponty's overriding concerns is to explore not primarily those inaugural choices by which our lives take on a decisively new historical direction, but the general practical orientations or dispositions that persist across, and in that sense are repeated through, an agent's episodic, moment-by-moment choices. Among other things, such general practical orientations account for how it is that there is a general, meaningful *continuity* across our various choices—an overall style of agency, if you will—and as such they involve not the ability to interrupt the past and inaugurate new directions, but rather the ability to maintain a single, coherent practical identity over time.

It is clear that, over time, agents betray a certain continuous style of approaching their practical situations, and are not making radically new and utterly autonomous decisions in each singular episode of action.[17] We gradually begin to notice recurring themes and tropes in a person's behavior—for instance, we see that a person is prone to shy away from situations involving adversity, or that she is the sort of person who typically seizes on opportunities to aggrandize herself before others—and this gives us reason to think that, beyond the sphere of self-conscious, *individuated* choice, there is another sort of enduring, more generalized form of agency at work in human life.

It seems clear that such general orientations are not like fixed essences or natural laws that determine an agent's behavior in a situation in the same way that, say, being a tomato plant determines the various ways in which an individual tomato plant will actualize itself in response to the determinacies of its physical environment. In the case of a plant, we are inclined to think that, if it meets with generally appropriate environmental conditions, and is not plagued by any disease or other abnormality, it *necessarily* and *automatically*—as by an inner law or essence—grows towards its proper form, that it has no possibility of its pursuing another, alternate route of its own accord. To ascribe a general orientation to an agent, in contrast, is to acknowledge that this agent can at times drift away from, or even expressly pose a challenge to, her characteristic patterns of action, even if this generally does not happen. A person whose overall way of being is shaped by an inferiority complex can, on occasion, summon the courage to speak confidently in a public setting, even if, for the most part, such an act is unimaginable to him.[18] Talk of general orientations thereby makes room for the sphere of indeterminacy that seems to lie at the heart of the human condition, and that we typically associate with freedom and with the possibility of substantial self-transformation: we are not *stuck* in an endless repetition of our already established ways of being, but always have some room to modify the direction of our lives, even if only in gradual increments.

Despite this indeterminacy, however, to have a general orientation towards a certain way of comporting *is* for this orientation to have an actual bearing on what directions and choices are experienced as compelling or as "live options" in a given situation, and thereby determine the basic contours of the present situation itself. And so, though orientations do not *determine* what an agent does, they do indeed make it *more likely* that an agent will engage in certain actions over others.[19] If my friend is generally self-disparaging in his overall orientation, and has a tendency to experience himself as unworthy of others' recognition, the chances are high that he will not perform very well at a job interview in which he will be called upon to give a compelling case for the significance and value of his accomplishments. For, despite his conscious resolve to impress the interviewers, singing his own praises does not come easily to him, and is a much more remote possibility in his practical orientation than that of selling himself short. An awkward performance seems both *more probable*, and *more fitting*, more immediately harmonious with the general way of being that he radiates in his overall demeanor, and that displays itself in so many other avenues of his life.

Given these considerations, it seems we can speak of there being a kind of general commitment at work at the heart of human agency, a commitment that occurs at the level of a person's overall and enduring character. This commitment is inherently general, I want to suggest, in the sense that it is not reducible to a particular choice made in a particular episode of action. Nor is its persistence a matter of being consciously re-chosen again and again in successive, distinct episodes of action. Nor is the content of this commitment peculiar to a particular situation and the concrete actions specifically relevant to it. Rather, it is a commitment to acting a certain way in general, or to a general way of approaching practical situations; it is a commitment that, in its generality, exists as such *only insofar as it cuts across otherwise distinct episodes and situations of action*, and sustains a kind of enduring presence. But how are we to understand such general commitments as playing a role in our lives as agents?

3. Two Levels of Agency and the Generality of the Body

To the extent that we find ourselves called upon to make particular choices in particular, relatively discrete temporal episodes, those more general commitments that constitute our overall style of being an agent in the world must be conceived as grounded in a relatively distinct level of agency, one that operates according to a different logic than our moment-by-moment choices. It is helpful to understand this general agency, not as directly involved in making such particular, moment-to-moment choices, but in terms of its role in opening up and shaping the general character of our practical situations themselves, such that the particular choices we as agents make *within* these situations are, so to speak, always already drawing upon the power and efficacy of this more general level of agency.

To illustrate how these two levels of agency operate together, we can turn to Merleau-Ponty's basic discussion of the ways in which the general features of our bodily life underlie and inform all of our particular practical situations, thus lending these situations a certain generic character that makes them open to a range of particular actions in the first place. For example, because I have hands that can take hold of things and wield them, the things I encounter in each situation I am in immediately present themselves to me, from the start, in terms of their possibilities of being wielded, or, conversely, in terms of their unwieldiness.[20] Thus, whenever I am, in some particular situation, consciously occupied by a particular decision to take hold of something or not, my particular decision is always nested within what Merleau-Ponty calls a general, non-representational bodily intentionality, by which the world is first opened up for me, in general, as a practical field populated with things that can be wielded in the first place. This intentionality is non-representational in the sense that it does not consciously picture its end to itself before resolving to enact it. Rather, the body's actuality consists precisely in the continual holding open of a world of interactional possibilities, and the perceptual appearing of these possibilities, *qua its* possibilities, *is* the concrete work of this bodily agency. In this sense, my active body is conscious of itself *only by way of the world and its possibilities*, not directly, via some sort of "internal" self-consciousness.

The working of this bodily intentionality is, of course, not something that depends upon my self-conscious, episodic agency; and whether things in the world appear to me as wieldy or not is not simply a function of whether I happen to have chosen to take on some particular project that involves the use of my hands. Rather, by the mere fact that I have hands, and that my hands exist for me as living portals of certain general kinds of motility involving gripping, holding, carrying, and so forth, all that is presented to me is *already* implicitly measured up in terms of their general potentialities for being wielded. In this respect, the practical outlines of my situation are not only generally the same from situation to situation, but they are also generally the same as those encountered by other agents with similar bodies. So it is not simply I, as a fully individuated, choosing agent, who perceives this chair as something that can be picked up and moved across the floor; it is the impersonal, generic agency of my body itself.[21] If we can speak of this bodily agency or intentionality as working, as being involved in generating something, what it generates is precisely a general and enduring practical terrain within which particular episodes of wielding things can unfold. Thus, though my act of picking up this pen in order to write a letter in which I finally confess my love may be a singular moment in my individual history, it draws on the power of a much more general and generic bodily agency, from the point of view of which this situation is only one among many other potential bodily performances I might undertake with arms and hands. And, insofar as this "natural" agency, which (like breathing) maintains its continuity across my various episodic engagements of it in practice, is nevertheless *my* agency, there is a sense in which I too (in my generality, in the generality of my experience) am at some level indifferent to the singularity of my particular act, a sense in which this act and this situation "dissolves" or is forgotten in its individuality. That is, insofar as I live my body in its general and continued openness to wielding handy things, in this "layer" of my experience I do not differentiate between the significance of this instance of picking up a pen and other comparable instances; from the point of view of my lived body, the particularly of this instance is a matter of indifference, for what I am most alive to is the continuity of a world that affords possibilities of wielding things.[22]

4. The Body as Basis of General Practical Orientations

Though the practical orientations that make up an individual agent's personal character are not as general and impersonal as the body's role in giving rise to the general practical terrain in which all of our actions take place, Merleau-Ponty argues that the generality that is at the core of such character traits is ultimately of a piece with the generality of the body's agency; and so, an agent's general manner of being is likewise conceived as a sort of impersonal agency that determines the general contours of her practical situation in advance of her conscious consideration of this or that particular line of action in a particular situation.

To highlight the way in which an enduring practical orientation takes shape by way of the body's power to generalize, Merleau-Ponty frequently appeals to examples of repression drawn from psychoanalytic thought.[23] For example, drawing from Binswanger's work, he discusses the case of a young woman who, as a response to her mother's forbidding her to see the young man with whom she is in love, cuts herself off from interpersonal interaction altogether (Merleau-Ponty 1962, 160ff.). She does this, not by consciously engaging in *particular* and *isolated* acts of refusal, but rather by "losing" her very ability to speak; that is, her refusal becomes a general manner of being in the world, and it comes about by way of a kind of impersonal and general agency that works, not at the level of episodic, self-conscious choice, but at the level of that general bodily function of speech that forms the underlying horizon of possibility for all of her more developed acts of interpersonal communication. Because her bodily organs of speech open up for her the general world of interpersonal communication in general, she is able, precisely by way of this general agency of her body, to shut down her very access to this world. This is something she simply could not pull off in her capacity as a self-conscious choosing agent. As Merleau-Ponty argues, she cannot be said to choose not to speak, for "will presupposes a field of possibilities among which [an agent] choose[s]" (1962, 162), whereas what collapses for her, in her aphonia, is precisely that general field of possibilities of speaking or remaining silent. Just as a man who loses his arms will eventually cease to experience things in the world in terms of their possibilities of being wielded, so too does she cease to experience a world populated by interlocutors, and has ceased to feel any sense of solicitation or motivation when someone poses a question to her.

Of course, her loss of speech is not paralysis, and she can regain the use of her voice through therapy or other events in her interpersonal situation, thus showing that this loss is, in some sense, a sort of activity that she herself is in the process of performing and maintaining. But it is not an activity that consciously puts before itself some specific, finite objective to be attained, at which point the action would end. It has the structure of an indefinite reiteration that is not leading anywhere in particular; that is, it has lost any trace of temporal individuality and historicality. Part of what this example shows, then, is that the body's generality—in this case, its general and enduring openness, through its organs of speech, to a world of interpersonal communication—becomes the vehicle through which some distinctively *personal, historical event* in an individual's life—here, the woman's being cut off from her lover—can itself become generalized into something more impersonal, something that comes to determine the contours of an agent's overall engagement with the world as a whole. This singular event, which concerned her contact with one particular person, lost its historical specificity and became merged with the general horizon of intersubjective contact that mediates all of her particular encounters with others. It comes to shape her experience of what is actually there, what is actually available and possible, in her world.

In the specific case of repression discussed, the woman continues to repeat the singular event of being cut off from her lover, by in effect transforming it into the general atmosphere of all of her present circumstances; as Merleau-Ponty writes (in describing the structure of repression generally), "one present among all presents thus acquires an exceptional value; it displaces the others and deprives them of their value as authentic presents" (1962, 424). In such cases "nothing further happens," and "there occurs only a recurrent and always identical 'now', life flows back on itself and history is dissolved in natural time" (1962, 164). As in the case of breathing, each new now, each new situation of having to take a breath, is immediately "dissolved" into the same humdrum drama, and it no longer presents itself in its historical distinctness and singularity, but rather in terms of its being yet another opportunity to affirm one's continuous, general way of being. It is as if, like the yoga practitioner who strives to "become his breathing," the woman recedes completely into her impersonal, generic orientation to the world, with the result that nothing eventful—that is, nothing historical, no new narratives—can occur.

5. Concluding Remarks: General Orientations and the Generalized Present

While we would call this sort of fixation a pathological *complex* in the psychoanalytic sense of that term, Merleau-Ponty would have us see that understanding the basic temporal structure of such complexes provides us with a better understanding of both the generalized agency that forms the core of our characters as agents, as well as of the relationship between this generalized agency and the episodic, more personal form of agency typically associated with the moment of self-conscious choice. One of Merleau-Ponty's main preoccupations, in his account of freedom at the end of the *Phenomenology of Perception*, concerns the manner in which our past as agents is never so thoroughly individuated and circumscribed that we can, in the present, distance ourselves from it completely, so as to be able to choose a radically new path at any individual instant. Rather, our present situation of action is quite literally *merged* into the past, to the point that we are literally not in a position to *distinguish* between present and past; or, in other words, our choices of the moment are inevitably embedded within a more extensive "moment," a *more general now*—one that bleeds off into (what, from the point of view of a narrower present, is) the past, and that our "current," but more general, commitments continue to keep alive into the future.

Thus, for instance, a man who is characterized by proneness to be adverse to risk, and who thus tends to engage in behaviors of avoidance and escape when confronted with what is unfamiliar or threatening, does not experience his current threatening situation as an original, self-contained whole unto itself, within which various possibilities of fleeing and standing one's ground are all equally present on the immediate horizon, waiting only for his current decision to determine which will win the day so as to become definitive of his next present. Rather, akin to the way in which our having bodies with hands opens up a general world already oriented in terms of the exigencies of wielding things, to have a risk-adverse character is to have an impersonal agency that opens up a world that is *oriented*, from the start, in terms of familiar paths of avoidance and fleeing. Indeed, it is as though an evasive manoeuvre were already underway when the risky situation presents itself: there is already a general atmosphere of evasion at work, such that the possibilities of standing his ground seem all the more distant. His decision, otherwise a singular, self-grounding moment, is, as it were, already in the process of being made *before* he even arrives in this particular now; and this "fore-decision" takes the form of the determinacy of the situation he finds himself in, the way the situation itself already presents itself as soliciting in his body evasive maneuvers. But this is only to say that his past, having ceased to take the form of discrete episodes of action, and having become a general now, has to some degree dissolved the distinctness of his present, in the sense that he is at some level of his experience incapable of distinguishing between the two—much as in the breathing case.[24]

The agent in this case is not *conscious* of himself *as willingly repeating* a past way of being. For, as I have already suggested, to be conscious of a repetition is at least to be alive to the actual distinctness of the present from a discrete past, even while affirming that they are substantially similar in content. But if we can say, drawing from Merleau-Ponty's discussions of natural temporality, that the past "displaces" or dissolves the present, or that the past is literally "re-opened" in the present, or that the past is a sort of enduring wound to which our agency in the present continues to dedicate its power (1962, 85), it must be that at some level we exist in such a way as to *blur* the past and the present, giving rise in effect to a continuous, "generalized" present that does not have determinate borders, and that thus does not easily fit into a register of circumscribed moments that clearly come before and after other moments as in a historical narrative. This collapsing of the distinction between present and past, this generalizing of the past into the general atmosphere of the present, is tantamount to a lack of consciousness of any significant difference between the present and the past. It is as though one's practical orientations give rise to a sort of eternal now, precisely because they in effect present the current situation as in no way dependent on, or as responding to, past reiterations. Rather, as in the case of breathing—where one is "back where one started," in the exact same situation one was in during one's last breath, and where one's last breath in no way contributes to one's ability to respond to the current respiratory exigency—one is, as it were, absorbed in the present, but not in the sense of having a heightened awareness of this present's historical momentousness, its singularity; on the contrary, one is, in one's general orientation, lost in what is in effect a generic present, the same one now that was present earlier and that will continue to be there as one goes on to face further, otherwise historically new situations with the same active orientation in place.

To have a practical orientation, I have been suggesting, involves operating with a substantially different way of experiencing time than is typical of our more familiar conception of time according to which time takes the form of individuated, historically datable episodes. Insofar as we have some experiential footing in this generalized, non-episodic temporality, there is a way in which we ourselves are not wholly "in" the distinct episodes of our practical lives, a way in which we are at some level indifferent to the unique demands of the individual situations we face. While this indifference might, from the point of view of historical time, seem like a kind of deficiency—akin to the indifference of nature to our personal projects and achievements—such indifference is nevertheless necessary for rendering the contours of our experiential situations similar to each other over time, necessary for making the experience of the continuity of our practical world, across the diverse situations we face, possible. On Merleau-Ponty's account, both ways of reckoning with time are necessary for understanding the fuller phenomenon of human agency, and so any privileging of historical time and its link with freedom—as in Sartre's thought—must be tempered by a phenomenology of what I have been calling the "general now" of practical life.

Notes

1. The links between temporal duration and agency are dealt with in Bergson (2001) and (1991). The latter book is focused especially on the way memory and anticipation shape the agent's perception of the present.
2. See Sartre (1993, pt. 1, ch. 1; pt. 2, ch. 2). For an especially helpful discussion of Sartre's conception of the link between freedom and time, see Bouton (2014, ch. 8). Bouton's chapter on Bergson (ch. 7) is also worth consulting.
3. For Husserl's link between the notion of an indeterminate experiential horizon and the experience of one's own freedom, see, for instance, Husserl (1964, § 19). It should be noted, however, that Husserl's own explicit discussions of the constitution of time and of the nature of protention do not expressly point to freedom as a condition of time.
4. I refrain from identifying them as unqualifiedly natural because even these processes can come to be (and perhaps always are) expressive of a person's individuality. Thus, for instance, the yoga master's breathing patterns, and Bette Davis's blinking patterns, carry traces of their respective personal identities and history of action. It seems that the conditions for natural processes like this taking on such expressive meaning are: (1) that we are not *totally* oblivious to them (even if they only appear to us as background to our determinate experiences, on the remote fringes of consciousness); and (2) that our personal actions and ways of being have some causal influence on how they unfold, even if only in a very indirect way. It is also worth noting here that any habitual, highly routine actions—for instance, showering, brushing teeth, making our morning coffee—are also characterized by a generic, repetitive character, and can be virtually automatic in their unfolding. Merleau-Ponty discusses the case of a worker (the patient, Schneider) for whom even the fairly complex activity of making a wallet has become automatic, to the point that he is "scarcely aware of any voluntary initiative" while in the midst of making the wallet; it is as though the sequence of events unfolds of its own accord (1962, 105).
5. The time characteristic of natural cycles, writes Merleau-Ponty, "continually erodes itself and undoes that which it has just done," and the "present which it brings to us is never a present for good, since it is already over when it appears" (1962, 453).
6. From the point of view of our lived experience of breathing (in contrast to our express, voluntary acts of attending to it, as when we set ourselves to make it an object of observation), we tend not to be alive to distinct breaths except in cases in which the continuous rhythm of breathing breaks down, as when we "lose our breath" or have the wind knocked of us and find ourselves gasping for air. Indeed, in such extreme cases it is as though our *whole consciousness* becomes absorbed in our breathing, and we are not able to concern ourselves with anything other than getting the next breath. But it does not seem right to say that we are at other times simply *unconscious* of our breathing. Rather, our continuous, "normal" breathing recedes to become part of the general background of our experiential field and, rather than something that we are (unconsciously) monitoring in a moment-by-moment manner, it is something the experience of which takes the form of a general aliveness to the world around us.

7. I have argued for this claim at more length in a forthcoming work (Ciavatta, forthcoming). See also Barbaras's insightful discussion of Merleau-Ponty's later challenge to the idea that nature, understood essentially as generality and totality, can be individuated into distinct temporal moments (2001, especially 32–36).

8. Compare Merleau-Ponty's discussion of the "time of the event" in contrast to natural time (1962, 423).

9. This proposed link between historical time and agency is perhaps more pronounced in Bergson's philosophy than in Merleau-Ponty's. See, for instance, Bergson (1991) for an account of how our indeterminacy as free agents comes hand in hand with what he calls "intensive durations," which give temporal individuation to an otherwise undifferentiated material world characterized by repetition. However, note the implicit link between agency and temporality discussed in Merleau-Ponty (1962, 83–85), where Merleau-Ponty links personal agency with the way in which the present is individuated, and note Merleau-Ponty's discussion of the idea that "time is the foundation and measure of our spontaneity" (1962, 428).

10. See Merleau-Ponty (1962, 453) for his identification of historical time with the notion of commitment. For a helpful elaboration of some of the complicated ways in which the temporal dimensions implicate each other—and, specifically, the way in which the past "becomes past" in virtue of the present that references it as past—see Mazis (1992).

11. On the specific issue of how an action can inaugurate a new present in terms of which the past must now be understood, compare the discussion of expressive actions in Merleau-Ponty (1964). That the meaning of the past can be reconfigured by present action is a theme central to Sartre's analysis of freedom; see Sartre (1993, 637–647).

12. On Merleau-Ponty's account,

> [e]very present grasps, by stages, through its horizon of immediate past and near future, the totality of possible time; thus does it overcome the dispersal of instants [characteristic of natural time], and manage to endow our past itself with its definitive meaning. (Merleau-Ponty 1962, 85)

See also his claim that

> [e]ach present may claim to solidify our life, and indeed that is what distinguishes it as the present. In so far as it presents itself as the totality of being and fills an instant of consciousness, we never extricate ourselves completely from it, time never completely closes over it. (Merleau-Ponty 1962, 85)

Compare also Merleau-Ponty's description of the historical present as an "indestructible individuality" that has a "once and for all" quality (1962, 424). I am suggesting that these features of the distinctive historical present are intimately bound up with our decisiveness as agents.

13. Compare Merleau-Ponty's notion of an "authentic present" and its link with "personal time" (in contrast to the impersonal time of nature) (1962, 83).

14. See Sartre (1993, 29–34, 439–453), among other places. Merleau-Ponty privileges the present—as the locus of the subject's self-presence—in a very similar way (1962, 424). Compare also Merleau-Ponty's claim that all our available past thoughts, which exist for us as a horizon surrounding our present thinking, "draw their sustenance from my present thought, they offer me a meaning, but I give it back to them" (1962, 130).

15. See Sartre (1993, 32–35), for a discussion of how freedom is, essentially, the same thing as that process by which the present introduces a discontinuity between itself and the past, precisely by "negating" the past, and thereby making it *past* (i.e., no longer present).

16. See Part 3, Chapter 3, titled "Freedom."
17. See Merleau-Ponty (1962, 451): "If the subject made a constant and at all times peculiar choice of himself, one might wonder why his experience always ties up with itself and presents him with objects and definite historical phases."
18. Though we should note that he is not, in this instance, free to be a *confident person*. He is rather, a characteristically self-doubting person who managed to put aside his self-doubts on a particular occasion. Compare Sartre's discussion of the way the past inevitably contextualizes our present decisions (1993, 496–504).
19. Compare Merleau-Ponty's discussion of the appropriateness of speaking of probability and likelihood in the prediction of human actions (1962, 442).
20. I am drawing here from Merleau-Ponty's influential discussion of the phantom limb (1962, 80–82).
21. See Merleau-Ponty (1962, 82–84), and in particular Merleau-Ponty's characterization of the body as an "inborn complex" (1962, 84).
22. Compare Merleau-Ponty's example of being locked up in one's personal grief, while one's eyes continue to be drawn in by the world's spectacle insofar as they, in their generic character, are relatively autonomous and indifferent to the singular event that absorbs one at the personal level (1962, 84).
23. For a description of repression as grounded in the character of the body, see Merleau-Ponty (1962, 82–86).
24. In having "committed myself to inferiority" in the past, having made it the abode of my life, this past "is not a set of [individuated] events over there, at a distance from me, but the [general] atmosphere of my present" (Merleau-Ponty 1962, 142).

References

Barbaras, R. (2001). Merleau-Ponty and nature. *Research in Phenomenology*, 31(1), 22–38.

Bergson, H. (1991). *Matter and memory*. Trans. Nancy Margaret Paul and W. Scott Palmer. New York: Zone Books.

——— (2001). *Time and free will*. Trans. F. L. Pogson. Mineola, NY: Dover.

Bouton, C. (2014). *Time and freedom*. Trans. Chris McCann. Evanston, IL: Northwestern University Press.

Ciavatta, D. (forthcoming). Merleau-Ponty and the phenomenology of natural time. In *Perception and its development in Merleau-Ponty's phenomenology*. Eds. Kirsten Jacobson and John Russon. Manuscript submitted for publication.

Husserl, E. (1964). *Cartesian meditations*. Trans. Dorian Cairns. The Hague: Martinus Nijhoff.

Mazis, G. A. (1992). Merleau-Ponty and the 'backward flow' of time: The reversibility of temporality and the temporality of reversibility. In *Merleau-Ponty, hermeneutics, and postmodernism* (pp. 53–68). Eds. Thomas W. Busch and Shaun Gallagher. Albany: State University of New York Press.

Merleau-Ponty, M. (1962). *Phenomenology of perception*. Trans. Colin Smith. London: Routledge, Humanities Press.

——— (1964). Indirect voices and the language of silence. In *Signs* (pp. 39–83). Ed. Merleau-Ponty. Trans. Richard C. McCleary. Evanston, IL: Northwestern University Press.

Sartre, J.-P. (1993). *Being and nothingness*. Trans. Hazel Barnes. New York: Washington Square Press.

16 Acts as Changes

A Metabolic Approach to the Philosophy of Action[1]

Micah D. Tillman

1. Introduction

"A popular approach both to the nature of action and the explanation of actions," writes Alfred Mele, "emphasizes causation" (Mele 1997, 2). Rather than simply being a popular alternative, however, the causal approach currently dominates the field (Frankfurt 1997, 42). Even debates between those who appeal to "agents" and those who appeal to "beliefs, desires, and intentions" are usually mere family feuds. Before I depart along a different route, therefore, a decent respect for the opinions of my colleagues requires me to explain the three difficulties I find blocking the causal path.[2]

First, a theory of agent causation might tell us that a person who raises her hand *caused* her hand to rise. However, I concur with those who claim, as Rowland Stout (2005, 56) reports, that "[i]n normal cases you do not *make* your hand rise; you just raise it." There is no room between agent and action—in Stout's words, "An action" simply "is an agent doing something" (2005, 3)—and thus I see no need to assert a causal relation between them.

Second, we might create room for a relation by saying that a person's *reasons* cause her actions. This approach, however, marginalizes the agent. If Sally raises her hand and I say, "Sally had some beliefs, desires, and intentions that caused her hand to move," I am looking through her rather than thematizing her as an acting person. It is as if her only role in the action were to own the reasons and body parts involved.

Third, whenever the notion of "cause" enters discussions of action, one finds freedom and determinism not far behind. With issues so important and complex crowding in, it grows difficult to focus on the comparatively mundane phenomenon of the action before me. I become lost in speculation, rather than coming to a clearer understanding of what acts really are.

I see no reason why the following theory might not be reframed in causal terms, and yet the three difficulties just mentioned encourage me at least to begin from a different direction. In Section 2, therefore, I introduce a three-category taxonomy of change, and argue that acts originate in changes of the third category. Section 3 will then describe the full extent of category 2c changes, Section 4 will provide a theory of acting from habit, and Section 5 will explore the nature of acting for reasons.

2. Three Categories of Change

For something to change (to undergo μεταβολή) is for it to be different at one point in time than at another, for it to move from one state to another. And yet there would seem to be at least nine ways in which a thing might be changed (Table 16.1).

Table 16.1 Nine types of changes

	. . . into other	*. . . into nothing*	*. . . into self*
Changed by other . . .	Category 1a: A thing is changed by something else into something else.	Category 1b: A thing is changed by something else into nothing.	Category 1c: A thing is changed by something else into itself.
Changed by nothing . . .	Category 2a: A thing is changed by nothing into something else.	Category 2b: A thing is changed by nothing into nothing.	Category 2c: A thing is changed by nothing into itself.
Changed by self . . .	Category 3a: A thing is changed by itself into something else.	Category 3b: A thing is changed by itself into nothing.	Category 3c: A thing is changed by itself into itself.

a. Category 1

The most obvious changes are those in which something is changed by something else (category 1a); a ball's location is changed by a person's throwing it or by another ball's striking it, for instance. In such events, the move is from one way of being to another. However, a change sometimes involves a thing's ceasing to be altogether (category 1b), as when a vase is destroyed by a sledgehammer. And occasionally, the change involves the thing's being changed into itself (category 1c), as when a painting is repaired by an art restorer.

b. Category 2

Category 2a changes would be those in which a thing changes, but nothing changes it. This category is controversial,[3] yet our common experience of aging and our secondhand understanding of particle decay and quantum fluctuations suggest that such changes occur. The result of deterioration, furthermore, is occasionally the ceasing-to-be of the thing changing (category 2b). Perhaps more common than a thing's declining into nonexistence, however, is the opposite of decay: the changes undergone by living beings—like healing, growth, and nutrition—which constitute their being alive (see Aristotle, *De Anima* II.1, 412a14). Aristotle attributes these to the "nutritive soul" (Aristotle, *De Anima*, II.4, 415a24–26; see also II.2, 413a25–30). believing they spring from the living being itself, while Karol Wojtyła (1979) attributes them to the natural "dynamism" inherent in living beings (60, 62). (That is, they derive from the living being itself, rather than from some outside source, or from chance [62].)

In essence, I concur with both Aristotle and Wojtyła.[4] The changes that constitute a thing's *being alive* are changes whereby a thing becomes itself more fully and maturely (growth), is maintained as itself and is able to express itself more fully (nutrition), or is restored to itself (healing). These changes, therefore, must belong to category 2c: a thing's changing into itself (see Aristotle, *De Anima* II.4, 415a24–b7). As an organism ages past a certain point, however, these changes become less and less successful. The thing begins to fail to change into itself, and thus slips further and further away from being fully what it is (at least in certain respects). Physical death then occurs when the being ceases to be capable of changing into itself at all, and changes of categories 1a/b and 2a/b take complete control.

c. Category 3

Here we have the changes at the heart of action. In category 3a, a thing changes itself. For example, if Bob kicks at a stool while waiting in a doctor's examination room, he is not only changing the location of a limb but is also a changing himself as a whole (Aristotle, *Physics* VIII.2, 252b17–25; VIII.4, 254b29–31, 257a31–258a26). He goes, to put it crudely,[5] from being a seated body whose right leg is not extended to being a seated body whose right leg is extended. In category 3a, therefore, there seems to be movement in two directions: a reflexive influence by the thing on itself, such that it is both the changer and the changed, and a shift by the thing away from itself into what it is not.

Actions, however, rarely involve the actor alone. If Bob becomes frustrated with waiting for his doctor, he may actually kick over the stool rather than playfully kicking at it. This is also an act, but one in which Bob changes both himself *and* the stool. Specifically, Bob changes the stool by changing himself.[6] Both the change to Bob (category 3a) and the change to the stool (category 1a) belong to Bob's whole act of kicking the stool over. Therefore, while an act may consist simply of a thing's changing itself, it may also consist of a thing's changing something else by changing itself.[7]

Category 3b changes, in contrast, are those in which a thing changes itself into nothing; that is, they are those in which a thing destroys itself. Some methods of suicide, like throwing oneself into a volcano, would seem to fall into this category (unless we join Plato [*Phaedo*, 78b–81a, 85e–86d, 115c–116a] in identifying the person with the soul and holding that the soul survives death). Category 3c changes, finally, would be those in which a thing changes itself into itself. Perhaps things like taking medicine, cleaning and bandaging one's own wounds, and even putting food into one's body, would belong to this type. In such changes, one is assisting (or stimulating) the ongoing category 2c changes that constitute one's being alive. One is changing oneself, but simply changing oneself into oneself, either by restoring oneself to a previous state or by assisting oneself in growth and maturation.

d. Conclusion

Though there are at least nine types of change, acts originate in category 3 (with a thing's changing itself) and can be extended to include category 1a changes (see Aristotle, *Physics*, VIII.5, 256a7–13). Furthermore, it would seem that category 2c changes are fundamental to acts, since only living beings have the ability to act (see Aristotle, *De Anima*, II.3, 414b31–415a11). Therefore, an act consists of a thing that naturally and persistently changes into itself either (1) changing itself or (2) changing one or more other things by changing itself. With this, we have the minimal beginnings of a theory of action, in need of significant elaboration.

3. Head and Heart: The Psychological Extent of Category 2c Changes

The stream of things running continually through our minds reveals, upon inspection, that most thinking is like digestion or blood circulation: it goes on incessantly without our actively doing it. Rather than a stream of acts (see Searle 1983, 3; Husserl 2001, vol. 2, 102), it constitutes our mental living— like nutrition, growth, and healing help constitute our physical living. The closest parallel between the lives of mind and body, however, may be that between thinking and breathing. Both are changes that go on constantly but are also changes with which we intermittently interfere. Our breathing, just like our thinking, occasionally shifts from being a set of category 2c changes to being a set of category 3 changes.

In his essay, "Does Consciousness Exist?", William James writes:

> I am as confident as I am of anything that, in myself, the stream of thinking . . . is only a careless name for what, when scrutinized reveals itself to consist chiefly of the stream of my breathing. . . . [B]reath, which was ever the original of "spirit," breath moving outwards, between the glottis and the nostrils, is, I am persuaded, the essence out of which philosophers have constructed the entity known to them as consciousness.
> (James 1912, 36–37)

James goes further than I would, finding an identity where I assert only a parallel. We cannot breathe a breath again, for example, but we can think a thought again. We cannot breathe the past or future, but we can think the past or future (in some sense). Breath, even when intentional, lacks intentionality, while thinking, even when unintentional, involves intentionality. However, the analogy between the "stream of thinking" and the "stream of breathing"—they are category 2c changes that (1) help constitute our "being alive" at different levels, and yet (2) can become category 3 changes—makes it easy to understand James's motivation for equating the two. If either were to cease altogether (rather than being transformed into category 3 changes), we would cease to be the kind of thing we are—thinking and animate beings—and thus would cease to be ourselves (see Descartes [1641] 1904, 27).

We can find category 2c changes, therefore, both on the level Aristotle's "plant" or "nutritive soul," and on the level of the "human" or "rational soul." But what of the middle, "animal" part of the soul, where we find sensation, emotion, and locomotion?[8] Can we understand these as category 2c changes as well? With the notion of telos in hand, I believe we can.

Aristotle's understanding of *telos* has to do not with human goals but with what things are ("essences") (Slade 2000, 58–59). A thing's telos is simply being, in the fullest, most excellent way, the kind of thing that it is (see Aristotle, *Politics*, I.2, 1252b30–1253a1), and this may include undergoing (and continuing to undergo) certain processes. Happiness, the human telos, for example, is "an activity of the soul in accordance with virtue" (Aristotle 1999, I.7, 1098a7–8, 15–16; I.8, 1098b30; I.9, 1099b25; I.10, 1100b10). That is, a human fulfills her telos by continually living and thinking excellently. Appealing to telos, however, does not mean we must also employ classical theories of final causality. Teleology, in its most basic form, is not etiology, and once divested of its causal baggage can be helpful even for us post-Cartesians (see Descartes [1641] 1904, 55).

Returning to the "middle" activities of the soul, we note that no animal, human or non-human, can be healthy without properly engaging in sensation, emotion, and locomotion. These are central to preserving and prolonging our physical lives by contributing to our ability to remain ourselves ("to survive") at the physical level (see Aristotle, *De Anima*, III.12, 434a32–b2). Furthermore, these processes constitute their own forms of life. They are part of the telos of animals—of what it means to be an animal—just as much as nutrition, growth, and healing. We, therefore, cannot become or be ourselves fully without undergoing them, and so they must be category 2c changes: instances of an animal changing into itself.

Nevertheless, many cases of locomotion would seem too active to fit category 2c. Though we fulfill our telos through them, they are not automatic; we are *changing ourselves* into ourselves, rather than simply changing into ourselves. Though instances of habitual movement may be category 2c changes, therefore, locomotion would often seem to fit category 3c best. Yet what, precisely, is the difference between habits and genuine actions? What does it mean to act from habit?

4. Acting from Habit

a. Initial Analysis

To have a habit is to engage in the same activity time and again—to "repeat oneself," returning to the same "place." Though it involves change, a habit is a process in which there is ultimately nothing new. Our habits define us, being our characteristic responses to, or ways of engaging with, the situations in which we find ourselves. In acting habitually, we are genuinely repeating ourselves.

"To act from habit," therefore, must be to undergo either a category 2c or 3c change. Two considerations speak in favor of assigning habitual action to 2c, however. First, there is the "automatic," "reactive," and "mindless" nature of acting from habit, which means such changes are not genuine acts. Second, you cannot have a "one-off" habit, nor indeed any "one-off" category 2c change—all such transformations are parts of larger, repetitious structures of change[9]—while acts seem more properly to be "one-offs."

Nevertheless, if we repeat an act, it has the tendency to form a habit (Aquinas, *Summa Theologica*, I-II, q. 51, aa. 2–3). The more repetition, the more the act belongs to an extended, cyclical pattern of changes, and falls into category 2c. We tend to understand a thing better the more instances of it we encounter, furthermore, gaining access to its essence through repetition.[10] In repeating an act to grasp its essence, however, we see a habit taking shape. Acts, therefore, are naturally habit-forming; this is essential to them and means that the intrinsic function or telos of acting (whether this is what we intend or not) is to pack down patterns into the category 2c changes that constitute our living. The telos of action is non-action (*qua* living, rather than *qua* inactivity). In acting, we are building a life.[11]

An act, therefore, is a whole consisting of a category 3 change, and perhaps certain category 1 changes, which has an intrinsic future-directed orientation to a repetitious structure of category 2c changes. A habit, on the other hand, is a part, with intrinsic reference to a whole that consists of other identical category 2c changes, spanning the past, present, and future. Thus, it would seem that acting is merely a step on the way to fulfilling our human telos, since acts, by their very nature, point ahead in time to habits.

b. Refining the Analysis

We might have seen this coming, given Aristotle's argument that (a) the human telos is happiness, (b) "happiness is an activity of the soul in accordance with virtue" (Aristotle, 1999, I.7, 1098a7–8, 15–16; I.9, 1099b25–27; slightly modified), (c) virtue is a "characteristic," or habit, of the soul (II.5, 1106a11–12; II.6, 1106b36–1107a1),[12] and (d) "we . . . are made perfect by habit" (Aristotle 2009, II.1, 1103a25). On Aristotle's account, our telos is the life of habitual excellence.

This cannot mean, however, that *eudaimonia* consists in mindless, automatic activity of the same basic type as instinctual reflex. This would shut the most properly human aspect of the soul (Aristotle, *Nicomachean Ethics*, 2009, I.7, 1097b33–1098a4) out of our telos, ruining the harmony with Aristotle[13] that I have found so helpful. Furthermore, this would contradict the phenomenon of discovering ourselves to have been engaged in a habitual activity while our minds were directed elsewhere. Such experiences feel like "waking up," and we would rather be mindful and awake with Plato (*Republic* V, 476c–d) than mindless and asleep to reality.[14]

Furthermore, category 3c changes sometimes depend on corresponding category 2c changes already in place, rather than merely pointing to future category 2c changes. We can only deliberately breathe, for example, because we were already breathing automatically, and we only begin to think purposefully because we were already "naturally" engaged in a stream of thought. Likewise, a category 3c change is occasionally meant to reinforce a category 2c change, as when an athlete runs through a drill to make it even more habitual.[15] In such cases, the act has an intrinsic reference to a past and present habit, rather than only to the future. There are habits not only "after" acts, then, but also "before" them.

How are we to deal with all of this? I believe we should say that acts not only point ahead to habits, but that habits point ahead to acts. The fact that the motor changes involved in walking become habitual frees us up to deliberately play football, or to purposefully ambulate from location to location. Likewise, the fact that typing has become habitual frees the writer's mind for complex reasoning as she composes. Perhaps these higher-level acts then also point to higher-level habits, such as when the once-purposeful act of walking home becomes habit, allowing us to use that time for examining the local flora. As we move through time, in other words, we transform category 3 changes into category 2c changes so as to engage in further category 3 changes that become further category 2c changes, and so on. Human nature, I suspect, is a matter of circulating from acts to habits to acts, such that we are "spiraling up" to ever-higher levels.

5. Acting for Reasons

Having examined the nature of acting from habit—highlighting once again the interconnectedness of acts and category 2c changes (see Section 2d)—we are now in a place to approach what would seem, at least at first, to be the very opposite of habitual action.

a. Reasons for Acting Are Category 2c Changes

One commonly hears that to understand the reasons for an act, we must examine the beliefs, desires, and intentions behind it.[16] Since only changes can be parts of acts on the aforementioned account, however, Searle's (1983) theory that intentional acts have reasons not only "behind" them but also "within" them (94) may lead us to ask whether reasons might be changes. The entrance of a belief, desire, or intention is clearly a change, but is "having" it, or its "being in effect," a change?

A belief or intention (in the Brentanian/Husserlian sense) is like a stance; it is a holding-oneself-in-position. In physical posture-holdings, it may appear that nothing is happening, but one is, at the very least, counteracting the force of gravity and the draining of energy. Similarly, the holding in consciousness of a belief or intention is at least a counteracting of distractions and the tendency of epistemic states to fade (that is, one is counteracting category 1a/b and 2a/b changes). To believe or intend, therefore, one must continually renew one's belief or intention.

I suggest that this "holding"—or "continual renewing"—is a type of category 2c or 3c change, and thus that both the onset of a belief or intention *and* its maintenance are changes.[17] The same must go for all epistemic or volitional "states," furthermore, since they are also subject to incessant category 1a/b and 2a/b changes and thus must be continually renewed. One might object, of course, that what we have here is simply the absurd proposition that even remaining the same is a kind of change, but the following three analogies seem to me to support the rationality of our position.

First, a note held constant in a musical piece against a changing background progression changes in relation to the background by not moving in parallel with it.[18] Something similar occurs when we hold a volitional state constant across changing time, instead of allowing it to die away as each moment does.[19]

Second, the circle seems the best symbol for category 2c changes, while a dot would better represent a single point in time, or state of being. In the limit, however, when a circle reaches a diameter of zero, it becomes a point. By saying that the holding of an epistemic or volitional "state" is a change, therefore, I am effectively asserting that we should visualize such states as circles of diameter zero. This amounts simply to interpreting stasis as the limit case of change.

Third, the operations of adding zero and multiplying by one are operations on a number, though they leave it as it was. Something similar can be said of identity functions and transformations. One is doing something, in one sense, though doing nothing, in another. Likewise, when we are consciously maintaining an epistemic or volitional "state," something is clearly going on or occurring, even though what we have is a "state." I can see no better way of describing this than saying that an epistemic or volitional state is a category 2c or 3c change. In it, a thing changes into itself.

b. Reasons and Their Expressions

Whether epistemic or volitional, reasons are often "for acting," and thus lead naturally to verbal and physical expressions. The belief, for example, that a bear is trying to eat you, combined with the desire to live, is naturally expressed by screaming and running. The one set of changes (the belief and desire) naturally forms a whole with the other (the screaming and running).

We might analyze the changes involved in acting for a reason at several levels. At the lowest, we have the change from not having a reason to having it, coupled with the change from not expressing that reason to expressing it. The beginning and ending states of the two, however, do not simply occur "side by side." At a higher level, we see that the beginning states cycle into each other, as do the ending states. Lacking a reason for acting, we tend not to; or failing to act, we tend to ignore whatever reasons we might have had. Likewise, having a reason for acting, we tend to express it; and if we act in a certain way, we tend to adopt the beliefs and desires this way of acting expresses.

At a third, still higher level, we see that the two beginning states are simply "sides" or "aspects" of a single, more inclusive state (not-having-a-reason-to-act-and-thus-not-acting), as are the two ending states (having-a-reason-to-act-and-thus-acting). We do not really have two changes, side by side, therefore, but a single unified change. Finally, at the highest level, we see that changing from not having a reason, to having it, and from not expressing the reason, to expressing it, means the person has moved from a unified state to a unified state. She has moved from being one with herself to being one with herself, and thus has simply changed into herself.

If you were to believe a bear is trying to consume you, were to desire to live, and yet were not to scream and run away, the changes that constitute the "expression" of your reasons would be missing and thus the whole set to which those reasons naturally belong would be incomplete. Having undergone only half the changes in the whole, we could only conclude you were somehow incomplete yourself, or "did not have it all together." You would have lost integrity in some sense, moving from a state in which you neither had a reason nor were engaged in an expression thereof, to a state in which you had a reason but engaged in no expression.

All of this leads to the following position. Epistemic and volitional "states" are full category 2c or 3c changes in themselves, but since they also form a whole with their "expressions," they are only half-changes with respect to that whole. This means that by maintaining an epistemic or volitional "state," a being may change into itself in part but not as a whole. If the appropriate expression does not step in to complete the change, the being loses cohesion (just as in the unsuccessful category 2c changes that occur in aging). The same is true, however, for the changes that "express" the category 2c changes begun by epistemic and volitional "states." Though they are complete changes in themselves, they also form a whole with certain reasons, and thus are only half-changes with respect to that whole. If they are not accompanied by the appropriate reasons, the being does not fully change into itself and thus loses cohesion. In both cases, either the first change must be undone (the reason or expression must be dropped), the change must be completed (the reason must be acted upon or adopted), or the being must live with the discomfort of internal conflict.[20]

c. Closer Analysis of the Reasons

When a person obtains a new belief because the object of that belief suddenly appears to him, it would seem the object is changing him. However, ongoing claims about "the theory-ladenness of perception," phenomenology's appeal to "intentionality" and "mental acts" in perception, and the assertion by many that perception involves the brain's "representing" its objects all speak against this. We cannot wholly blame our perceptual beliefs on the objects thereof.

In view of this, I would suggest the following. A charging bear changes the state of our "perceptual apparatus" by reflecting light, which changes the state of cones and rods in our eyes. The person seeing the bear, therefore, has been changed by something else (category 1a), like a person whose flesh is cut by a sword-wielding opponent, or metaphorically like a king whose castle's wall is breached by an invading army. This change is not enough, however, to constitute the person's actually perceiving the charging bear as such. Though it provides her with the "sensory givens" that would support such an experience, to have a full perceptual belief the viewer must respond to the change in her "perceptual apparatus" through an intention, automatically maintaining herself (her "integrity") through certain epistemological or intentional changes. This is like the skin's healing itself from the sword wound, or the king's stepping into the hole in the wall so as to use it as a window from which to better survey his enemy.

To analyze perception, then, we would place the changes wrought in the viewer's perceptual apparatus "beside" the perceiver's epistemological response to these changes. The viewer moves from not having been changed and not having responded to this change, to having been changed and having responded. The beginning states belong together, as do the ending states, with each pair simply being two sides of a single state, and the two changes being two aspects of a larger change. This larger change, then, turns out to be the viewer's changing into herself, as she moves from one unified state to another.[21] At the highest level, finally, perception belongs to category 2c (or perhaps 3c), though it contains a category 1a change as a part. This is the same as when we heal from a wound. If such a category 2c change is "unpacked," we may find the 1a change that was the original being-changed-by-reflected-light, or the original wounding, but the whole is not our being-changed-by-something-else. It is, instead, our changing-into-ourselves.

What, however, of epistemic states that are more purely mental, or conceptual? For such to function as reasons for acting here and now, they must be tied to other "reasons" about the perceptual environment. "It is good to help the elderly cross the street" becomes a reason for acting only when coupled with, "That is an elderly person trying to cross the street."[22] The state of not seeing an elderly person crossing the street, after all, flows naturally into not thinking of the fact that the elderly should be helped to cross the street, while the state of seeing an elderly person trying to cross the street flows naturally into actively maintaining the belief that the elderly should be helped. To some extent, likewise, the reverse is also true. To ignore our belief that elderly people should be helped, we have to ignore any elderly people in our environment who need help, and thinking about elderly people needing help naturally flows into seeing an elderly person needing help—even if only "in our mind's eye."[23]

What, then, of volitional states? Such category 2c changes usually seem more urgent than simple beliefs. In fact, beliefs often come to be expressed only by being coupled with volitional states. We flee from the bear not only because we see it and believe it is trying to eat us, but because we wish to live. We help the elderly person not only because we believe that such actions are good, and that there is an elderly person who needs help, but also because we want to help, to be the kind of person who helps, or to achieve some consequence of helping.[24] There are cases, however, in which the volition is primary. The sudden urge for chocolate often comes over us on its own, and the perceptual and conceptual reasons that we express by getting up and navigating to wherever the chocolate is arise only because the urge was first experienced.

In all such cases, the volitional, perceptual, and conceptual changes unite as parts within the general change that is the reason for acting. The state of not having one reason tends to cycle naturally into not having the others, and the state of having one tends to cycle naturally into having the others. We are, at least at the highest level of description, simply changing into ourselves by undergoing all these linked changes.

d. Closer Analysis of Expressions of Reasons

But what of the other part of acting for a reason: the expression? If it is habitual, as when we automatically brake to avoid a squirrel, then nothing needs to change in our previous analysis. However, if the expression is a genuine act, we require a few nuances.

When we have a reason to act, but no habitual expression ready-to-hand, we must instead act "deliberately," and the whole to which our two changes belong loses some cohesion. First, having the reason does not flow automatically into expressing the reason; we remain for a split second in the state of not expressing the reason, even though we now have it. Second, the unity between the act that expresses the reason and the reason itself is only accomplished through a second reason. For example, given that we believe a bear is trying to eat us, we may freeze, and then have to force ourselves to think more clearly about the object of our belief in order to come to a practical conclusion. Sometimes, of course, we simply have to wait for all our reasons to make themselves felt. We see the bear, believe it is trying to eat us, and yet take a moment to respond because we are groggy with sleep (see Aristotle, *Nicomachean Ethics*, 2009, VII.3, 1147a10–b17); but once the full complement of reasons registers, the expression thereof is automatic. At other times, however, we must actually make a choice that produces (or *is*) a new reason, and this, plus our expression of this new reason, is the genuine act that expresses our original reason.

So long as we do eventually find connecting reasons that allow us to express our original reason, we maintain our cohesion as beings who act for reasons. However, this cohesion will be "loose" (since the original states naturally flow into each other, while the final states do not) until we have had enough experience with the reason in question to develop a characteristic response. In many cases, this will be a long process; in others, the reason may be too complex to admit of a habitual expression (see Dreyfus 2005, 57); and in still others, the reason may forever remain unexpressed, since we simply do not know what to do.

e. Conclusion

It would seem, then, that humans are capable of "acting for reasons" without doing so on purpose. In such "acts," they are simply being themselves: acting habitually, undergoing the changes that are characteristic of them, and changing into themselves. In other cases, however, we act for reasons in the genuine sense of acting. We change ourselves, which involves the introduction of a new reason in order to bring the original reason to expression.

We are forever acting for reasons, of course, and thus the question of whether we are acting (1) out of habit or (2) in the genuine sense is one that it would be reasonable to ask for much of what we do. This is not, however, because our moral judgments depend upon knowing whether we were actually acting, or only acting from habit, when an event occurred. When acting from habit we are being ourselves, changing in either good or bad ways, and thus can be praised or reprimanded. We do not have to act in the genuine sense to be moral subjects.

However, as time progresses, and thus as we encounter new reasons for which we do not yet have characteristic responses, we are continually forced to act for reasons in the genuine sense of acting. Likewise, as we become more mature intellectually and emotionally, and more capable of having more complex reasons, we find ourselves continually in need of acting if we are going to express our reasons. Our growth through time enables us to recognize complexities in our reasons that make reliance on previous habits impossible. Furthermore, as we noted earlier, the development of habits often serves to enable us to act in higher, or more complex ways. We not only progress through acts to the habits that are their *tele*, but depend upon temporally prior habits in order to engage in further acts.

It is not just our initial definition of acts as category 3 changes that ties agency to time, therefore. If acts *are* changes, there could be no acts without time, since there can be no change without time. However, we have also seen that (1) acts refer to future habits, just as habits prepare us for future acts, (2) our maturation in time breaks the circle of habit by allowing us to encounter novel nuances in our reasons for acting, and (3) our progression through time introduces us to completely new reasons for which we have not even had a chance to develop habitual expressions. Time, in other words, makes agency not only possible, but—for those who make a habit of growth and attentiveness—also necessary.

Notes

1. My thanks to all who attended the discussion following the presentation of this chapter's first draft, but most especially to Lior Levy and Roman Altshuler. Their comments led to what I think are a number of significant improvements in the text. Any remaining confusions or mistakes are, of course, on me.
2. A fuller allusion would reveal even Thomas Jefferson, whose words I borrow here, to have been an adherent of the causal approach.
3. Aristotle, for example, rejects it (*Phys.*, VIII.4, 254b25, 256a2; *Met.*, XII.3, 1069b36–1070a1).
4. Wojtyła's (1979) text convinced me that these changes needed to be treated as belonging to a separate category, and his tripartite distinction between "man-acts," "something happens in man," and "something happens with man" (61–63) parallels my own distinction between category 3a, category 2c, and category 1 changes, respectively. It should be instructive, furthermore, to compare and contrast the theory expounded here with that offered by Jonas (2001), of which I only became aware after completing this chapter's analysis.
5. Leaving out reference to his mind, familial relationships, vocation, and so forth.
6. See Searle's (1983) distinction between actions "by means of" and actions "by way of" (98–99).
7. This does not mean, however, that the agent *consciously* experiences changing himself as a means to an end; in this case, for example, kicking over the stool would be a "basic action" (see Searle 1983, 100). See also Heidegger (1962), who contrasts the "withdrawnness" of a tool that is ready-to-hand with "the work" at which we aim in using it (99). What is ready-to-hand only "becomes conspicuous" when we find it unusable (102–103). Dreyfus (2005) develops Heidegger's ideas in terms of "coping" (55–56, 60).
8. See Aristotle, *DA*, II.2, 413b1–4, II.3, 414a32–b5, II.4, 415b22–23, and III.9 (see also III.10, 433a9–12, 22, and 433b27–30).
9. See Derrida (1973) on the repetitive nature of signs (41, 49–52, 57; see also 6, 9–10).
10. On the (Husserlian) phenomenology of intuiting essences, see Sokolowski (1974, ch. 3).
11. See Sartre's (1975) insistence that human beings are nothing, and proceed to make themselves through their actions and choices (349).
12. "Characteristic" is Ostwald's translation of *hexis*. See Aquinas, *Summa Theologica*, I-II, q. 55.
13. Given his identification of the highest happiness with the contemplative life (Aristotle, *Nicomachean Ethics*, 2009, X.7–8).
14. On the health benefits of mindfulness (as understood from a Western psychological point of view), see Langer (1989).
15. There are cases, of course, of an attempted category 3c change turning counterproductive, as with the centipede in the poem.

16. On beliefs and desires, see, for example, Davidson (1997, 27–41). For Searle's argument that it is better to understand intentions, not simply beliefs and desires, as the reasons for (intentional) action, see Searle (1983, 103–106).
17. See Aristotle, *DA*, II.4, 415b25; but also see II.5, 417b2–16, III.3, 427a20, and 429a2.
18. See the line, "I changed by not changing at all," in Vedder (1993).
19. See Descartes's [1641] (1904) argument that sustaining is a continual recreation—a continual change from non-being to being (48–49).
20. Here, we might compare the James-Lange theory of emotion. See Golightly (1953, 286–299).
21. If the reason for an act is a perceptual belief, therefore, we have a quasi-fractal structure, with the whole set of changes that makes up the reason having the same basic structure as acting for a reason.
22. See Aristotle on practical syllogisms (for example, *Nicomachean Ethics*, 2009, VII.3, 1146b30–1147b17).
23. This has to do with the Husserlian distinction between empty and fulfilled intentions, the ordering of the former to the latter, and the intuitive nature of imagination. See Husserl (2001, vol. 1, 192; vol. 2, 222). On the ordering of intention to fulfillment, see esp. Sokolowski (2002, 175–176).
24. See Kant [1785] (1993, 10–14 [Ak. 397–402], and 14n14 [Ak. 401n]).

References

Aristotle (1999). *Nicomachean Ethics* (M. Ostwald, Trans.). Upper Saddle River, NJ: Prentice Hall.

——— (2009). *Nicomachean Ethics* (D. Ross & L. Brown, Trans.). New York: Oxford University Press.

Davidson, D. (1997). Actions, reasons, and causes. In A. Mele (Ed.). *The Philosophy of Action* (pp. 27–41). New York: Oxford University.

Derrida, J. (1973). *Speech and Phenomena: And Other Essays on Husserl's Theory of Signs* (D. Allison, Trans.). Evanston, IL: Northwestern University Press.

Descartes, R. [1641] (1904). *Meditationes de Prima Philosophia. Oeuvres de Descartes*, VII (C. Adam & P. Tannery, Eds.). Paris: Ministère de l'Instruction Publique/Léopold Cerf.

Dreyfus, H. (2005). Overcoming the myth of the mental: How philosophers can profit from the phenomenology of everyday expertise. *Proceedings and Addresses of the American Philosophical Association*, 79(2), 47–65.

Frankfurt, H. (1997). The problem of action. In A. Mele (Ed.). *The Philosophy of Action* (pp. 42–52). New York: Oxford University.

Golightly, C. (1953). The James-Lange theory: A logical post-mortem. *Philosophy of Science*, 20(4), 286–99.

Heidegger, M. (1962). *Being and Time* (J. Macquarrie & E. Robinson, Trans.). San Francisco: HarperSanFrancisco.

Husserl, E. (2001). *Logical Investigations* (2 volumes) (J.N. Findlay, Trans.). New York: Routledge.

James, W. (1912). Does consciousness exist? In *Essays in Radical Empiricism* (pp. 1–38). New York: Longmans, Green and Co.

Jonas, H. (2001). *The Phenomenon of Life: Toward a Philosophical Biology*. Evanston, IL: Northwestern University Press.

Kant, I. [1785] (1993). *Grounding for the Metaphysics of Morals* (3rd ed.) (J.W. Ellington, Trans.). Indianapolis: Hackett.

Langer, E. J. (1989). *Mindfulness* Cambridge, MA: Perseus Books, 1989.

Mele, A. (1997). Introduction. In *The Philosophy of Action* (pp. 1–26). New York: Oxford University.

Sartre, J.-P. (1975). Existentialism is a humanism. In W. Kaufman (Ed.), *Existentialism from Dostoevsky to Sartre* (rev. & exp. ed.) (pp. 287–311). New York: Plume.

Searle, J. (1983). *Intentionality: An Essay in the Philosophy of Mind*. Cambridge: Cambridge University Press, 1983.

Slade, F. (2000). On the ontological priority of ends and its relevance to the narrative arts. In A. Ramos (Ed.), *Beauty, Art, and the Polis* (pp. 58–69). Washington, DC: Catholic University of America Press/American Maritain Association.

Sokolowski, R. (1974). *Husserlian meditations: How Words Present Things*. Evanston, IL: Northwestern University Press.

——— (2002). Semiotics in Husserl's *Logical Investigations*. In D. Zahavi & F. Stjernfelt (Eds.). *One Hundred Years of Phenomenology: Husserl's 'Logical Investigations' revisited* (pp. 171–184). Dordrecht: Kluwer.

Stout, R. (2005). *Action*. Ithaca, NY: McGill-Queen's University Press.

Vedder, E. (1993). Elderly woman behind the counter in a small town [Recorded by Pearl Jam]. On *Vs.* New York: Epic Records.

Wojtyła, K. (1979). *The Acting Person* (A. Potocki, Trans.; A.-T. Tymieniecka, Ed.). Boston, MA: D. Reidel.

17 Hamlet and the Time of Action

Henry Somers-Hall

1. Introduction

In this chapter I want to explore a comment made by the French philosopher Gilles Deleuze that presents a connection between two figures: Kant and Hamlet.[1] In his most important early work, *Difference and Repetition*, Deleuze writes, "the Northern Prince says 'time is out of joint'. Can it be that the Northern philosopher says the same thing?" (Deleuze 2004, 111). In this chapter, I want to look at the question of drama and see how different conceptions of drama allow us to understand action, or more precisely inaction, in Shakespeare's *Hamlet*. I want to show how these different conceptions of dramatic action tell us something about the nature of temporality. I will begin by reversing Hamlet's claim, and discussing what time "in joint" would look like, tracing it back to Aristotle's conception of drama, before moving on to Plato's characterization of temporality.

2. Aristotle's Conception of Drama

Aristotle claims that tragedy is

> the imitation of an action that is serious and also, as having magnitude, complete in itself; in language with pleasurable accessories, each kind brought in separately in the parts of the work; in a dramatic, not in a narrative form; with incidents arousing pity and fear, wherewith to accomplish its catharsis of such emotions.
>
> (Aristotle 1995, 1449b25–29)

For our purposes, what is important to note is that in Greek tragedy, what is central is action ("they do not act in order to portray the characters; they include the characters for the sake of the action" [Aristotle 1995, 1450a20–21]). Tragedy cannot occur simply through a misfortune on the part of good characters, as this is "not fear-inspiring or piteous, but simply odious to us" (Aristotle 1995, 1452b36). Likewise, it cannot involve evil characters suffering misfortune, as "pity is occasioned by undeserved misfortune"

(Aristotle 1995, 1453a4). In this sense, the plot of tragedy involves "a man not pre-eminently virtuous and just, whose misfortune is, however, brought about not by vice and depravity but by some fault" (Aristotle 1995, 1453a6–10). As Michael Davis points out, "the *Poetics* is about two things: *poiêsis* understood as poetry, or imitation of action, and *poiêsis* understood as action, which is also imitation of action" (Davis 1999, 9).

If we look at a play such as *Antigone*, we can see that the drama arises from the actions of its characters, and traces their consequences. The conflict at the center of the drama emerges through King Creon's prohibition on the burial within the city of Antigone's brother, the traitor Polyneices, and Antigone's transgression of this prohibition. Here the conflict occurs between the laws "of the gods, that are unwritten and unfailing" (Sophocles 2003, 501), the right to bury the dead, and the laws of the state, exemplified by Creon's prohibition. Each character chooses to act solely according to one determining principle, and does not recognize the rights of the other. The one-sidedness of their opposed actions means that a resolution can only be achieved by the tragic death of Antigone, and Creon's loss of his wife and son. In order to explore how drama represents action, I want to bring in here the analysis of the art critic, Harold Rosenberg. In what Rosenberg calls "old drama," the basis of tragedy must involve some kind of transgression for which the agent is responsible:

> [T]he concepts of morality or social law, applying exclusively to human beings and ignoring possible analogies with other living creatures, tend to define the individual not as an entity enduring in time but by what he has done in particular instances. A given sequence of acts provokes a judgement, and this judgement is an inseparable part of the recognition of the individual.
>
> (Rosenberg 1994, 136)

When we look at the legal conception of the person, it isn't the case that the unity of the individual can be given in terms of their acts themselves. Rather, when someone comes before a judge, what the judge relates to is not a unity governed by personality, but rather a series of acts that are unified by the last act's relationship to the law. As Rosenberg notes, the acts of a murderer are in large part no different from the acts of anyone else, and are only made criminal by the fact that they precede the murder itself: "entering an automobile, stepping on the gas, obeying the traffic lights" (Rosenberg 1994, 138). These actions are all perfectly innocent, but gain their criminal character by the fact that they make possible the transgression of the law. When we look at a criminal act, it is the law that provides a framework for the analysis of action, and which imposes a structure of artifice that unifies the conduct of the perpetrator. This is not merely the case in the sense that it is the transgression of the law that defines the act as criminal, but also that the law provides a way of synthesizing certain other acts that, without reference to the crime, would be indifferent in relation to the law. We can see that there is an artificiality to such an analysis as becomes clear if it is suddenly discovered that the alleged perpetrator did not commit the crime. The entire identity of series of actions before the law then disintegrates. The actions of "stepping on the gas" and "obeying the traffic lights" now take on an entirely innocent aspect. Law is therefore what draws together a series of acts that in themselves are indifferent to one another.

Now, obviously in the case of the law, the problem is determining whether the structures of the law apply to the actions or not (whether the person is guilty or innocent). In the case of "old drama," we do not have the difficulty of determining whether acts properly accord with the structure of the law. Rather, characters in "old drama" are constituted to be in accordance with their fate from the outset:

> The dramatist's definition of the character was not an arbitrary super-imposition that exchanged the emotional, intellectual, and mechanical characteristics of a biological and social organism for some one deed that concerned the court; it constituted instead the entire reality of a character, avoiding the ruinous abstraction of the law by determining in advance that his emotions, his thoughts and his gestures should correspond with and earn in every respect the fate prepared for him.
>
> (Rosenberg 1994, 139)

In "old drama," therefore, character is seen simply as a manifestation of the underlying unity provided by the actions' relations to the law. Once character has been subordinated to the structure of the law through action, the possibility of an exploration of the change or development of character is ruled out. Thus, while Creon's relations to those around him—Haemon, for instance—may change during the course of the play, this is simply as a result of the expression of the same moral identity. This is explicit, for instance, in Hegel's reading of tragedy, where "the truly inviolable law is the unity of the action" (Hegel 1975, 1166). Hegel's understanding of tragedy sees it emerging through each character's adoption of a partial aspect of the ethical as their guiding principle. Thus Antigone is justified in appealing to the rights of the family in seeking to bury Polyneices, just as Creon is justified in upholding the laws of the state. They are both also culpable in that their action fails to take into account the complete structure of the ethical. The resolution of tragedy is the restoration of balance to the ethical, but as "the *individuals* have more or less put their whole will and being into the undertaking they are pursuing" (Hegel 1975, 1166), the resolution of the tragedy can only occur with the dissolution of these partial views, and with them the characters who embody them. Thus by focusing on action, we rule out an account of character, or at least of a change of character. Similarly, the singularly human nature of action as relation to ethical structure reflects that what is being played out is something primarily atemporal. Here, the phenomenal manifestations of characters in classical drama are merely manifestations of an underlying law, or an underlying judgment: the fate of the character. Hence, "psychology can establish the plausibility of Macbeth's or Lear's behavior, but for the sufficiency of his motivation, we must not refer to a possible Macbeth or Lear 'in real life' but to the laws of the Shakespearean universe" (Rosenberg 1994, 140).

3. Plato's Metaphysics of Time

I want to now introduce Deleuze's insights on this point by bringing in a distinction he draws between time in joint and time out of joint. In *Difference and Repetition*, he writes:

> The joint, *cardo*, is what ensures the subordination of time to those properly cardinal points through which pass the periodic movements which it measures (time, number of the movement, for the soul as much as for the world). By contrast, time out of joint means demented time or time outside the curve which gave it a god, liberated from its overly simple circular figure, freed from the events which made up its content, its relation to movement overturned; in short, time presenting itself as an empty and pure form.
> (Deleuze 2004, 111)

In order to develop an understanding of what Deleuze means by time out of joint, I want to begin by looking at what "jointed" time might look like. The key to this is the notion of the joint itself, which Deleuze presents in more detail in his 1978 lecture, "Synthesis and Time":

> Cardinal comes from cardo; cardo is precisely the hinge, the hinge around which the sphere of celestial bodies turns, and which makes them pass time and again through the so-called cardinal points, and we note their return: ah, there's the star again, it's time to move my sheep!
> (Deleuze 1978)

Deleuze is here referring to Plato's conception of time, presented in the *Timaeus*, which gives a mythical account of the creation of the world by a demiurge, who seeks to create the universe by imposing form on chaotic matter. The demiurge "took over all that was visible—not at rest but in discordant and disorderly motion—and brought it from a state of disorder to a state of order, because he believed that order was in every way better than disorder" (Plato 1997, 30a). As the creator himself is perfect, he desires to create the universe as far as possible as an image of himself. Given the world is already in motion, he can only create the world as a likeness to eternity:

> Now it was the Living Thing's nature to be eternal, but it isn't possible to bestow eternity fully upon anything that is begotten. And so he began to think of making a moving image of eternity, at the same time as he brought order to the universe, he would make an eternal image, moving

according to number, of eternity remaining in unity. This number, of course, is what we now call "time."

(Plato 1997, 37d–e)

Now, at this point, we can already note the first important point in this myth. Time is here seen as a structure of the realm of appearance. While the world itself is a copy of the creator, the creator himself is eternal, and hence has no relationship to time. So how does time come about? Before the universe is organized according to time, it is still in motion, although this motion is "disorderly." Thus, for Timaeus, motion is not dependent on time, as motion is prior to the imposition of time. Time is not that which allows movement to take place, but that which allows movement to become rational. In fact, Timaeus believes that time is grounded in the elements that are most perfect in the universe, the celestial bodies:

> In this way and for these reasons night-and-day, the period of a single circling, the wisest one, came to be. A month has passed when the Moon has completed its own cycle and overtaken the Sun; a year when the Sun has completed its own cycle. As for the periods of the other bodies, all but a scattered few have failed to take any note of them. Nobody has given them names or investigated their numerical measurements relative to each other. And so people are all but ignorant of the fact that time really is the wanderings of these bodies, bewilderingly numerous as they are and astonishingly variegated.

> (Plato 1997, 39c–d)

On the Platonic model, time is simply an imperfect way in which the eternal patterns of the world present themselves. It is always an ancillary time premised on a logically prior movement. In this regard, we can see the *Timaeus* as, with some reservations,[2] providing an account of how an essentially rational structure becomes exemplified in the world as a whole.[3] We can see that the subordination of time to an eternal, intelligible, and also, representational, model of time is central not just to Plato's conception, but also to a more general trend in pre-Kantian philosophy. Leibniz, for instance, argues that the notions of space and time are simply ways in which we perceive what are essentially a series of intelligible relations between things:

> As for my own opinion, I have said more than once that I hold space to be something purely relative, as time is—that I hold it to be an order of coexistences, as time is an order of successions. For space denotes, in terms of possibility, an order of things that exist at the same time, considered as existing together, without entering into their particular manners of existing. And when many things are seen together, one consciously perceives this order of things among themselves.

> (Leibniz 2000, 15)

Space and time are simply "well-founded phenomena" by which we inadequately perceive the true "conceptual" order of things. As such, time is really a mode in which the essential structure of succession appears to us. In this case too, therefore, time is secondary to a rational, conceptual, and representational way of ordering things. Time is predicated on a prior rational structure. Number is expressed in the celestial movement of the heavens. To be "in joint" is therefore to be hinged, tied to cardinal numbers, and tied to a prior rational order. We can further note that Aristotle's model of drama mirrors this notion of a metaphysics of time. Rather than seeing the character as an entity who develops in time, we have a drama where the time of the play is an expression of the movement of the action. Just as in Plato and Leibniz's metaphysics, time is the mode of expression of a prior rational structure in the world of appearances, so for Aristotle, the time of a tragedy is that by which an action is expressed in the world. How do these two notions of jointed time relate to *Hamlet*?

4. Hamlet

The focus of Deleuze's philosophical interest in *Hamlet* is on Hamlet's hesitation. In *Hamlet's* opening scene, Marcellus sets up the opposition to the Platonic scheme by noting that war has severed the connection between action and the celestial bodies in the Kingdom of Denmark:

> Why such impress of shipwrights, whose sore task
> Does not divide the Sunday from the week.
> What might be toward, that this sweaty haste
> Doth make the night joint-labourer with the day?
> (Shakespeare 2003, I.i.75–78)

As Spencer notes (Spencer 1943, 11), the view of the celestial spheres as a rational realm mirroring the rational faculty of man was widespread at the time that Shakespeare was writing, and views such as that of the humanist G.-B. Gelli—that "man is made of two natures, one corporeal and terrestrial, the other divine and celestial; in the one he resembles beasts, in the other those immaterial substances which turn the heavens"—were the standard fare of renaissance cosmology. Hamlet relates his own inability to act to something like the Platonic cosmology:

> And indeed it goes so heavily with my disposition that this goodly frame, the earth, seems to me a sterile promontory; this most excellent canopy the air, look you, this brave o'erhanging firmament, this majestical roof fretted with golden fire—why, it appeareth no other thing to me but a foul and pestilent congregation of vapours.
> (Shakespeare 2003, II.ii.280–286)

In this sense, Hamlet's concerns are not simply psychological, but are also related to a feeling of incongruence with the kind of rational cosmology put forward by the *Timaeus*. *Hamlet* sees Hamlet himself not as an identity in the legal sense, or in the dramatic senses that we find in Aristotle. He is not defined by action as, for instance Sophocles's own model of the vengeful son Orestes is, who claims that "talk is expensive," and is instead concerned with "practical details": "where we should hide, where we can leap out and push that enemy laughter right back down their throats!" (Sophocles 2001, 1720–1725). We do not need to turn to Greek tragedy for the figure of an identity, however, as Hamlet's character is paralleled by Laertes's need for revenge, which plays out in the traditional manner. The drama prior to Hamlet's return from England precisely concerns this inability to act:

> I do not know
> Why yet I live to say "This thing's to do";
> Sith I have cause and will and strength and means
> To do't
>
> (Shakespeare 2003, IV.iv.43–46)

Now, as this quote makes clear, Hamlet is very much aware of what he should do, but he is simply not able to do it. To this extent, we have an odd dramatic structure, since, if characters are understood in terms of the relations of acts to the judgment of the law, then Hamlet's various speeches, and use of speech in the first half of the play, are simply irrelevant to the structure of his role. On this basis, Deleuze writes that "Hamlet is the first hero who truly needed time in order to act, whereas earlier heroes were subject to time as the consequence of an original movement (Aeschylus) or aberrant action (Sophocles)" (Deleuze 1998, 28). Rather than action determining the time of the drama, Hamlet experiences time itself as being the ground of action. In this sense, he operates within a time out of joint where his relation to the world, and to himself,[4] is governed by a fundamental passivity. His actions unfold in a time that is not his own. Rosenberg's interpretation is precisely this, that Hamlet's character cannot be understood as purely derived from the structure of the action he is to undertake, unlike in the case of the kind of identity central to Aristotle's theory, and hence he exists outside of the role that the play assigns him. The task of taking on the role of avenging his father is simply too big for him. The sea voyage is therefore necessary to the structure of *Hamlet*, as it represents the break in the structure of the play whereby Hamlet becomes equal to the task allotted to him. In this sense, we can note that the first half of *Hamlet* can only be characterized negatively by an Aristotelian account of drama, as it is precisely governed by the absence of action. In Aristotle's terms, *Hamlet* is a dramatic failure. If we accept that Hamlet's hesitation is more than simply a negative moment or a failure on Shakespeare's part, then we also have to recognize that the time of Hamlet's action is not simply an expression of that action itself. The Aristotelian model of drama functioned analogously to the Platonic conception of time, and so if *Hamlet* breaks with Aristotle's conception of drama, we might expect it to also require a new metaphysics of time. In order to see what this new metaphysics might look like, I now want to introduce Kant's critique of the Leibnizian model of time.

5. Kant's Characterization of Time

For both Leibniz and Plato, what is given to thought is a confused image of the true nature of things; that is, a copy that differs in degree from it. For Plato, what is presented is an image of eternity in time, but time is simply a manifestation of the rational movement of the celestial bodies. Similarly, for Leibniz, space and time are simply confused perceptions of a properly conceptual reality. Kant's claim in the *Critique of Pure Reason* is essentially that rather than regarding perception as merely a confused form of conceptual thought (what Kant calls in relation to Leibniz "intellectualised appearances" [Kant 1929, A271/B327]), there is a difference in kind between perception and concepts. In his article, "Concerning the Ultimate Foundation for the Differentiation of Regions in Space" (1968), Kant presents this difference explicitly as the grounds for a refutation of Leibniz. If we take Leibniz's position as being that space and time are simply confused presentations of conceptual relations, then we would expect all of the properties of spatial and temporal entities to be ultimately explicable in conceptual terms. If we consider a hypothetical universe containing simply one glove, however, then even if we were to completely specify the conceptual relations between the parts of a glove, we will not be able tell whether the glove is left- or right-handed, as the mirror image of the glove will have exactly the same relations between the parts that compose it:

> However, since there is no difference in the relations of the parts to each other, whether right hand or left, the hand would be completely indeterminate with respect to such a quality, that is, it would fit on either side of the human body. But this is impossible.
>
> (Kant 1968, 42–43)

Given that the glove must be left- or right-handed, it therefore possesses a determination (its "handedness") that is not reducible to conceptual relations. If this is the case, then spatial entities cannot be seen merely as expressions of a non-spatial rational nature (we might think of this claim that space is also out of joint as a parallel thesis to the one we are considering), but also contain something in excess of the rational. The argument from incongruent counterparts shows that space cannot be seen as a mode of presentation of conceptual relations in the way that Leibniz proposed, and Kant makes similar arguments in the *Critique of Pure Reason* for the difference between time and succession. Kant's distinction of concepts and intuition therefore has the consequence that time cannot be seen as the moving image of eternity, as it is no longer the expression of an underlying representational structure, whether the "number of movement" or the true "order of things." That is, that rather than time being a mode of succession, succession is rather a mode in which time appears to us. In fact, succession is simply a way in which *we* organize time for Kant. Deleuze puts this point forward as follows:

> Time cannot be defined by succession because succession is only a mode of time, coexistence is itself another mode of time. You can see that he arranged things to make the simple distribution: space-coexistence, and time-succession. Time, he tells us, has three modes: duration or permanence, coexistence and succession. But time cannot be defined by any of the three because you cannot define a thing through its modes.
>
> (Deleuze 1978)

Once we recognize a difference in kind between concepts and intuitions (or between time and measure), then our understanding of time is no longer "joined" to, and dependent upon, an understanding of rational movement.

Does this mean that Deleuze's "time out of joint" is an essentially Kantian form of temporality? In fact, Kant's understanding of time is limited by his commitments to an essentially epistemic and subjectival project. That is, Kant is concerned with the conditions under which a subject can attain *a priori* knowledge of a world of objects. On this basis, he claims that whereas

> hitherto it has been assumed that all our knowledge must conform to objects, . . . we must . . . make trial whether we may have more success in the tasks of metaphysics, if we suppose objects must conform to our knowledge.
>
> (Kant 1929, Bxvi)

In working out the implications of this assumption, Kant takes time and space to be faculties of the subject. That is, they are ways in which that which is to be thought by the subject is given to the subject. As space and time are faculties through which the subject apprehends objects, then those formal properties of the object that derive from its spatio-temporal nature can be known *a priori*. Given that we are interested in knowledge, which involves concepts, and Kant has posited a radical split between concepts and intuitions, the major project of the *Critique of Pure Reason* is to show how the faculty of the understanding can be related to the faculty of intuition. In Deleuze's words, "does he not once again come up with the idea of harmony, simply transposed to the level of faculties of the subject which differ in nature?" (Deleuze 2003, 19). Kant himself recognizes that this is the central problem of the *Critique*, and it is the role of the transcendental deduction to show that it is impossible that "appearances might very well be so constituted that the understanding should not find them to be in accordance with the conditions of its unity" (Kant 1929, A90/B123).

This is not the place to give a full account of Kant's account of how intuition and the understanding are related, but Kant's essential claim is that the understanding relates to intuition through the notion of synthesis. Kant defines synthesis as "the act of putting different representations together, and of grasping what is manifold in them in one act of knowledge" (Kant 1929, A77/B109). In this sense, it functions like Rosenberg's law, which draws together a series of acts that in themselves are indifferent to one another. Now conscious empirical synthesis takes the form of making judgments. Deleuze's claim is that Kant has essentially taken a psychological account of how we synthesize judgments and reiterated it at a transcendental level. "The same function that gives unity to the different representations in a judgment also gives unity to the mere synthesis of different representations in an intuition, which, expressed generally, is called the pure concept of the understanding" (Kant 1929, A79/B104). By using this model, Kant ties the notion of synthesis to the notion of consciousness, and hence debars any form of synthesis that is not ultimately governed by judgment. This leads to a further sharp dichotomy between passivity and spontaneity. If the activity of synthesis is tied to the subject, then there is no possibility of a form of synthesis that does not involve a subject—intuition becomes a passive material to be taken up and organized by the conceptual understanding. Time remains a faculty of a subject rather than something capable of imposing itself upon the subject. If we do not see synthesis as tied to the subject in this way, we open up the possibility of seeing time as capable of exhibiting organization in its own right. It is in this sense that time can be seen as escaping from the subject. We might say that as well as the order of action, Hamlet must also operate according to an order of time. In this sense, Bergson's example of waiting for sugar to dissolve is instructive:

> For here the time I have to wait is not that mathematical time which would apply equally well to the entire history of the material world, even if that history were spread out instantaneously in space. It coincides with my impatience, that is to say, with a certain portion of my own duration, which I cannot protract or contract as I like. It is no longer something *thought*, it is something *lived*.
>
> (Bergson 2002, 176)

The question is not how this time of duration can be overcome, but of how it can be related to the time of action. We can therefore see "old drama" as dramatizing an ontology of movement, whereas Hamlet dramatizes an ontology of time as out of joint or freed from movement. Hamlet's question is how he is able to reconcile his personality with the identity of action. For Deleuze, it is only against the future that these two moments can be related. It is only the future that allows the self of the past to be brought into a "secret coherence" with the present, as action precisely is this relation of past and present towards a future. There are thus three moments to Hamlet. The first is that of passivity, where he is incapable of acting. The second is the *caesura*, the voyage to England, which Deleuze represents as the present, and the final moment, the future, the horizon against which the action takes place. What Bergson brings to the fore is that the difficulty in understanding Hamlet's inability to act can be related to the question of the metaphysical account of time as the expression of rational structure. Hamlet's task in the play is to reconcile his acts with a temporality that is of a different order. That is, to bring his actions into accord with the temporality in which he is operating. In this sense, for Deleuze, Hamlet mirrors the central problem of Kant's transcendental philosophy: how can two orders, which are different in kind, be brought into alignment with one another? For Kant, this problem is posed in terms of the incongruence of intuition and the understanding. For Rosenberg and Deleuze, the problem is posed in terms of the incongruence of personality as "an organic coherence intuitively based on the real world of sensation" (Rosenberg 1994, 135), and legal identity as the basis for action. To properly understand what it is to act therefore involves a proper understanding of the nature of time, and a deduction of how the structure of the act can be incorporated into the order of time. *Hamlet* can be seen as an attempt to provide such an understanding, offering an enquiry into the difference in kind between thinking and temporality.

Notes

1. This chapter includes some material previously published in Somers-Hall (2011). I am grateful to Edinburgh University Press for granting permission to reprint this material here.
2. As Francis Cornford notes, the forms have no generative power, and so cannot replace the role of the demiurge in *creating* a moving image of eternity (Cornford 1997, 28).
3. Cornford gives the following account of the relationship of myth to the rational structure of the world:

    It is to be taken, not literally, but as a poetical figure. The whole subsequent account of the world is cast in a mould which this figure dictates. What is really an analysis of the elements of rational order in the visible universe and of those other elements on which order is imposed, is presented in mythical form as the story of a creation in time. (Cornford 1997, 27)

4. Deleuze cites Kant's paralogisms as providing a model for Hamlet's hesitation, where Kant argues that even the self is given under the form of time.

References

Aristotle (1995) Poetics, trans. I. Bywater, in Jonathan Barnes (ed.), *The Complete Works of Aristotle* (pp. 2316–2340), Princeton, NJ: Princeton University Press.

Bergson, Henri (2002) Creative Evolution, in Keith Ansell Pearson and John Mullarky (eds.), *Bergson: Key Writings* (pp. 169–201), London: Continuum Press.

Cornford, Francis MacDonald (1997) *Plato's Cosmology: The Timaeus of Plato*, Cambridge: Hackett.

Davis, M. (1999). *The Poetry of Philosophy: On Aristotle's Poetics*, South Bend, Indiana: St. Augustine's Press.

Deleuze, Gilles (1978) 'Time and Synthesis', trans. Melissa McMahon, available at http://www.webdeleuze.com/php/texte.php?cle=66&groupe=Kant&langue=2 (accessed 25/05/11).

——— (1998) On Four Poetic Formulas that Might Summarize the Kantian Philosophy, trans. Daniel W. Smith and Michael A. Greco, in *Essays Critical and Clinical* (pp. 27–35), London: Verso Books.

——— (2003). *Kant's Critical Philosophy*, trans. H. Tomlinson and B. Habberjam, Minneapolis: University of Minnesota Press.

——— (2004) *Difference and Repetition*, trans. Paul Patton, London: Continuum Books.

Hegel, Georg (1975) *Aesthetics: Lectures on Fine Arts*, trans. T.M. Knox, Oxford: Clarendon Press, vol. 2.

Kant, Immanuel (1929) *Critique of Pure Reason*, trans. Norman Kemp Smith, London: Macmillan Press.

——— (1968) Concerning the Ultimate Foundation for the Differentiation of Regions in Space, in G.B. Kerferd and D.E. Walford (trans. and eds.), *Selected Pre-Critical Writings* (pp. 36–43). Manchester: Manchester University Press.

Leibniz, G.W. and Clarke, S (2000) *Correspondence*, ed. Roger Ariew, Cambridge: Hackett.

Plato (1997) Timaeus, trans. Donald J. Zeyl, in John M. Cooper (ed.), *Plato: Complete Works* (pp. 1224–1291), Cambridge: Hackett.

Rosenberg, Harold (1994) *The Tradition of the New*, New York: De Capo Press.

Shakespeare (2003) *Hamlet*, Cambridge: Cambridge University Press.

Somers-Hall, Henry (2011) "Time Out of Joint: Hamlet and the Pure Form of Time," *Deleuze Studies*, 5.4: 56–76.

Sophocles (2001) *Electra*, trans. Anne Carson, Oxford: Oxford University Press.

——— (2003) *Antigone*, trans. Reginald Gibbons and Charles Segal, Oxford: Oxford University Press.

Spencer, Theodore (1943) *Shakespeare and the Nature of Man*, Cambridge: Cambridge University Press.

Contributors

Roman Altshuler is Assistant Professor of Philosophy at Kutztown University, USA.

Santiago Amaya is Assistant Professor of Philosophy at the Universidad de los Andes.

Monika Betzler is Chair for Practical Philosophy and Ethics at Ludwig-Maximilians-Universität Munich.

David Ciavatta is Associate Professor of Philosophy at Ryerson University.

John J. Drummond is the Robert Southwell S.J. Distinguished Professor in the Humanities and Professor of Philosophy at Fordham University.

Luca Ferrero is Associate Professor of Philosophy at the University of Wisconsin–Milwaukee.

Kim Frost is Assistant Professor of Philosophy at Syracuse University.

Shaun Gallagher is the Lillian and Morrie Moss Professor of Philosophy at the University of Memphis.

Edward S. Hinchman is Associate Professor of Philosophy at the University of Wisconsin–Milwaukee.

Daniel D. Hutto is Professor of Philosophical Psychology at the University of Wollongong.

Patrick McGivern is Senior Lecturer in Philosophy at the University of Wollongong.

B. Scot Rousse is a visiting scholar in the Department of Philosophy at the University of California, Berkeley.

Michael J. Sigrist is Professorial Lecturer in Philosophy at George Washington University.

Henry Somers-Hall is Senior Lecturer in Philosophy at Royal Holloway, University of London.

Helen Steward is Professor of Philosophy of Mind and Action in the School of Philosophy, Religion and History of Science at the University of Leeds.

Micah D. Tillman is an adjunct lecturer at McDaniel College.

J. David Velleman is Professor of Philosophy and Bioethics at New York University.

Jonathan Webber is Reader in Philosophy at Cardiff University.

Ben Wolfson, Ph.D. Stanford, is an independent researcher.

Index